Scientific Insights and Technological Advances in Gluten Free Products Development

Scientific Insights and Technological Advances in Gluten Free Products Development

Editors

Maria Papageorgiou
Theodoros Varzakas

MDPI • Basel • Beijing • Wuhan • Barcelona • Belgrade • Manchester • Tokyo • Cluj • Tianjin

Editors
Maria Papageorgiou
Department of Food Science and Technology
International Hellenic University
Thessaloniki
Greece

Theodoros Varzakas
Food Science and Technology
Universtity of the Peloponnese
Kalamata
Greece

Editorial Office
MDPI
St. Alban-Anlage 66
4052 Basel, Switzerland

This is a reprint of articles from the Special Issue published online in the open access journal *Foods* (ISSN 2304-8158) (available at: www.mdpi.com/journal/foods/special_issues/Scientific_Insights_Technological_Advances_Gluten_Free_Products_Development).

For citation purposes, cite each article independently as indicated on the article page online and as indicated below:

LastName, A.A.; LastName, B.B.; LastName, C.C. Article Title. *Journal Name* **Year**, *Volume Number*, Page Range.

ISBN 978-3-0365-6565-1 (Hbk)
ISBN 978-3-0365-6564-4 (PDF)

Contents

About the Editors

Maria Papageorgiou

Maria Papageorgiou, Professor. Specialised in the field of polysaccharide rheology and mixed biopolymer systems (Ph.D. in Cranfield University, UK). Her current job position is Professor at the Food Science and Technology Department of the International Hellenic University (IHU). She is being involved in several international initiatives. Currently, she is serving as Secretary-Treasurer of the European Section of the Cereals and Grains, Cereals and Europe, with ca 400 members from all over Europe (http://www.aaccnet.org), she is elected Executive Committee member and member of the Technical Committee of the International Association of Cereal Chemists (ICC, http://www.icc.or.at) having also served as National Delegate of Greece (2002-2010) in that organization. In 2020 she was elected as Board member of the ISEKI FOOD Association (IFA) (http://www.iseki-food.net/). She has coordinated bilateral research projects, national research activities and she has being a partner of EU-funded projects and COST actions. She is also a referee at 23 scientific journals. Her main research interests are the tailoring of functional properties of cereal grains and their components in view of developing specific cereal healthy foods and ingredients and their structural characterization. She has more than 2000 citations to her research work in scopus (https://orcid.org/0000-0001-7009-846X) and has participated in various conferences. Currently she is a member of the Quality Assurance Unit of IHU and since 2016 member of the Greek National Rular Network (http://www.ead.gr/index.php/en/)

Theodoros Varzakas

Theodoros Varzakas has a Bachelor (Honours) in Microbiology and Biochemistry (1992), a Ph.D. in Food Science and Technology, and an MBA in Food and Agricultural Management from Reading University, UK (1998). He has also worked as a postdoctoral research staff at the same university. He has worked in large pharmaceutical and multinational food companies in Greece for 5 years and has at least 22 years of experience in the public sector. Since 2005, he has served as Assistant, Associate, and Full Professor at the Department of Food Science and Technology, University of Peloponnese, ex Technological Educational Institute of Peloponnese, Greece, specializing in issues of food technology, food processing/engineering, and food quality and safety. He is also a Section Editor in Chief for the Journal *Foods* in Food Security and *Sustainability* (2020-), was an ex Editor in Chief for Current Research in *Nutrition and Food Science* (2015-2019), and is a reviewer and member of the editorial board in many international journals. He has written more than 250 research papers and chapters in books and has presented more than 160 papers and posters at national and international conferences. He has written and edited 15 books in Greek and 11 in English on sweeteners, biosensors, food engineering, and food processing, chemometrics and authenticity published by CRC. He has participated in many European and national research programs as a coordinator or scientific member. His work has been cited in over 4000 citations with an h-index of 31.

Ex EFSA advisor/scientific expert to panel of biological hazards, visiting professor at Ghent University global campus, South Korea, Research Fellow at University Technology Malaysia.

According to September 2022 data-update for "Updated science-wide author databases of standardized citation indicators"Varzakas Theodoros is in the 2% of citations worldwide.

Preface to "Scientific Insights and Technological Advances in Gluten Free Products Development"

The prevalence of autoimmune disorders along with intolerance toward gluten and lifestyle trends have led to increased consumption of gluten-free products in the last two decades. The above has been accompanied with a steep rise in scientific publications on the topic of gluten-free, since both academia and the R&D sector of the food industry have been faced with the challenge to eliminate gluten and seek for substitutes. The prognosis for the global gluten-free products market is that it will continue to grow at a compound annual growth rate of 8.5% from 2020 to 2027.

Exploring alternative ingredients for the development of gluten-free products aiming to mimic the unique viscoelastic properties of gluten as a gas retention ingredient during fermentation, water binding and enabler of starch gelatinization on baking, and a bread flavor enhancer via gluten-related proteases is still a major field of research. Low specific volume, rapid staling, crumble and crumb structure, dry crumb, taste, and unbalanced nutritional profile are among the common defects in the final products associated with gluten absence.

This Special Issue addresses both new scientific insights and technological advances of gluten-free product development aiming to tackle the aforementioned issues. A selection of 12 peer reviewed papers covering this broad topic have been assembled. These will be of interest to anyone in the gluten free supply chain, researchers, students, and the farming community.

Maria Papageorgiou and Theodoros Varzakas
Editors

foods MDPI

Editorial

Scientific Insights and Technological Advances in Gluten-Free Product Development

Maria Papageorgiou [1] and Theodoros Varzakas [2,*]

[1] Department of Food Science and Technology, International Hellenic University, P.O. Box 141, GR-57400 Thessaloniki, Greece
[2] Department of Food Science and Technology, University of the Peloponnese, GR-24100 Kalamata, Greece
* Correspondence: t.varzakas@uop.gr; Tel.: +30-2721045279

This Special Issue addresses new scientific insights and technological advances in the area of gluten-free product development with the aim of controlling gluten intolerance and autoimmune diseases.

This Special Issue publishes seven research papers and four review articles. In the paper by Gasparre et al. [1], the authors focused on the nutritional qualities of gluten-free bakery products labeled ketogenic and/or low-carb and compared them to standard gluten-free products. All of the ketogenic and/or low-carb products showed lower carbohydrate and had higher protein contents ($p < 0.05$) compared to standard products, as well as higher ($p < 0.05$) fat contents. Bokic et al. [2] investigated the effect of the addition of chicory root (20–40%) and extrusion conditions (moisture content from 16.3 to 22.5%, and screw speed from 500 to 900 rpm) on the contents of bioactive compounds (inulin, sesquiterpene lactones, and polyphenols) of gluten-free rice snacks, and found an improvement in bioactive compounds and mineral contents, as well as antioxidative activities in all extrudates compared to the pure rice control sample. The paper by Rados et al. [3] aimed to develop crackers with high fiber and low fermentable oligosaccharides, disaccharides, monosaccharides, and polyols (FODMAP) content. They found that crackers made from maize and millet flour mixtures with sourdough and chia or flax seeds were rated highest for overall impression, including improvement in taste and appearance. However, according to the authors, soluble fiber content should also be taken into account as it confines undesirable descriptive texture attributes and bitter taste.

Chochkov et al. [4] explored the effect of sourdoughs on the quality traits of gluten-free dough (composed of teff, rice, corn, and sorghum flours) and GF bread. They found that sourdough-fermented doughs were softer and more elastic compared to control dough and yielded reduced baking losses. Moreover, the most pronounced positive effect on aroma, taste, and aftertaste was attributed to the *Pediococcus pentosaceus* strain.

Dos Reis Gallo et al. [5] determined the chemical composition, antioxidant activity and capacity, and the glycemic as well as insulinemic responses of gluten-free (GF) sorghum bread. They conducted a randomized clinical trial and found that brown sorghum was superior to other genotypes.

Gazikalović et al. [6] produced gluten hydrolysates through suitable combinations of partial enzymatic hydrolysis and microwave pretreatment parameters with the aim of reduced allergenicity and the preservation of technofunctional features for food applications. Microwave treatment yielded protein hydrolysates with enhanced antioxidant and functional properties.

Laignier et al. [7] developed gluten-free bread samples using different concentrations of *Amorphophallus konjac* (a perennial plant from the subtropical regions of Southeast Asia and Africa) flour. The bread samples with konjac showed a high fiber content and lower levels of carbohydrates, hence lower calories.

Citation: Papageorgiou, M.; Varzakas, T. Scientific Insights and Technological Advances in Gluten-Free Product Development. *Foods* **2023**, *12*, 250. https://doi.org/10.3390/foods12020250

Received: 23 December 2022
Revised: 29 December 2022
Accepted: 1 January 2023
Published: 5 January 2023

The review papers deal with the use of additives to replace gluten and ensure the stability and elasticity of the dough, hence improving the nutritional quality and sensory properties of gluten-free bread [8]. The application of hydrocolloids in GF bread and pasta, affecting dough rheology, bread hardness, specific volume, staling, and the glycemic index, is discussed in [9], whereas plant-based gluten-free proteins as well as high-protein sources of animal origin, sea-microorganism- and insect-based proteins, are illustrated in [10]. Finally, sourdough biotechnology based on an ecosystem of lactic acid bacteria (LAB) and yeasts to facilitate gluten-free products is described in [11].

Acknowledgments: The editors would like to thank all of the contributors who made this Special Issue a success and also *Foods* for this great accomplishment, which will become an e-book.

Conflicts of Interest: The authors declare no conflict of interest.

References

1. Gasparre, N.; Pasqualone, A.; Mefleh, M.; Boukid, F. Nutritional Quality of Gluten-Free Bakery Products Labeled Ketogenic and/or Low-Carb Sold in the Global Market. *Foods* **2022**, *11*, 4095. [CrossRef] [PubMed]
2. Bokić, J.; Kojić, J.; Krulj, J.; Pezo, L.; Banjac, V.; TumbasŠaponjac, V.; Travičić, V.; Moreno, D.A.; Bodroža-Solarov, M. Bioactive, Mineral and Antioxidative Properties of Gluten-Free Chicory Supplemented Snack: Impact of Processing Conditions. *Foods* **2022**, *11*, 3692. [CrossRef] [PubMed]
3. Radoš, K.; ČukeljMustač, N.; Varga, K.; Drakula, S.; Voučko, B.; Ćurić, D.; Novotni, D. Development of High-Fibre and Low-FODMAP Crackers. *Foods* **2022**, *11*, 2577. [CrossRef] [PubMed]
4. Chochkov, R.; Savova-Stoyanova, D.; Papageorgiou, M.; Rocha, J.M.; Gotcheva, V.; Angelov, A. Effects of Teff-Based Sourdoughs on Dough Rheology and Gluten-Free Bread Quality. *Foods* **2022**, *11*, 1012. [CrossRef] [PubMed]
5. Dos Reis Gallo, L.R.; Reis, C.E.G.; Mendonça, M.A.; da Silva, V.S.N.; Pacheco, M.T.B.; Botelho, R.B.A. Impact of Gluten-Free Sorghum Bread Genotypes on Glycemic and Antioxidant Responses in Healthy Adults. *Foods* **2021**, *10*, 2256. [CrossRef]
6. Gazikalović, I.; Mijalković, J.; Šekuljica, N.; Jakovetić-Tanasković, S.; Đukić-Vuković, A.; Mojović, L.; Knežević-Jugović, Z. Synergistic Effect of Enzyme Hydrolysis and Microwave Reactor Pretreatment as an Efficient Procedure for Gluten Content Reduction. *Foods* **2021**, *10*, 2214. [CrossRef] [PubMed]
7. Laignier, F.; Akutsu, R.d.C.C.d.A.; Maldonade, I.R.; Bertoldo Pacheco, M.T.; Silva, V.S.N.; Mendonça, M.A.; Zandonadi, R.P.; Raposo, A.; Botelho, R.B.A. *Amorphophallus konjac*: A Novel Alternative Flour on Gluten-Free Bread. *Foods* **2021**, *10*, 1206. [CrossRef] [PubMed]
8. Šmídová, Z.; Rysová, J. Gluten-Free Bread and Bakery Products Technology. *Foods* **2022**, *11*, 480. [CrossRef] [PubMed]
9. Culetu, A.; Duta, D.E.; Papageorgiou, M.; Varzakas, T. The Role of Hydrocolloids in Gluten-Free Bread and Pasta; Rheology, Characteristics, Staling and Glycemic Index. *Foods* **2021**, *10*, 3121. [CrossRef] [PubMed]
10. Skendi, A.; Papageorgiou, M.; Varzakas, T. High Protein Substitutes for Gluten in Gluten-Free Bread. *Foods* **2021**, *10*, 1997. [CrossRef] [PubMed]
11. Ramos, L.; Alonso-Hernando, A.; Martínez-Castro, M.; Morán-Pérez, J.A.; Cabrero-Lobato, P.; Pascual-Maté, A.; Téllez-Jiménez, E.; Mujico, J.R. Sourdough Biotechnology Applied to Gluten-Free Baked Goods: Rescuing the Tradition. *Foods* **2021**, *10*, 1498. [CrossRef] [PubMed]

foods

MDPI

Article

Nutritional Quality of Gluten-Free Bakery Products Labeled Ketogenic and/or Low-Carb Sold in the Global Market

Nicola Gasparre [1], Antonella Pasqualone [2], Marina Mefleh [2] and Fatma Boukid [3,*]

1 Department of Food and Human Nutritional Sciences, University of Manitoba, Winnipeg, MB R3T 2N2, Canada
2 Department of Soil, Plant and Food Science (DISSPA), University of Bari Aldo Moro, 70121 Bari, Italy
3 ClonBio Group LTD, 6 Fitzwilliam Pl, D02 XE61 Dublin, Ireland
* Correspondence: fboukid@clonbioeng.com

Abstract: Gluten-free and ketogenic bakery products are gaining momentum. This study aims to develop a better understanding of the nutritional quality of gluten-free bakery products labeled ketogenic and/or low-carb. For this reason, the products available on the global market that were labeled ketogenic and/or low-carb ($n = 757$) were retrieved and compared to standard gluten-free products ($n = 509$). Overall, nutritionally, no significant differences were found among ketogenic and/or low-carb products due the high intra-variability of each type, but they differed from standard products. Compared to standard products, all ketogenic and/or low carb, irrespective of categories, showed lower carbohydrates that derived chiefly from fibers and, to a lesser extent, from sugars. They also had higher protein contents ($p < 0.05$) compared to standard products. Fats was higher ($p < 0.05$) in ketogenic and/or low-carb baking mixes, savory biscuits, and sweet biscuits than in their standard counterparts. Saturated fats were higher ($p < 0.05$) in low-carb savory biscuits and breads, as well as in ketogenic sweet biscuits than in the same standard products. Overall, median values of the nutrients align with the definition of the ketogenic diet. Nevertheless, several products did not align with any of the ketogenic definitions. Therefore, consumers need to carefully read the nutritional facts and not rely on mentions such as low-cab and ketogenic to make their decision of purchase/consumption.

Keywords: low carb; high protein; high fiber; bread; cake; biscuits; flour mixes

Citation: Gasparre, N.; Pasqualone, A.; Mefleh, M.; Boukid, F. Nutritional Quality of Gluten-Free Bakery Products Labeled Ketogenic and/or Low-Carb Sold in the Global Market. *Foods* **2022**, *11*, 4095. https://doi.org/10.3390/foods11244095

Academic Editor: Claudia Monika Haros

Received: 16 October 2022
Accepted: 16 December 2022
Published: 18 December 2022

1. Introduction

Bakery products are staple foods worldwide, made basically from wheat flour, salt, and/or sugar. Gluten is a protein complex that is key for the development of bakery products such as bread and cakes, owing to its viscoelastic features [1,2]. Nevertheless, gluten intake might trigger adverse reactions in individuals genetically predisposed to gluten-related allergies and intolerances, and thus they must follow a lifetime gluten-free diet [3,4]. From a technological viewpoint, producing gluten-free products with an equivalent quality to that of gluten-containing counterparts is challenging due to the pivotal role played by gluten in forming a strong protein network that provides structure and allows for gas retention in bread and bakery products [5–7]. Gluten-free bakery products are mainly made using flours and starches that have a low functionality compared to wheat flour and thus other ingredients (e.g., hydrocolloids and crosslinking enzymes) are added to create a network similar to gluten [8–10]. The main sources of gluten-free flours/starches are rice, corn, potato, and tapioca. Nutritionally, gluten-free products are generally characterized by a high content of carbohydrates (due to starchy ingredients), low protein content, and high calorie content [11,12]. Additionally, gluten-free products are often associated with a high predicted glycemic index owing to their high glycemic load due their starch-based composition, which might be related to serious metabolic issues such as obesity and diabetes [13–15]. In recent years, significant research and development have been undertaken to enhance the technological and nutritional quality of gluten-free bakery products by increasingly using wholegrains, pseudocereals, and pulses to raise their protein and fiber contents and decrease that of easily digested carbohydrates [16,17].

Overall, it is assumed that all low carbohydrate strategies aim to reduce the overall intake of carbohydrates, but there is no clear consensus on the definition of a low-carb diet. The ketogenic diet, a specific low-carb diet, has gained immense popularity during the last decade [18]. The ketogenic diet market was valued at USD 10,221.40 million in 2019 and is projected to reach USD 15,266.36 million by 2027 [19]. There are several versions of the ketogenic diet [20]. Classic ketogenic diets are defined as high in fat (90%) and low in carbohydrates (restricting daily carbs to 4%) and proteins (6%) [20]. A modified Atkins diet does not restrict energy content and allows 65% fat, 5% carbohydrate, and 30% protein [21]. Another ketogenic diet is the very low-energy ketogenic diet which allows 13% carbohydrates, 44% fat, and 43% proteins and provides a total energy intake of <800 kcal/day [22]. The consumption of carbohydrate-rich foods, mostly cereal-based foods, fruits, and vegetables is limited during a ketogenic diet, yet it is not a carbohydrate-free diet [20]. Food companies have therefore started to develop, and commercialize, several food products specific for such a diet. The health benefits of the ketogenic diet are speculated to be in association with glycogen depletion and fat mobilization that might result in reducing blood glucose and improving fat burning [23–25]. The ketogenic diet has been used against neurologic conditions including autism, dementia, epilepsy, and nerve regeneration [26,27]. However, the long-term effects of a ketogenic diet are not yet fully understood.

From a market perspective, the gluten-free diet moved from a specialty diet to a mainstream market due to consumers associating a gluten-free diet with a healthy lifestyle [28,29]. In fact, the gluten-free bakery market is projected to reach USD 1819.4 million by the end of 2022 and is expected to expand at a compound annual growth rate of 8.2% by 2030 [30]. The demand for gluten-free ketogenic bakery products is increasing exponentially as a "high-fat, low-carbohydrate, adequate-protein" diet strategy is being adopted for weight loss, and treating/preventing diabetes, and neurological disorders [31,32]. Gluten-free ketogenic bakery products have not been researched extensively. For consumers, a deeper understanding of the nutritional facts of bakery products labeled gluten-free, ketogenic, and/or low carbohydrate could help in making a conscious and suitable decision of purchase/consumption. Therefore, the objective of this study was to evaluate the nutritional facts of commercial gluten-free ketogenic and/or low carb bakery products and compare them to standard gluten-free products to identify similarities/dissimilarities. For this reason, this study relied on a market database, Mintel, to be as exhaustive as possible by enabling a concrete illustration of the nutritional quality of products available on the supermarket shelves.

2. Materials and Methods

2.1. Data Collection

The market search of commercial gluten-free ketogenic and/or low carb bakery products was carried out by consulting the Mintel Global New Product Database (Mintel GNPD-Mintel Group Ltd., London, UK). The Mintel GNPD tracks packaged food and beverage launches in 86 markets worldwide. Each item has detailed product information, such as manufacturer, brand, price, ingredients, claims, and nutritional facts. The search was conducted on 16 September 2022.

The search considered the sub-category of "Bakery". The inclusion criteria were the date of product launches (1 January 1996 to 16 September 2022), the region (the global market), the presence of gluten-free claim, and the presence of the nutritional facts (i.e., energy, carbohydrates, sugar, protein, fat, saturated fat acids (SFA), fiber, and sodium). Using these settings, three searches were conducted with specific keywords:

- Search 1 was conducted with the inclusion of "Keto/ketogenic" and "Low/No/Reduced Carb" to retrieve ketogenic and low carb products (K-LC).
- Search 2 was conducted with the inclusion of "Keto/ketogenic" and the exclusion of products labeled "Low/No/Reduced Carb" to retrieve ketogenic products (K).
- Search 3 was conducted with the inclusion "Low/No/Reduced Carb" and the exclusion of products labeled "Keto/ketogenic" to retrieve low carb products (LC).
- Search 4 was conducted with the exclusion of "Keto/ketogenic" and "Low/No/Reduced Carb" mentions to retrieve standard gluten-free bakery products (STD) to be used in the nutritional comparison. This search considered the products launched in the last six months (16 March–16 September 2022) owing to the high number of standard gluten-free products and to capture the most recent launches.

2.2. *Data Extraction*

For all searches, nutritional facts and nutrition claims were collected. The results of all searches were exported to Microsoft Excel (Microsoft Office, Washington, WA, USA). Nutritional facts related to energy (kcal/100 g), total fat (g/100 g), saturated fatty acids-SFA (g/100 g), carbohydrates (g/100 g), sugars (g/100 g), protein (g/100 g), fiber (g/100 g), and sodium (mg/100 g) were retrieved, as well as nutrition claims.

2.3. *Statistical Analysis*

The statistical analysis was carried out using the Statistical Package for Social Sciences software (IBM SPSS Statistics, Version 25.0, IBM Corp., Chicago, IL, USA). Based on Kolmogorov–Smirnov test, the normality of data distribution was rejected, and therefore data were expressed as median values with inter-quartile ranges 25th–75th percentile. Energy and nutrient contents of products were analyzed using Kruskal–Wallis non-parametric one-way ANOVA for independent samples with multiple pairwise comparisons.

3. Results and Discussion

3.1. *Description Analysis*

Figure 1 shows the evolution of the number of K-LC, K, and LC bakery products in the global market. In total, the number of products retrieved was higher in LC, followed by K and K-LC. LC products have a longer history. The first low products (low-carb cookies) were launched in 2003 in the US market. Products showed a small peak in 2004 due to new launches in both US and Canadian markets. This aligns with a market report underlining that the low-carbohydrate movement started in North America in 2003–2004, where low-carb products ranked fifth among the most desirable new foods [33]. From 2004 to 2014, fluctuations in the number of launches were observed, to start increasing in a steady way in 2015. K-LC and K products appeared in 2016 and 2017, respectively. These products start to grow exponentially from 2018. This can be attributed to carb-watcher consumers increasing across the globe considering ketogenic, and/or views of low-carb diets as healthier diets and responsible for rapid weight loss. However, the main concern remains in the quality of the ingredients added, and, especially, the fat sources of low carb/ketogenic products. On the other hand, special caution should be taken by consumers who decide to adopt such diets because a high fat and protein diet could induce renal dysfunction on the long term [34,35]. A notable peak was observed in all gluten-free products in 2020 which can be attributed to Coronavirus 2019 pandemic [36,37]. The main boosters can be the health halo surrounding gluten-free diet, as well as the mounting number of consumers choosing this diet for weight management.

With 334 products, LC were the most abundant products (Table 1) due to their established history in the market, followed by K and K-LC. The examined bakery products were classified into five categories: baking flour mixes; bread products; cakes, pastries, and sweet goods; savory biscuits and crackers; and sweet biscuits and cookies. Baking flour mixes markedly prevailed over the other product categories, irrespective of being K, K-LC, or LC. The need for excluding or limiting the introduction of carbohydrates obviously excludes, or limits, all standard bakery products. Baking mixes have therefore been formulated by food companies to substitute regular grain flours, allowing for the domestic production of baked goods. In their K version, these mixes are basically composed of vegetable oils (sunflower or palm oil) absorbed on a cellulose substrate and powdered and added to protein from various sources [32]. In decreasing order, the second category most often labelled K or LC was represented by sweet biscuits and cookies, while bread products were the second category for K-LC. Categories such as sweet biscuits and cookies are widely appreciated by consumers because are very palatable and ready-to-eat, helping to improve diet adherence [32]. Regarding the geographical distribution of the examined products, our results confirmed that North America is the leader in K, K-LC, and LC gluten-free bakery products. The low-carb and ketogenic diets have been promoted for weight loss due to the numerous low-carbohydrate diet books, the over-sensationalism of these diets in the media and by celebrities, and the promotion of these diets in fitness centers and health clubs [38]. The high rates of obesity in North America explain the strong interest of the US consumers, who associated the low-carb diet with health and wellness [39], and thus moved toward K, K-LC, and LC products. The typical US diet is high in carbohydrates, mostly simple, accounting overall for approximately more than 65% of caloric intake [39,40]. Low-carbohydrate diets such the Atkins diet (a version of the ketogenic diet) has fueled the diet industry in North America for years, despite limited scientific evidence about its health benefits and risks [18]. During the last 20 years, weight loss and weight management in the US market witnessed a steady growth and was valued at USD 78 billion in 2019 [41]. It should be also underlined that a gluten-free diet is sometimes adopted, and perceived as an effective diet for weight loss, by consumers without gluten related intolerances/allergies [3,42]. The gluten-free diet has sometimes been associated with weight loss in non-gluten intolerant/allergic people because of improvements in insulin resistance and lipolysis [43,44].

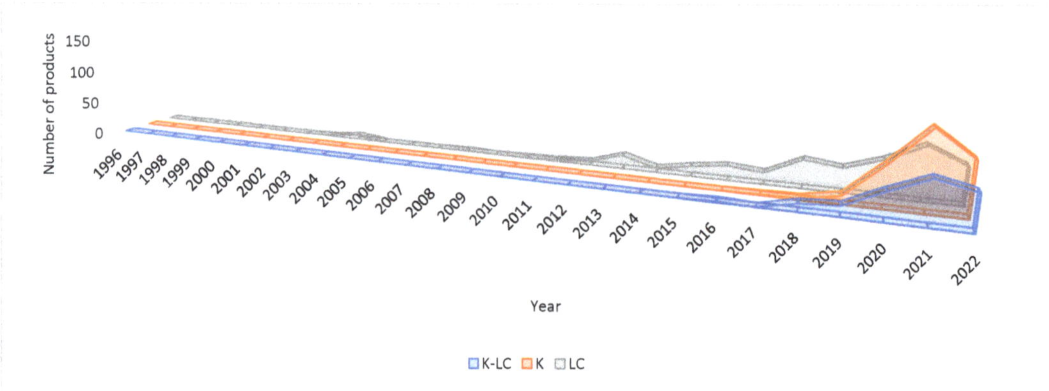

Figure 1. Evolution of the number of gluten-free bakery products labeled ketogenic low carb (K-LC), ketogenic (K), and/or low carb (LC).

LC and K products are also popular in Latin America as this geographical area is facing major diet-related health problems linked to overweight and obesity among all ages, accompanied by enormous social costs [45]. Therefore, in Latin America consumers are also shifting to these diets with the objective to manage their weight. The traditional Latin American cuisine is rich in complex carbohydrates, micronutrients, fiber, and phytochemicals [46,47]. However, during the last 40 years, Latin American countries have been experiencing a nutrition transition, moving from under- to overweight due to excessive consumption of refined carbohydrates and added sugars [47].

Table 1. Categorization of gluten-free bakery products labeled ketogenic (K) and/or low-carb (K-LC, LC).

Criteria	Segmentation	K-LC	K	LC
Type of product	Baking flour mixes	89	93	173
	Bread products	30	53	36
	Cakes, pastries, and sweet goods	15	18	28
	Savory biscuits and crackers	11	10	35
	Sweet biscuits and cookies	27	74	62
	Total	172	248	334
Region	North America	101	186	161
	Latin America	23	42	45
	Asia Pacific	33	13	80
	Middle East and Africa	15	5	33
	Europe	0	2	15
	Total	172	248	334
Nutrition claim *	Low/reduced/no added/free sugar	121	123	219
	High/added fiber	36	41	105
	Low/no/reduced trans fat	11	22	25
	High/added protein	14	12	77
	Low/no/reduced fat	2	11	25
	Low/no/reduced calories	7	11	30
	Low/no/reduced saturated fat	3	4	6
	Low/no/reduced sodium	1	6	20
	Total	195	230	507

* More than one can apply.

In the Asia-Pacific, a similar pattern was observed, especially for LC products. Over the last 20 years, Asian countries decreased the average carbohydrate intake due to increased prevalence of diabetes [48,49]. The traditional Asian diet is characterized by a high intake of rice, soy-based foods, and fish [50]. The Japanese Diabetes Society recommended a caloric reduction of 25–35 kcal/kg ideal body weight with carbohydrates constituting 50–60% of total energy consumption [51]. A recent review underlines that available evidence suggests there is a strong physiological rationale supporting the role of carbohydrate restriction for the management of Type 2 diabetes without inducing an increased risk of cardiovascular disease [52].

The Middle East and Africa ranked fourth, with a total of 53 products (as the sum of K, K-LC, and LC). Traditionally, the Mediterranean diet adopted in the Middle East and North Africa was one of the healthiest diets, as it is rich in vegetable proteins, fibers, minerals, and vitamins [53,54]. However, due to urbanization and changes in lifestyle, this geographical area too experienced a relatively recent nutrition transition to a diet rich in added sugars, and often lacking in vegetables, fruits, and whole grains [55,56]. This transition, associated with an increased burden of non-communicable diseases [57,58], is expected to strengthen the market of K, K-C, and LC products in the Middle East and Africa.

The European market had the fewest launches of products labeled K, K-LC, and LC. Indeed, low carb and ketogenic claims are not among the permitted claims in Europe (Regulation (EU) No 1047/2012). The retrieved products were mostly marketed in the UK, which has different legislation than the European union. The UK government's dietary guidelines recommend no more than 55% carbohydrate intake per day [59]. Mostly LC products were observed, only two K and no K-LC.

Additional nutrition claims may apply to K, K-LC, and LC products. LC products—the "historical" and most numerous category—were those labelled with the highest number of additional claims (507 claims for a total of 334 products, e.g., 1.5 claims per product). On the contrary, not all K products were additionally labelled (230 claims found for 248 products). Regarding the type of claim, K-LC chiefly had products with sugar reduction claims, followed by fiber claims. K products were mainly related to sugar, fiber, protein, and trans-fat reduction claims. LC had claims mostly related to sugar reduction, followed by fiber and protein enrichment. The reduction of sugar aligns with the fact that a large number of products were sweet and agrees with the ongoing trend in bakery products, especially in gluten-free products, to reduce sugar as reported in previous market surveys [11]. The focus was to reduce starchy ingredients by substituting them with fiber, protein, and sweeteners to preserve product structure and organoleptic features. For K and K-LC, few products had claims related to fat reduction, as the ketogenic diet was typically defined as high in fat [60]. For all categories, few products had claims related to calories, saturated fat, and sodium reduction.

3.2. *Nutritional Quality*

3.2.1. Bakery Flour Mixes

The nutritional composition of gluten-free flour mixes labeled ketogenic and/or low carb compared to their standard counterparts is displayed in the Figure 2. In terms of energy value, the median of all the product categories stayed around 400 kcal/100 g, and non-significant differences were reported ($p > 0.05$). K, LC, and K-LC flour mixes offered the highest fat amounts (below 30 g/100 g), while the lowest values belonged to the standard ones ($p < 0.05$). Regarding saturated fatty acids, the same trend was shown where standard products contained significantly ($p < 0.05$) less with respect to the gluten-free baking K and/or K-LC flour mixes. To better understand the nutritional variations, valuable help is provided by Table S1 (supplementary material), which contains all the ingredients present in the product ingredient lists. According to Table S1, high-fat flours from almond, coconut, soybean, and tiger nut appeared most frequently among the ingredients of the K, K-LC, and LC products due to their high content of protein and fat [61,62], unlike the standard bakery flour mixes that, in addition, included also flours from cassava, corn, potato, and rice. Oils from coconut, sunflower, and palm were the most employed fat for these product categories, with butter appearing only in the ingredient lists of the LC products. No significance ($p > 0.05$) was found among the different sodium contents, the value of which was below 500 mg/100 g. The median values of the carbohydrate contents confirmed what was reported on the product packages, standard bakery flour mixes had significant ($p < 0.05$) higher contents (around 65 g/100 g), as opposed to the K, K-LC, and LC products (below 40 g/100 g). The sugar content followed the same tendency, in which amounts below 5 g/100 g characterized the gluten-free K, K-LC, and LC products, whereas significantly ($p < 0.05$) slightly higher contents were found in the standard products. This could be explained by looking at the Table S1, which shows that ingredients with a high glucose content, such as cane sugar, coconut palm sugar, brown sugar, glucose, and maltodextrin were specially employed for the production of the standard products; no added sugars were reported on the ingredient list of the K products. On the other hand, sweeteners, such as erythritol, stevia, and xylitol were largely employed in the K, K-LC, and LC products to substitute sugars; allulose was the only sweetener used for the production of standard products. The situation changed when products were compared in terms of fiber content. Standard products were significantly ($p < 0.05$) the lowest, while K products occupied the middle position, and K-LC with LC bakery flour mixes held the highest fiber content (around 20 g/100 g). The number of fiber ingredients present in the ingredient list of the K, K-LC, and LC products was much higher compared to that of the standard products. Xanthan gum, psyllium seed husks, inulin, oat fiber, resistant dextrin, guar gum, apple fiber, and cellulose were the most utilized fibers. The adoption of the aforementioned ingredients not only serves for nutritional improvements, making the gluten-free products healthier, but it is also intended for a structuring function within the food matrix, while adhering to the low net carbohydrate requirements [23,63,64].

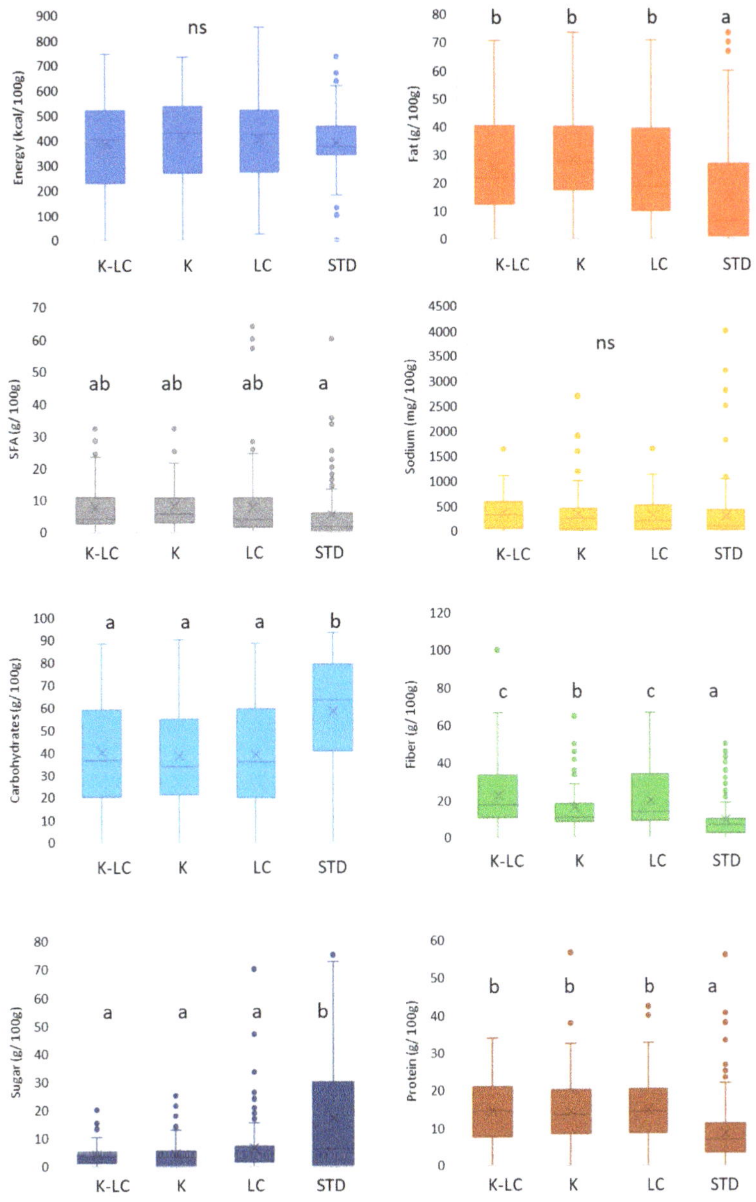

Figure 2. Nutritional composition of gluten-free flour mixes labeled ketogenic and/or low carb. K-LC: ketogenic and low carb; K: ketogenic; LC: low carb; STD: standard; different letters indicate significant difference at $p < 0.05$; ns: non-significant; the box-plot legend: the box is limited by the lower (Q1 = 25th) and upper (Q3 = 75th) quartile; the median is the horizontal line dividing the box; whiskers above and below the box indicate the 10th and 90th percentiles; outliers: are the points outside the quartile 10–90th percentiles.

From a technological standpoint, the gluten absence and the reduced presence of starchy raw materials mandates the use of hydrocolloids, gums, and fiber to recreate a pseudo gluten network with the aim of increasing the gas retention and creating a well-defined crumb structure [8,65]. All the K, K-LC, and LC gluten-free products were significantly ($p < 0.05$) higher in protein content (around 15 g/100 g) compared to their standard equivalents (below 7 g/100 g). Egg white proteins were the only protein used for the manufacture of the standard products. Typically, the protein source in gluten-free products derived from animal sources [23]. However, for the other product categories, beside proteins from animal sources (egg, whey, and milk), plant-based proteins were used, such as pea and rice (Table S1). Pea and rice proteins are increasingly used due their gluten-free nature and good functionality. Nutritionally, both have limitations in terms of amino acids compared to egg, but the blended cereal–legumes proteins could offer a balanced amino acid profile [66,67]. This can be attributed to the increased interest towards vegan products, and thus K, K-LC, and LC products are considered to fit within this trend in the bakery sector [68,69]. Decreasing carbohydrates, fat, and protein contents resulted increases in K, K-LC, and LC products. By resorting the use of proteins, food manufacturers principally wanted to meet nutritional enhancement and strengthening of the structure, which allows the obtention of gluten-free products with better properties in terms of texture and mouthfeel [70]. Fats rich in saturated fatty acids are increasingly being avoided due to their undesirable health effects, such as increasing the risk of cardiovascular disease and metabolic syndrome [71,72]. Fats play different roles in bakery products, including the promotion of moistness, mouthfeel, and soft texture [73].

3.2.2. Bread Products

Figure 3 displays the nutritional analysis of the gluten-free ketogenic and/or low carb and standard breads. Standard breads had significantly ($p < 0.05$) the highest energy content (Figure 3). No significance ($p > 0.05$) was found among the different fat contents in all the product categories, but LC breads showed the highest saturated fatty acid amounts, significantly higher than in their standard gluten-free counterparts, maybe because of the presence of coconut oil and full fat milk in their formulations (Table S2). This observation has negative health implications, especially considering that the overall nutritional quality of the fatty fraction of standard gluten-free products is not particularly high, as highlighted by several studies [74–76], so it should not be worsened further. For the specific purpose of improving nutritional quality, the use of extra virgin olive oil in gluten-free bread-making has been proposed [77]. For sodium content, statistical analysis showed no significant difference among categories ($p > 0.05$). The analysis of the carbohydrate contents related to the K, K-LC, and LC products highlighted that their median values were below 30 g/100 g, making them significantly ($p < 0.05$) lower than the standard products (above 45 g/100 g). Concerning the fiber content, K-LC and K products reached the highest values (between 12 and 18 g/100 g), followed by LC (10 g/100 g), with the standard gluten-free breads that contained the significantly ($p < 0.05$) lowest fiber content. Fibers from bamboo, oat, as well as rice bran, cellulose, and carob bran were found in the ingredient lists of LC, K, and K-LC breads (Table S2) and this underlies the versatility of fiber enrichment. Increasing fiber is a proven strategy to increase gluten-free bread's nutritional features and improve its sensory properties [78]. Small differences were found in sugar content, attributed to the high intra-variability of each group, especially the standard products. This is due to the general tendency of reducing sugar in gluten-free baked goods for health motives [79,80]. White sugar, cane sugar, brown sugar, glucose, rice syrup, and agave syrup were the sugars employed for the standard gluten-free breads, as opposed to maple syrup and glycerol, which were only utilized in the K and LC formulations, respectively (Table S2). A protein content around of 5 g/100 g positioned the gluten-free standard breads in the last position, whereas K-LC and K breads reached the highest values (above 12 g/100 g). As shown by the Table S2, proteins from pumpkin seeds and hemp were incorporated in the recipes for the ketogenic and/or low carb breadmaking in addition to protein from eggs, soybean, and pea that were present also in the standard formulations. These proteins are added to raise the nutritional value and to substitute gluten functionality, improving bread properties, such as crumb structure and volume [73,81].

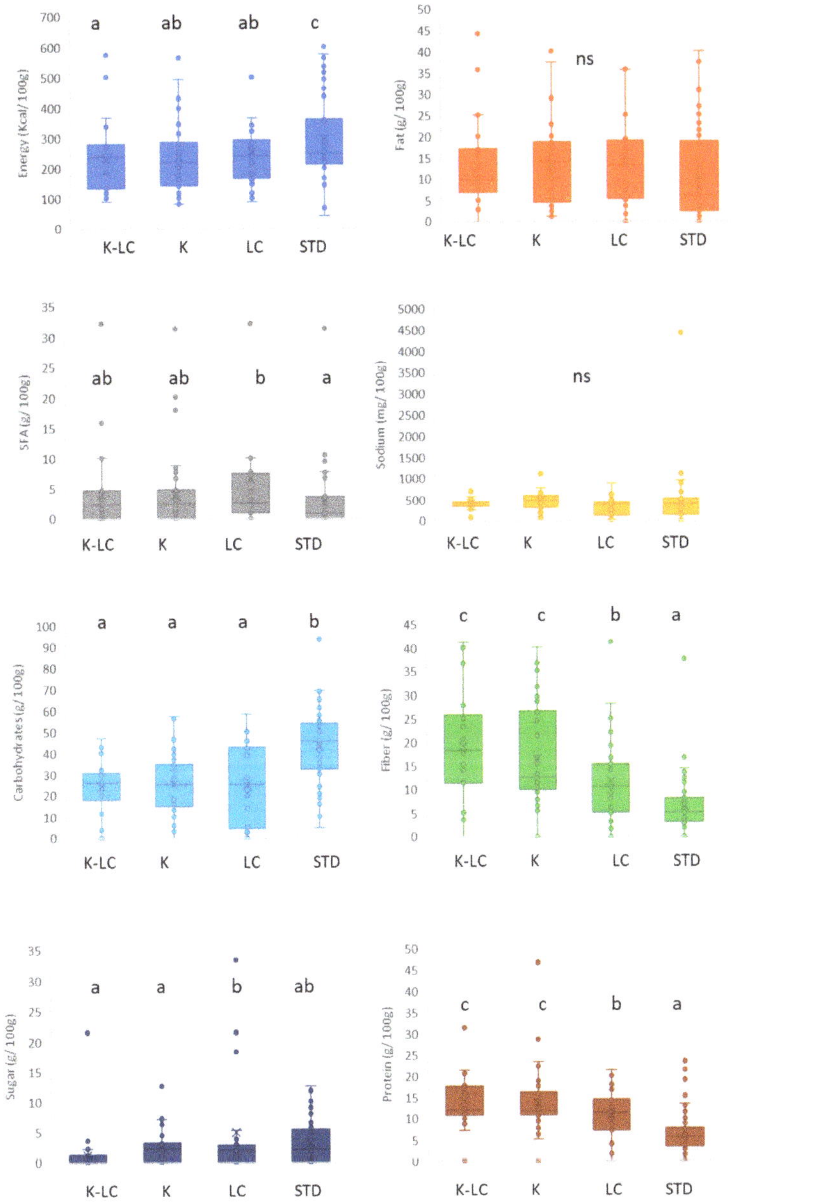

Figure 3. Nutritional composition of gluten-free bread products labeled ketogenic and/or low carb. K-LC: ketogenic and low carb; K: ketogenic; LC: low carb; STD: standard; different letters indicate significant difference at $p < 0.05$; ns: non-significant; the box-plot legend: the box is limited by the lower (Q1 = 25th) and upper (Q3 = 75th) quartile; the median is the horizontal line dividing the box; whiskers above and below the box indicate the 10th and 90th percentiles; outliers: are the points outside the quartile 10–90th percentiles.

3.2.3. Cakes, Pastries, and Sweet Goods

The nutritional features of the K, K-LC, and LC gluten-free cakes, pastries, and sweet goods are reported in the Figure 4, as well as their standard homologues.

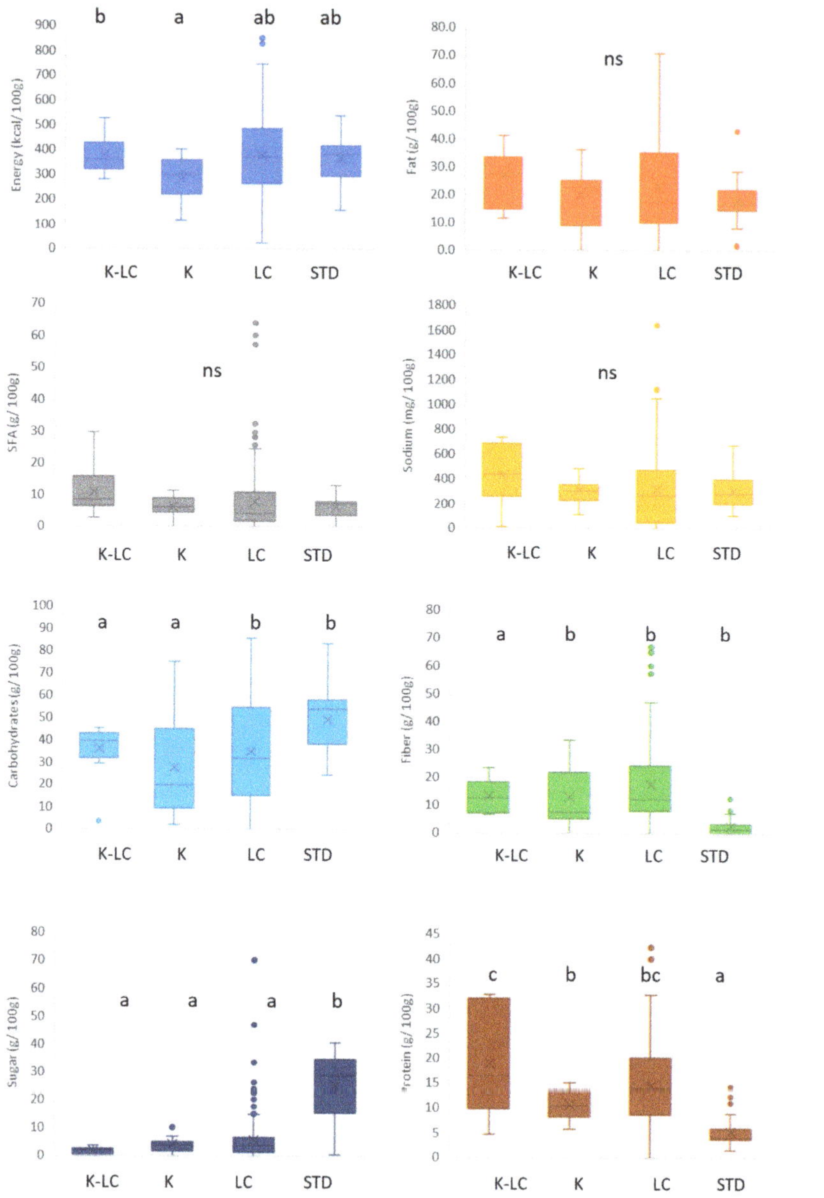

Figure 4. Nutritional composition of gluten-free cakes, pastries and sweet goods labeled ketogenic and/or low carb. K-LC: ketogenic and low carb; K: ketogenic; LC: low carb; STD: standard; different letters indicate significant difference at $p < 0.05$; ns: non-significant; the box-plot legend: the box is limited by the lower (Q1 = 25th) and upper (Q3 = 75th) quartile; the median is the horizontal line dividing the box; whiskers above and below the box indicate the 10th and 90th percentiles; outliers: are the points outside the quartile 10–90th percentiles.

K cakes offered the lowest energy content (around 300 kcal/100 g) compared to K-LC products with the highest energy density (350 kcal/100 g). Statistical analysis exposed no significant ($p > 0.05$) differences among the fat, saturated fatty acids, and sodium contents of the products considered for this study. Concerning the carbohydrate contents, they were significantly ($p < 0.05$) lower in K-LC and K products than in LC and standard products. These latter showed the highest sugar content (around 30 g/100 g), as opposed to their K, K-LC, and LC equivalents, which stayed below 5 g/100 g. Regarding the fiber content, LC cakes, pastries, and sweet goods significantly ($p < 0.05$) contained the highest quantity (above 10 g/100 g), as opposed to the other products included in the study. Reducing sugar in K and/or LC induced the increase of fiber to mimic the functionality of sugar, while sweeteners were added to preserve the taste [82]. This is a general strategy in reduced sugar/sugar-free bakery owing to rising health concerns over the high consumption of sugar [83,84]. These changes explain, in part, the reduction of calories in LC and/or K products. Therefore, these products might fit the requirement not only of celiacs or people following keto and low carb diets, but also of health-conscious consumers looking for a tasty and low caloric cake. According to the Table S3, only eggs and spirulina extract were used as protein ingredients in the gluten-free standard product formulations, reaching a median value of 5 g/100 g (Figure 4). Products belonging to the K-LC category showed the highest protein content (above 15 g/100 g), followed by LC and K products, respectively, with egg proteins and casein or milk proteins being the main sources of protein.

3.2.4. Savory Biscuits and Crackers

Figure 5 shows that the energy value provided by K, LC, and K-LC products was significantly ($p < 0.05$) higher than that of the standard products. Considering the energy value of fat (9 kcal/g), the fat content boxplots confirmed the previous outcomes, as they showed a similar pattern to energy content. With a fat amount above 40 g/100 g, K products led the group, while the gluten-free standard savory biscuits and crackers provided the lowest values (slightly above 5 g/100 g). The use of fats, mostly from vegetable sources (Table S4), allowed the obtainment of the significantly ($p < 0.05$) lowest content of saturated fatty acids (around 1 g/100 g) in the regular gluten-free products, as opposed to LC products, in which the median content was around 5 g/100 g, with the upper quartile touching values up to 40 g/100 g. Concerning the carbohydrate content, K-LC, K, and LC products reached significantly ($p < 0.05$) higher median values, around 15, 25, and 35 g/100 g, respectively; standard gluten-free products continued showing significantly ($p < 0.05$) higher carbohydrate contents (around 75 g/100 g). Shifting to the fiber contents, outcomes highlighted that gluten-free products labeled K, K-LC, and LC provided significantly ($p < 0.05$) higher fiber, with K-LC products leading the group. Xanthan and guar gums, as well as inulin, represented the only fiber sources utilized in the formulations of the gluten-free standard savory biscuits and crackers (Table S4); this could explain their lowest median values (around 5 g/100 g). Dried eggs, whey protein concentrate, milk protein, and hemp protein (Table S4) were the protein ingredients that most frequently appeared on the ingredient lists of K-LC, K, and LC products, giving them median values ranging from 15 to 20 g/100 g. On the other hand, gluten-free standard products contained just eggs, with a protein content of 5 g/100 g. No significant ($p > 0.05$) differences were found among the different products regarding the amounts of sodium (around 500 mg/100 g) and sugar (about 3 g/100 g).

3.2.5. Sweet Biscuits and Cookies

Outcomes from the nutritional analysis of the gluten-free sweet biscuits and cookies labeled ketogenic and/or low carb, as well as gluten-free standard products, are summarized in Figure 6.

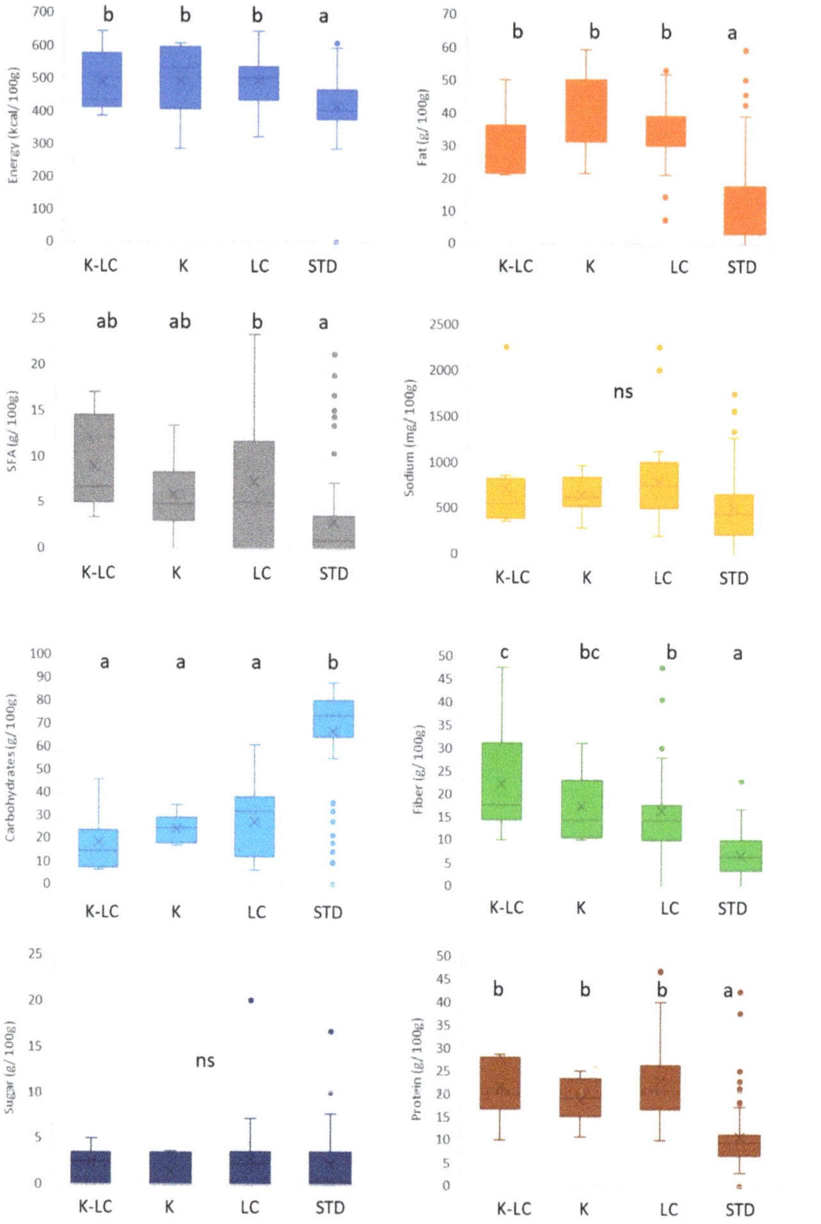

Figure 5. Nutritional composition of gluten-free savory biscuits and crackers labeled ketogenic and/or low carb. K-LC: ketogenic and low carb; K: ketogenic; LC: low carb; STD: standard; different letters indicate significant difference at $p < 0.05$; ns: non-significant; the box-plot legend: the box is limited by the lower (Q1 = 25th) and upper (Q3 = 75th) quartile; the median is the horizontal line dividing the box; whiskers above and below the box indicate the 10th and 90th percentiles; outliers: are the points outside the quartile 10–90th percentiles.

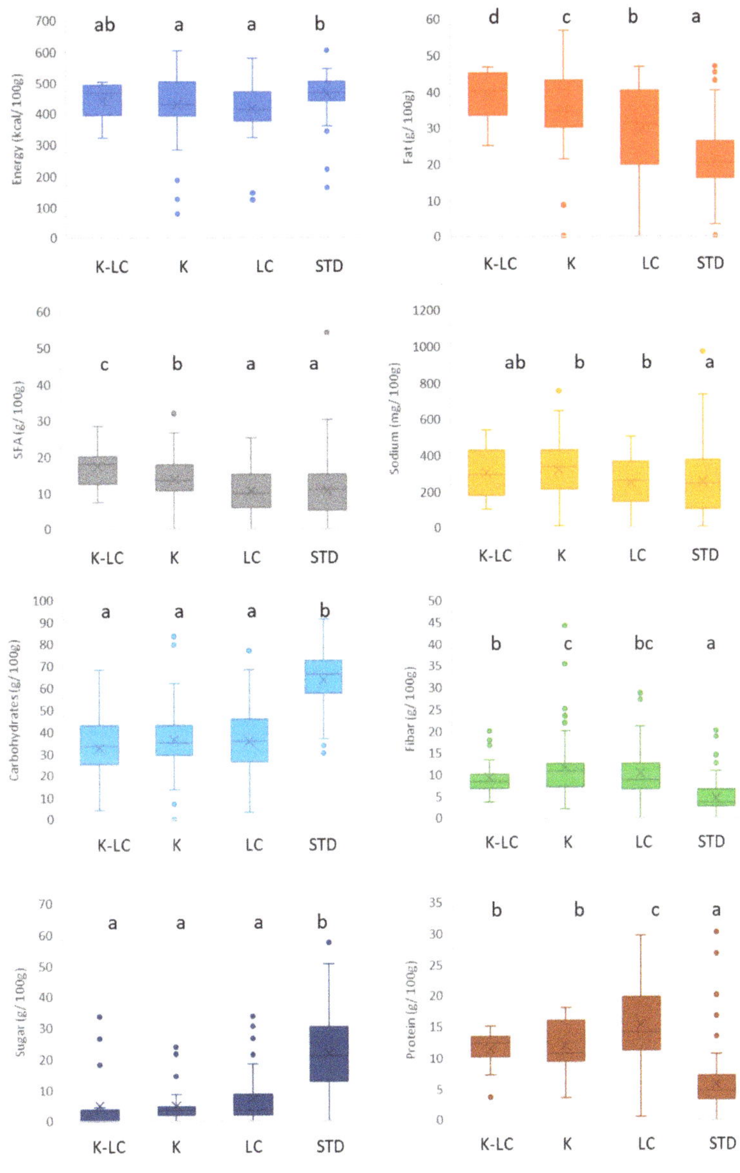

Figure 6. Nutritional composition of gluten-free sweet biscuits and cookies labeled ketogenic and/or low carb. K-LC: ketogenic and low carb; K: ketogenic; LC: low carb; STD: standard; different letters indicate significant difference at $p < 0.05$; ns: non-significant; the box-plot legend: the box is limited by the lower (Q1 = 25th) and upper (Q3 = 75th) quartile; the median is the horizontal line dividing the box; whiskers above and below the box indicate the 10th and 90th percentiles; outliers: are the points outside the quartile 10–90th percentiles.

The energy value box plots revealed a significantly ($p < 0.05$) little higher value offered by the standard products, as compared to those of K-LC, K, and LC. Despite this, standard products (~20 g/100 g) contained significantly ($p < 0.05$) less fat than LC (~20 g/100 g), K (~30 g/100 g), and K-LC (40 g/100 g), with these last two product categories showing the highest content of saturated fatty acids. In all the analyzed products, the median values for the sodium content were below 400 mg/100 g, with K and LC products having the highest amounts. These results were dependent on the only additive present in the ingredient lists, sodium hydrogen carbonate (Table S5). In the presence of some acids, this commonly used chemical leavening agent reacts with them, releasing carbon dioxide [64] and ensuring proper biscuit porosity. K-LC, K, and LC gluten-free products presented a median value of ~35 g/100 g of carbohydrates and 5 g/100 g of sugar. The total energy coming from sugar is lower than 5%, confirming that K-LC, K, and LC gluten-free sweet biscuits and cakes were suitable for the low carb dietary regimes, unlike gluten-free regular products that significantly ($p < 0.05$) contained the highest amounts of carbohydrates (~65 g/100 g) and sugars (20 g/100 g). A different pattern was presented when fiber content was analyzed. In particular, standard products were the poorest (~5 g/100 g), while the K, K-LC, and LC products significantly ($p < 0.05$) included up to twice the amount of fiber. The resorting to the use of eggs, egg whites isolate, soy protein, whey protein concentrate, milk protein, and whey protein isolate (Table S5) contributed to increase the protein content in the LC products. Hence, LC products significantly ($p < 0.05$) had the highest content (15 g/100 g), while gluten-free standard sweet biscuits and cookies only reached ~5 g/100 g.

4. Conclusions

Consumers' dietary patterns are changing and the food industry is trying to meet their expectations by launching new products. In the field of gluten-free foods, K, LC, and K-LC bakery products are conquering an important slice of the market. Up to now, no research has been carried out to analyze the global market of gluten-free bakery products labelled as ketogenic and/or low carb. The broad view, offered by this study, pointed out the pivotal role of North America in driving the global market for these food products. Baking flour mixes, bread products, cakes, pastries and sweet goods, savory biscuits and crackers, as well as sweet biscuits were the main categories forming the world market of gluten-free products labelled K, LC, and K-LC.

Overall, nutritionally, no significant differences were found among K, K-LC, and LC products due the high intra-variability of each type, but they differed from the standard products. A common trend was observed in the majority of the product categories analyzed: compared to their standard counterparts, gluten-free K, LC, and K-LC products contained higher levels of fiber and protein, while carbohydrate and sugar contents were lower. The fat content was significantly higher in K, LC, and K-LC baking mixes, savory biscuits, and sweet biscuits than in their regular gluten-free homologous products. Moreover, saturated fatty acids were significantly more abundant in LC savory biscuits and LC breads, as well as in K and K-LC sweet biscuits, compared to gluten free regular products of the same categories. While the higher fiber content is an obviously positive nutritional feature, the higher amount of saturated fatty acids constitutes a potential red flag for human health, especially when consumed for extended periods of time.

These findings suggest that the prolonged consumption of these new product categories always requires prior approval by health specialists. On the other hand, this research will open up a new scenario, which could be valuable in order to intensify the collaboration between researchers and the food industry with the aim of improving the nutritional quality of gluten-free ketogenic and/or low carb bakery products. At the same time, the nutritionally questionable composition observed in some products raises the need for special attention to the content of nutrients whose excessive intake has a negative effect on health, such as saturated fatty acids, in foods that are commonly perceived as healthy by the consumers. Consumers are invited to thoroughly read the ingredients and nutritional facts of these products before purchase.

Supplementary Materials: The following supporting information can be downloaded at: https://www.mdpi.com/article/10.3390/foods11244095/s1, Table S1: List of ingredients of baking flour mixes; Table S2: List of ingredients of bread products; Table S3: List of ingredients of cakes, pastries, and sweet goods; Table S4: List of ingredients of savory biscuits and crackers; Table S5: List of ingredients of sweet biscuits and cookies.

Author Contributions: Conceptualization, F.B.; methodology, F.B.; software, F.B.; validation, N.G., A.P., M.M. and F.B.; formal analysis, N.G., A.P., M.M. and F.B; investigation, N.G., A.P., M.M. and F.B.; resources, F.B.; data curation, F.B.; writing—original draft preparation, N.G. and F.B.; writing—review and editing, N.G., A.P., M.M. and F.B.; project administration, F.B. All authors have read and agreed to the published version of the manuscript.

Funding: This research received no external funding.

Data Availability Statement: The data presented in this study are available on request from the corresponding author.

Conflicts of Interest: The authors declare that they have no known competing financial interests or personal relationships that could have appeared to influence the objectivity and authenticity of the experimental results and discussion.

References

1. Boukid, F. *Cereal-Based Foodstuffs: The Backbone of Mediterranean Cuisine*; Springer: Cham, Switzerland, 2021.
2. Mefleh, M. Cereals of the Mediterranean Region: Their Origin, Breeding History and Grain Quality Traits. In *Cereal-Based Foodstuffs: The Backbone of Mediterranean Cuisine*; Springer International Publishing: Cham, Switzerland, 2021; pp. 1–18.
3. Boukid, F.; Mejri, M.; Pellegrini, N.; Sforza, S.; Prandi, B. How Looking for Celiac-Safe Wheat Can Influence Its Technological Properties. *Compr. Rev. Food Sci. Food Saf.* **2017**, *16*, 797–807. [CrossRef] [PubMed]
4. Morreale, F.; Boukid, F.; Carini, E.; Federici, E.; Vittadini, E.; Pellegrini, N. An Overview of the Italian Market for 2015: Cooking Quality and Nutritional Value of Gluten-Free Pasta. *Int. J. Food Sci. Technol.* **2019**, *54*, 780–786. [CrossRef]
5. Pico, J.; Reguilón, M.P.; Bernal, J.; Gómez, M. Effect of Rice, Pea, Egg White and Whey Proteins on Crust Quality of Rice Flour-Corn Starch Based Gluten-Free Breads. *J. Cereal Sci.* **2019**, *86*, 92–101. [CrossRef]
6. Kowalczewski, P.Ł.; Walkowiak, K.; Masewicz, Ł.; Bartczak, O.; Lewandowicz, J.; Kubiak, P.; Baranowska, H.M. Gluten-Free Bread with Cricket Powder—Mechanical Properties and Molecular Water Dynamics in Dough and Ready Product. *Foods* **2019**, *8*, 240. [CrossRef]
7. Mefleh, M.; Boukid, F.; Fadda, C. Suitability of Improved and Ancient Italian Wheat for Bread-Making: A Holistic Approach. *Life* **2022**, *12*, 1613. [CrossRef]
8. Belorio, M.; Gómez, M. Psyllium: A Useful Functional Ingredient in Food Systems. *Crit. Rev. Food Sci. Nutr.* **2021**, *62*, 527–538. [CrossRef]
9. Gao, Y.; Janes, M.E.; Chaiya, B.; Brennan, M.A.; Brennan, C.S.; Prinyawiwatkul, W. Gluten-Free Bakery and Pasta Products: Prevalence and Quality Improvement. *Int. J. Food Sci. Technol.* **2018**, *53*, 19–32. [CrossRef]
10. Xu, J.; Zhang, Y.; Wang, W.; Li, Y. Advanced Properties of Gluten-Free Cookies, Cakes, and Crackers: A Review. *Trends Food Sci. Technol.* **2020**, *103*, 200–213. [CrossRef]
11. Myhrstad, M.C.W.; Slydahl, M.; Hellmann, M.; Garnweidner-Holme, L.; Lundin, K.E.A.; Henriksen, C.; Telle-Hansen, V.H. Nutritional Quality and Costs of Gluten-Free Products: A Case-Control Study of Food Products on the Norwegian Marked. *Food Nutr. Res.* **2021**, 65. [CrossRef]
12. Jnawali, P.; Kumar, V.; Tanwar, B. Celiac Disease: Overview and Considerations for Development of Gluten-Free Foods. *Food Sci. Hum. Wellness* **2016**, *5*, 169–176. [CrossRef]
13. Di Cairano, M.; Galgano, F.; Tolve, R.; Caruso, M.C.; Condelli, N. Focus on Gluten Free Biscuits: Ingredients and Issues. *Trends Food Sci. Technol.* **2018**, *81*, 203–212. [CrossRef]
14. Romão, B.; Falcomer, A.L.; Palos, G.; Cavalcante, S.; Botelho, R.B.A.; Nakano, E.Y.; Raposo, A.; Shakeel, F.; Alshehri, S.; Mahdi, W.A.; et al. Glycemic Index of Gluten-Free Bread and Their Main Ingredients: A Systematic Review and Meta-Analysis. *Foods* **2021**, *10*, 506. [CrossRef]
15. Krupa-Kozak, U.; Lange, E. The Gluten-Free Diet and Glycaemic Index in the Management of Coeliac Disease Associated with Type 1 Diabetes. *Food Rev. Int.* **2019**, *35*, 587–608. [CrossRef]
16. Boukid, F.; Vittadini, E.; Lusuardi, F.; Ganino, T.; Carini, E.; Morreale, F.; Pellegrini, N. Does Cell Wall Integrity in Legumes Flours Modulate Physiochemical Quality and in Vitro Starch Hydrolysis of Gluten-Free Bread? *J. Funct. Foods* **2019**, *59*, 110–118. [CrossRef]
17. Boukid, F.; Rosell, C.M. The Nutritional Quality of Wholegrain and Multigrain Breads Is Not Necessarily Better than White Breads: The Case of Gluten-Free and Gluten-Containing Breads. *Int. J. Food Sci. Nutr.* **2022**, *73*, 902–914. [CrossRef]

18. Batch, J.T.; Lamsal, S.P.; Adkins, M.; Sultan, S.; Ramirez, M.N. Advantages and Disadvantages of the Ketogenic Diet: A Review Article. *Cureus* **2020**, *12*, e9639. [CrossRef]
19. Market Research Ketogenic Diet Market Forecast to 2027—COVID-19 Impact and Global Analysis by Product Type (Supplements, Beverages, Snacks, Dairy, and Others); and Distribution Channel (Hypermarket & Supermarket, Specialty Stores, Convenience Stores, and Others); and G. Available online: https://www.marketresearch.com/TIP-Knowledge-Services-v4095/Ketogenic-Diet-Forecast-COVID-Impact-14031593/ (accessed on 18 September 2022).
20. Ashtary-Larky, D.; Bagheri, R.; Bavi, H.; Baker, J.S.; Moro, T.; Mancin, L.; Paoli, A. Ketogenic Diets, Physical Activity and Body Composition: A Review. *Br. J. Nutr.* **2022**, *127*, 1898. [CrossRef]
21. Ratia, N.; Pietiläinen, K.H.; Auranen, M.; Saksa, L.; Luukkonen, R.; Suomalainen, A.; Piirilä, P. Modified Atkins Diet Modifies Cardiopulmonary Exercise Characteristics and Promotes Hyperventilation in Healthy Subjects. *J. Funct. Foods* **2021**, *81*, 104459. [CrossRef]
22. Caprio, M.; Infante, M.; Moriconi, E.; Armani, A.; Fabbri, A.; Mantovani, G.; Mariani, S.; Lubrano, C.; Poggiogalle, E.; Migliaccio, S.; et al. Very-Low-Calorie Ketogenic Diet (VLCKD) in the Management of Metabolic Diseases: Systematic Review and Consensus Statement from the Italian Society of Endocrinology (SIE). *J. Endocrinol. Investig.* **2019**, *42*, 1365–1386. [CrossRef]
23. Gillespie, R.; Ahlborn, G.J. Mechanical, Sensory, and Consumer Evaluation of Ketogenic, Gluten-Free Breads. *Food Sci. Nutr.* **2021**, *9*, 3327–3335. [CrossRef]
24. Alarim, R.A.; Alasmre, F.A.; Alotaibi, H.A.; Alshehri, M.A.; Hussain, S.A. Effects of the Ketogenic Diet on Glycemic Control in Diabetic Patients: Meta-Analysis of Clinical Trials. *Cureus* **2020**, *12*, e10796. [CrossRef]
25. Paoli, A. Ketogenic Diet for Obesity: Friend or Foe? *Int. J. Environ. Res. Public Health* **2014**, *11*, 2092–2107. [CrossRef]
26. Davis, J.J.; Fournakis, N.; Ellison, J. Ketogenic Diet for the Treatment and Prevention of Dementia: A Review. *J. Geriatr. Psychiatry Neurol.* **2021**, *34*, 3–10. [CrossRef]
27. Li, R.J.; Liu, Y.; Liu, H.Q.; Li, J. Ketogenic Diets and Protective Mechanisms in Epilepsy, Metabolic Disorders, Cancer, Neuronal Loss, and Muscle and Nerve Degeneration. *J. Food Biochem.* **2020**, *44*, e13140. [CrossRef]
28. Wójcik, M.; Różyło, R.; Schönlechner, R.; Berger, M.V. Physico-Chemical Properties of an Innovative Gluten-Free, Low-Carbohydrate and High Protein-Bread Enriched with Pea Protein Powder. *Sci. Rep.* **2021**, *11*, 14498. [CrossRef]
29. Caponio, G.R.; Difonzo, G.; de Gennaro, G.; Calasso, M.; De Angelis, M.; Pasqualone, A. Nutritional Improvement of Gluten-Free Breadsticks by Olive Cake Addition and Sourdough Fermentation: How Texture, Sensory, and Aromatic Profile Were Affected? *Front. Nutr.* **2022**, *9*, 75. [CrossRef]
30. Market Search Future Gluten Free Bakery Market Size, Share, and Industry Analysis by 2030. Available online: https://www.marketresearchfuture.com/reports/gluten-free-bakery-market-3228 (accessed on 18 September 2022).
31. Hopkin, L.; Broadbent, H.; Ahlborn, G.J. Influence of Almond and Coconut Flours on Ketogenic, Gluten-Free Cupcakes. *Food Chem. X* **2022**, *13*, 100182. [CrossRef]
32. Leone, A.; De Amicis, R.; Lessa, C.; Tagliabue, A.; Trentani, C.; Ferraris, C.; Battezzati, A.; Veggiotti, P.; Foppiani, A.; Ravella, S.; et al. Food and Food Products on the Italian Market for Ketogenic Dietary Treatment of Neurological Diseases. *Nutrients* **2019**, *11*, 1104. [CrossRef]
33. Sloan, A.E. The Top 10 Functional Food Trends 2004. *Food Technol. Mag.* **2004**, *58*, 28–51.
34. Ebbeling, C.B.; Feldman, H.A.; Klein, G.L.; Wong, J.M.W.; Bielak, L.; Steltz, S.K.; Luoto, P.K.; Wolfe, R.R.; Wong, W.W.; Ludwig, D.S. Effects of a Low Carbohydrate Diet on Energy Expenditure during Weight Loss Maintenance: Randomized Trial. *BMJ* **2018**, 363. [CrossRef]
35. Schutz, Y.; Montani, J.P.; Dulloo, A.G. Low-Carbohydrate Ketogenic Diets in Body Weight Control: A Recurrent Plaguing Issue of Fad Diets? *Obes. Rev.* **2021**, *22* (Suppl. 2), e13195. [CrossRef]
36. Demasi, M. COVID-19 and Metabolic Syndrome: Could Diet Be the Key? *BMJ Evid.-Based Med.* **2021**, *26*, 1–2. [CrossRef]
37. Gangitano, E.; Tozzi, R.; Mariani, S.; Lenzi, A.; Gnessi, L.; Lubrano, C. Ketogenic Diet for Obese COVID-19 Patients: Is Respiratory Disease a Contraindication? A Narrative Review of the Literature on Ketogenic Diet and Respiratory Function. *Front. Nutr.* **2021**, *8*, 1078. [CrossRef]
38. Bilsborough, S.; Crowe, T. Low-Carbohydrate Diets: What Are the Potential Short- and Long-Term Health Implications? *Asia Pac. J. Clin. Nutr.* **2003**, *12*, 396–404.
39. O'Driscoll, T.; Minty, R.; Poirier, D.; Poirier, J.; Hopman, W.; Willms, H.; Goertzen, A.; Madden, S.; Kelly, L. New Obesity Treatment: Fasting, Exercise and Low Carb Diet—The NOT-FED Study. *Can. J. Rural. Med.* **2021**, *26*, 55–60. [CrossRef]
40. Norwood, F.B. Perceived Impact of Information Signals on Opinions about Gluten-Free Diets. *PLoS ONE* **2021**, *16*, e0248570. [CrossRef]
41. Research and Markets. The U.S. Weight Loss & Diet Control Market. Available online: https://www.researchandmarkets.com/reports/5313560/the-u-s-weight-loss-and-diet-control-market (accessed on 22 September 2022).
42. Jones, A.L. The Gluten-Free Diet: Fad or Necessity? *Diabetes Spectr.* **2017**, *30*, 118. [CrossRef]
43. Soares, F.L.P.; de Oliveira Matoso, R.; Teixeira, L.G.; Menezes, Z.; Pereira, S.S.; Alves, A.C.; Batista, N.V.; de Faria, A.M.C.; Cara, D.C.; Ferreira, A.V.M.; et al. Gluten-Free Diet Reduces Adiposity, Inflammation and Insulin Resistance Associated with the Induction of PPAR-Alpha and PPAR-Gamma Expression. *J. Nutr. Biochem.* **2013**, *24*, 1105–1111. [CrossRef]

44. Ehteshami, M.; Shakerhosseini, R.; Sedaghat, F.; Hedayati, M.; Eini-Zinab, H.; Hekmatdoost, A. The Effect of Gluten Free Diet on Components of Metabolic Syndrome: A Randomized Clinical Trial. *Asian Pac. J. Cancer Prev.* **2018**, *19*, 2979. [CrossRef]
45. Popkin, B.M.; Reardon, T. Obesity and the Food System Transformation in Latin America. *Obes. Rev.* **2018**, *19*, 1028–1064. [CrossRef]
46. Fisberg, R.M.; Fontanelli, M.M.; Kowalskys, I.; Gómez, G.; Rigotti, A.; Cortés, L.Y.; García, M.Y.; Pareja, R.G.; Herrera-Cuenca, M.; Fisberg, M.; et al. Total and Whole Grain Intake in Latin America: Findings from the Multicenter Cross-Sectional Latin American Study of Health and Nutrition (ELANS). *Eur. J. Nutr.* **2022**, *61*, 489. [CrossRef]
47. Kovalskys, I.; Fisberg, M.; Gómez, G.; Pareja, R.G.; Yépez García, M.C.; Cortés Sanabria, L.Y.; Herrera-Cuenca, M.; Rigotti, A.; Guajardo, V.; Zalcman Zimberg, I.; et al. Energy Intake and Food Sources of Eight Latin American Countries: Results from the Latin American Study of Nutrition and Health (ELANS). *Public Health Nutr.* **2018**, *21*, 2535–2547. [CrossRef]
48. Mohan, V.; Unnikrishnan, R.; Shobana, S.; Malavika, M.; Anjana, R.M.; Sudha, V. Are Excess Carbohydrates the Main Link to Diabetes & Its Complications in Asians? *Indian J. Med. Res.* **2018**, *148*, 531. [CrossRef]
49. Rhee, E.J. Diabetes in Asians. *Endocrinol. Metab.* **2015**, *30*, 263–269. [CrossRef]
50. Sakurai, M.; Nakamura, K.; Miura, K.; Takamura, T.; Yoshita, K.; Nagasawa, S.Y.; Morikawa, Y.; Ishizaki, M.; Kido, T.; Naruse, Y.; et al. Dietary Carbohydrate Intake, Presence of Obesity and the Incident Risk of Type 2 Diabetes in Japanese Men. *J. Diabetes Investig.* **2016**, *7*, 343. [CrossRef]
51. Sanada, M.; Kabe, C.; Hata, H.; Uchida, J.; Inoue, G.; Tsukamoto, Y.; Yamada, Y.; Irie, J.; Tabata, S.; Tabata, M.; et al. Efficacy of a Moderately Low Carbohydrate Diet in a 36-Month Observational Study of Japanese Patients with Type 2 Diabetes. *Nutrients* **2018**, *10*, 528. [CrossRef]
52. Wheatley, S.D.; Deakin, T.A.; Arjomandkhah, N.C.; Hollinrake, P.B.; Reeves, T.E. Low Carbohydrate Dietary Approaches for People With Type 2 Diabetes—A Narrative Review. *Front. Nutr.* **2021**, *8*, 687658. [CrossRef]
53. Castro-Quezada, I.; Román-Viñas, B.; Serra-Majem, L. The Mediterranean Diet and Nutritional Adequacy: A Review. *Nutrients* **2014**, *6*, 231. [CrossRef]
54. Huang, C.L.; Sumpio, B.E. Olive Oil, the Mediterranean Diet, and Cardiovascular Health. *J. Am. Coll. Surg.* **2008**, *207*, 407–416. [CrossRef]
55. Azizi, F.; Hadaegh, F.; Hosseinpanah, F.; Mirmiran, P.; Amouzegar, A.; Abdi, H.; Asghari, G.; Parizadeh, D.; Montazeri, S.A.; Lotfaliany, M.; et al. Metabolic Health in the Middle East and North Africa. *Lancet Diabetes Endocrinol.* **2019**, *7*, 866–879. [CrossRef]
56. Afshin, A.; Micha, R.; Khatibzadeh, S.; Fahimi, S.; Shi, P.; Powles, J.; Singh, G.; Yakoob, M.Y.; Abdollahi, M.; Al-Hooti, S.; et al. The Impact of Dietary Habits and Metabolic Risk Factors on Cardiovascular and Diabetes Mortality in Countries of the Middle East and North Africa in 2010: A Comparative Risk Assessment Analysis. *BMJ Open* **2015**, *5*, e006385. [CrossRef] [PubMed]
57. Golzarand, M.; Mirmiran, P.; Jessri, M.; Toolabi, K.; Mojarrad, M.; Azizi, F. Dietary Trends in the Middle East and North Africa: An Ecological Study (1961 to 2007). *Public Health Nutr.* **2012**, *15*, 1835–1844. [CrossRef] [PubMed]
58. Sali, S.; Farhadnejad, H.; Asghari, G.; Teymoori, F.; Mirmiran, P.; Djazayeri, A.; Azizi, F. Animal Based Low Carbohydrate Diet Is Associated with Increased Risk of Type 2 Diabetes in Tehranian Adults. *Diabetol. Metab. Syndr.* **2020**, *12*, 87. [CrossRef] [PubMed]
59. Public Health England. In *Government Dietary Recommendations: Government Recommendations for Energy and Nutrients for Males and Females Aged 1–18 Years and 19+ Years*; PHE: London, UK, 2016. Available online: https://assets.publishing.service.gov.uk/government/uploads/system/uploads/attachment_data/file/618167/government_dietary_recommendations.pdf (accessed on 15 October 2022).
60. Crosby, L.; Davis, B.; Joshi, S.; Jardine, M.; Paul, J.; Neola, M.; Barnard, N.D. Ketogenic Diets and Chronic Disease: Weighing the Benefits Against the Risks. *Front. Nutr.* **2021**, *8*, 702802. [CrossRef] [PubMed]
61. Fernández-Cuesta, Á.; Fernández-Martínez, J.M.; Socias i Company, R.; Velasco, L. Near-Infrared Spectroscopy for Analysis of Oil Content and Fatty Acid Profile in Almond Flour. *Eur. J. Lipid Sci. Technol.* **2013**, *115*, 211–216. [CrossRef]
62. Gasparre, N.; Rosell, C.M. Role of Hydrocolloids in Gluten Free Noodles Made with Tiger Nut Flour as Non-Conventional Powder. *Food Hydrocoll.* **2019**, *97*, 105194. [CrossRef]
63. Anton, A.A.; Artfield, S.D. Hydrocolloids in Gluten-Free Breads: A Review. *Int. J. Food Sci. Nutr.* **2008**, *59*, 11–23. [CrossRef]
64. Roman, L.; Belorio, M.; Gomez, M. Gluten-Free Breads: The Gap Between Research and Commercial Reality. *Compr. Rev. Food Sci. Food Saf.* **2019**, *18*, 690–702. [CrossRef]
65. Matos, M.E.; Rosell, C.M. Understanding Gluten-Free Dough for Reaching Breads with Physical Quality and Nutritional Balance. *J. Sci. Food Agric.* **2015**, *95*, 653–661. [CrossRef]
66. Boukid, F. The Realm of Plant Proteins with Focus on Their Application in Developing New Bakery Products. In *Advances in Food and Nutrition Research*; Zhou, W., Gao, J., Eds.; Academic Press: Cambridge, UK, 2021; Volume 99, pp. 101–136.
67. Boukid, F.; Rosell, C.M.; Castellari, M. Pea Protein Ingredients: A Mainstream Ingredient to (Re) Formulate Innovative Foods and Beverages. *Trends Food Sci. Technol.* **2021**, *110*, 729–742. [CrossRef]
68. He, Y.; Meda, V.; Reaney, M.J.T.; Mustafa, R. Aquafaba, a New Plant-Based Rheological Additive for Food Applications. *Trends Food Sci. Technol.* **2021**, *111*, 27–42. [CrossRef]
69. Boukid, F.; Gagaoua, M. Vegan Egg: A Future-Proof Food Ingredient? *Foods* **2022**, *11*, 161. [CrossRef] [PubMed]
70. Ziobro, R.; Juszczak, L.; Witczak, M.; Korus, J. Non-Gluten Proteins as Structure Forming Agents in Gluten Free Bread. *J. Food Sci. Technol.* **2016**, *53*, 571–580. [CrossRef] [PubMed]

71. Astrup, A.; Teicholz, N.; Magkos, F.; Bier, D.M.; Thomas Brenna, J.; King, J.C.; Mente, A.; Ordovas, J.M.; Volek, J.S.; Yusuf, S.; et al. Dietary Saturated Fats and Health: Are the U. S. *Guidelines Evidence-Based? Nutrients* **2021**, *13*, 3305. [CrossRef] [PubMed]
72. Lawrence, G.D. Perspective: The Saturated Fat–Unsaturated Oil Dilemma: Relations of Dietary Fatty Acids and Serum Cholesterol, Atherosclerosis, Inflammation, Cancer, and All-Cause Mortality. *Adv. Nutr.* **2021**, *12*, 647–656. [CrossRef] [PubMed]
73. Waziiroh, E.; Schoenlechner, R.; Jaeger, H.; Brusadelli, G.; Bender, D. Understanding Gluten-Free Bread Ingredients during Ohmic Heating: Function, Effect and Potential Application for Breadmaking. *Eur. Food Res. Technol.* **2022**, *248*, 1021–1034. [CrossRef]
74. Pellegrini, N.; Agostoni, C. Nutritional Aspects of Gluten-Free Products. *J. Sci. Food Agric.* **2015**, *95*, 2380–2385. [CrossRef]
75. Miranda, J.; Lasa, A.; Bustamante, M.A.; Churruca, I.; Simon, E. Nutritional Differences Between a Gluten-Free Diet and a Diet Containing Equivalent Products with Gluten. *Plant Foods Hum. Nutr.* **2014**, *69*, 182–187. [CrossRef]
76. Caponio, F.; Summo, C.; Clodoveo, M.L.; Pasqualone, A. Evaluation of the Nutritional Quality of the Lipid Fraction of Gluten-Free Biscuits. *Eur. Food Res. Technol.* **2008**, *227*, 135–139. [CrossRef]
77. Pasqualone, A.; Caponio, F.; Summo, C.; Paradiso, V.M.; Bottega, G.; Pagani, M.A. Gluten-Free Bread Making Trials from Cassava (Manihot Esculenta Crantz) Flour and Sensory Evaluation of the Final Product. *Int. J. Food Prop.* **2010**, *13*, 562–573. [CrossRef]
78. Sciarini, L.S.; Bustos, M.C.; Vignola, M.B.; Paesani, C.; Salinas, C.N.; Pérez, G.T. A Study on Fibre Addition to Gluten Free Bread: Its Effects on Bread Quality and in Vitro Digestibility. *J. Food Sci. Technol.* **2017**, *54*, 244–252. [CrossRef] [PubMed]
79. Allen, B.; Orfila, C. The Availability and Nutritional Adequacy of Gluten-Free Bread and Pasta. *Nutrients* **2018**, *10*, 1370. [CrossRef] [PubMed]
80. Fry, L.; Madden, A.M.; Fallaize, R. An Investigation into the Nutritional Composition and Cost of Gluten-Free versus Regular Food Products in the UK. *J. Hum. Nutr. Diet.* **2018**, *31*, 108–120. [CrossRef] [PubMed]
81. Boukid, F.; Rosell, C.M.; Rosene, S.; Bover-Cid, S.; Castellari, M. Non-Animal Proteins as Cutting-Edge Ingredients to Reformulate Animal-Free Foodstuffs: Present Status and Future Perspectives. *Crit. Rev. Food Sci. Nutr.* 2021. [CrossRef]
82. Gao, J.; Brennan, M.A.; Mason, S.L.; Brennan, C.S. Effect of Sugar Replacement with Stevianna and Inulin on the Texture and Predictive Glycaemic Response of Muffins. *Int. J. Food Sci. Technol.* **2016**, *51*, 1979–1987. [CrossRef]
83. Krupa-Kozak, U.; Drabińska, N.; Rosell, C.M.; Piłat, B.; Starowicz, M.; Jeliński, T.; Szmatowicz, B. High-Quality Gluten-Free Sponge Cakes without Sucrose: Inulin-Type Fructans as Sugar Alternatives. *Foods* **2020**, *9*, 1735. [CrossRef]
84. Garvey, E.C.; O'Sullivan, M.G.; Kerry, J.P.; Milner, L.; Gallagher, E.; Kilcawley, K.N. Characterising the Sensory Quality and Volatile Aroma Profile of Clean-Label Sucrose Reduced Sponge Cakes. *Food Chem.* **2021**, *342*, 128124. [CrossRef]

foods

Article

Bioactive, Mineral and Antioxidative Properties of Gluten-Free Chicory Supplemented Snack: Impact of Processing Conditions

Jelena Bokić [1,*], Jovana Kojić [1], Jelena Krulj [1], Lato Pezo [2], Vojislav Banjac [1], Vesna Tumbas Šaponjac [3], Vanja Travičić [3], Diego A. Moreno [4] and Marija Bodroža-Solarov [1]

1 Institute of Food Technology, University of Novi Sad, Bul. Cara Lazara 1, 21000 Novi Sad, Serbia
2 Institute of General and Physical Chemistry, University of Belgrade, Studentski Trg 12–16, 11000 Beograd, Serbia
3 Faculty of Technology, University of Novi Sad, Bul. Cara Lazara 1, 21000 Novi Sad, Serbia
4 Phytochemistry and Healthy Food Lab (LabFAS), Department of Food Science Technology, Centro de Edafología y Biología Aplicada del Segura (CEBAS-CSIC), University Campus of Espinardo—25, Murcia, E-30100 Murcia, Spain
* Correspondence: jelena.perovic@fins.uns.ac.rs

Abstract: This study aimed to investigate the impact of chicory root addition (20–40%) and extrusion conditions (moisture content from 16.3 to 22.5%, and screw speed from 500 to 900 rpm) on bioactive compounds content (inulin, sesquiterpene lactones, and polyphenols) of gluten-free rice snacks. Chicory root is considered a potential carrier of food bioactives, while extrusion may produce a wide range of functional snack products. The mineral profiles were determined in all of the obtained extrudates in terms of Na, K, Ca, Mg, Fe, Mn, Zn, and Cu contents, while antioxidative activity was established through reducing capacity, DPPH (2,2-diphenyl-1-picrylhydrazyl) and ABTS (2,2-azino-bis(3-ethylbenzothiazoline-6-sulfonic acid) tests. Chicory root addition contributed to the improvement of bioactive compounds and mineral contents, as well as antioxidative activities in all of the investigated extrudates in comparison to the pure-rice control sample. An increase in moisture content raised sesquiterpene lactones and minerals, while high screw speeds positively affected polyphenols content. The achieved results showed the important impact of the extrusion conditions on the investigated parameters and promoted chicory root as an attractive food ingredient in gluten-free snack products with high bioactive value.

Keywords: chicory root; bioactive compounds; antioxidative activity; gluten free snack; extrusion

Citation: Bokić, J.; Kojić, J.; Krulj, J.; Pezo, L.; Banjac, V.; Tumbas Šaponjac, V.; Travičić, V.; Moreno, D.A.; Bodroža-Solarov, M. Bioactive, Mineral and Antioxidative Properties of Gluten-Free Chicory Supplemented Snack: Impact of Processing Conditions. *Foods* **2022**, *11*, 3692. https://doi.org/10.3390/foods11223692

Academic Editor: Laura Gazza

Received: 8 October 2022
Accepted: 9 November 2022
Published: 17 November 2022

1. Introduction

Human nutrition habits have recently been focused on innovative healthy food. The development of gluten-free foods intended for people suffering from celiac disease, vegetarians, or people practicing a gluten-free diet based on rice, buckwheat, amaranth, and quinoa for some other reason is a technological challenge nowadays. Celiac disease is an autoimmune disorder that affects about 1% of the world's population, but a strict gluten-free diet is practiced by about 10% of the population [1]. Celiac disease might be alleviated only by adhering to a gluten-free diet [2]. Among gluten-free materials, rice possesses some advantages: easy digestion, hypoallergenicity, mild taste, and colorlessness [3]. In addition to the gluten-free composition, rice is an important material in the production of extruded foods due to its high starch content. However, to improve the nutritional composition of rice extrudates, which are often enriched with protein or dietary fiber-rich raw materials [4]. Therefore, gluten-free snacks appear to be a promising basis for fortification. The roots and rhizomes of plants, representing rich sources of bioactive and nutritional valuable compounds, are attracting the interest of food technologists with the goal of developing novel functional products.

Chicory (*Cichorium intybus* L.) root is a rich source of inulin, oligofructose, polyphenols, and sesquiterpene lactones, observed as potential carriers of food functionality [5,6]. Inulin, as non-digestible dietary fiber, is a linear fructose polymer with β (2→1) glycosidic linkage, potentially causing antihyperglycemic and antidyslipidemic effects and improving bowel movements [7]. Chicory root flour causes different health effects than inulin alone, affecting cancer prevention, antibacterial and antiviral defense, hypoglycemic and hypolipidemic response, and antioxidant activity [6]. This statement may be related to the presence of other bioactive compounds in chicory root, such as sesquiterpene lactones (SLs). SLs, the bitter compounds from chicory, including dihydrolactucin, lactucin, 8-deoxylactucin, jacquinelin, dihydrolactucopicrin, and others, are C15 terpenoids that show primarily anti-inflammatory health benefits [8]. In addition, Perović et al. [5] highlighted chicory as a rich source of a variety of phenolic compounds managing antioxidative properties, including mono- and dicaffeoylquinic acids, chicoric acid, chlorogenic acid, caffeic acid, and others. Minerals are essential components of the human diet that serve as cofactors in the various enzyme-controlled reactions that enable the normal functioning of the human body [9]. Among selected pasture plants, such as dandelion and white clover, chicory root showed an abundant mineral content [10]. Product enrichment with whole chicory root flour instead of isolation and implementation of individual bioactive compounds may prevent the accumulation of by-products. Aiming to produce a rice-based snack with an improved bioactive profile, chicory root was chosen as an extrudate supplement.

Extrusion is flexible food technology that ensures improved digestibility and high sensory quality of final products [11]. A combination of high temperature and mechanical shear under pressure during extrusion cooking changes raw materials and modifies the functional properties, nutrient, and phytochemical composition of the food [12]. Several studies investigated the impact of the extrusion process on the nutritional composition, bioactive constituents, antioxidant potential, and the physicochemical and functional characteristics of rice-based extrudates [4,13]. Extrusion affected the investigated phytochemicals in different manners in the study of Arribas et al. (phenols increased while lectins and protease inhibitors were eliminated during extrusion) [4]. Dilrukshi et al. noted that extrusion also increased the oligosaccharides, while the insoluble fiber content was not significantly affected during the production of gluten-free extruded snacks fortified with cowpea and whey protein concentrate [13].

The objective of this study was to define the impact of the extrusion process variables (moisture content, screw speed, and chicory root flour addition) on the contents of bioactive compounds (inulin, sesquiterpene lactones, and polyphenols) and minerals, as well as the antioxidant activity of the obtained snacks. An artificial neural network (ANN) was employed to generate the dependence on the final product characteristics concerning the input extrusion process parameters.

2. Materials and Methods

The details of the raw materials' characteristics, the preparations of the blends, the applied instruments, and the production steps of the rice-based extrudates enriched with chicory root are presented in a previously published paper [14]. Briefly, a twin-screw extruder, Bühler BTSK 30/28D (Bühler, Uzwil, Switzerland), was employed to produce novel gluten-free rice snacks with varying feed moisture (M, 16.3–22.5%), screw speed (SS, 500–900 rpm), and chicory root flour content (P, 20–40%), according to the central composite design (CCD). The rice-based extrudates enriched with chicory root flour were generated from five different blends (with 20, 24.1, 30, 35.9, or 40% chicory root), while the control sample (CS) was obtained from pure rice flour.

2.1. Inulin Determination

Then, 1 g of the ground sample was extracted with 10 mL of distilled water for 1 h in an agitation water bath (80 °C), which continued with 30 min of ultrasonication (ATU Ultrasonidos, Valencia, Spain). After centrifugation (10,000 rpm, 15 min; Eppendorf Centrifuge

5804R, Eppendorf, Wien, Austria), the supernatant was precipitated with four volumes of pure acetone (Avantor®, Radnor Township, PA, USA), overnight at 4 °C. The precipitates were redissolved in a water bath (80 °C) for 1 h, with an additional 30 min sonication. Appropriate dilutions of the samples were filtrated (0.45 μmØ pore size, regenerated cellulose from Agilent Technologies, Santa Clara, USA) prior to chromatographic analysis. The detection and quantification of inulin were determined by applying a high-performance liquid chromatographic method with the evaporative light-scattering detection (HPLC-ELSD) method developed and validated by Perović et al. [15]. This HPLC-ELSD analysis (16 min) has been performed by isocratic elution with water as a mobile phase and optimized detector parameters (the temperature of the evaporator was 80 °C, the temperature of the nebulizer was 80 °C, the gas flow rate was 1.3 standard liters per minute, and the detector gain was 1). The proposed method was carried out using a cation exchange Rezex™ RSO-Oligossacharide Ag+ (Phenomenex, Aschaffenburg, Germany) column (4%, 12 μm particle size, 200 × 10 mm).

2.2. Sesquiterpene Lactones (SLs) Content

The most represented SLs from chicory root: lactucin, lactucopicrin, di-hydrolactucin, and di-hydrolactucopicrin were extracted according to the procedure of Schmidt et al. [16]. The extraction was carried out by shaking with 95% Ethanol (Zorka Pharma-Hemija DOO, Šabac, Serbia) for 24 h at room temperature, with a ratio of sample:extraction agent = 1:10. The defatting of ethanolic extract with the n-heptane was carried out in a separatory funnel. The defatted ethanolic extract was further extracted with ethyl acetate to obtain a concentrated extract of sesquiterpene lactones. The evaporated extracts (under a stream of nitrogen) were dissolved in 2 mL of extractant (25:1:24 = methanol:formic acid:distilled water), placed in an ultrasonic bath for 10 minutes, filtered through 0.22 μm filters (PVDF, Millipore, Burlington, US), and subjected to chromatographic analysis.

The chromatographic analyses were carried out on a Luna C18 column (250 × 4.6 mm, 5 mm particle size; Phenomenex, Macclesfield, UK). Water/formic acid (99:1, *v/v*) and acetonitrile were used as the mobile phases A and B, respectively, with a flow rate of 1 mL/min. The linear gradient started at 0 min–5% B; 20 min–80% B, for separation; at 25 min–95% B, 30 min–95% B; wash; and back to initial 35 min–5% B, 40 min–5% B. The injection volume was 20 μL. Chromatograms were recorded at 254 and 320 nm.

The HPLC–DAD–ESI/MSn analyses were carried out using an Agilent HPLC 1200 system (Agilent Technologies, Waldbronn, Germany) coupled to a mass detector in series. The HPLC system consisted of a binary capillary pump (model G1376A), an autosampler (model G1377A), a degasser (model G1379B), a sample cooler (model G1330B), and a photodiode array detector (model G1315D) was controlled by ChemStation software (v.B.0103-SR2). The mass detector was a Bruker model UltraHCT (Bremen, Germany) ion trap spectrometer equipped with an electrospray ionization interface (ESI) and controlled by Bruker Daltonik Esquire software (v.6.1) (Bruker Daltonik GmbH, Bremen, Germany). The ionization conditions were 350 °C and 4 kV, for capillary temperature and voltage, respectively. The nebulizer pressure and nitrogen flow rate were 65.0 psi and 11 L/min, respectively. The full-scan mass covered the range of *m/z* from 100 to 1000. Collision-induced fragmentation experiments were performed in the ion trap using helium as the collision gas, with voltage ramping cycles from 0.3 to 2 V. The mass spectrometry data were acquired in the negative ionization. MS^n was carried out in the automatic mode on the more abundant fragment ion in $MS^{(n-1)}$.

2.3. Mineral Elements

Mineral content analysis was performed on the Varian spectra AA 10 (Varian Techtron Pty Limited, Australia). Na^+, K^+, Ca^{2+}, Mg^{2+}, Fe^{3+}, Mn^{2+}, Zn^{2+}, and Cu^{2+} content in the extrudates and control samples were determined following the standard SRPS EN ISO 6869/2008 [17]. A mixture of air and acetylene gas was used in all experiments. Cathode lamps were used for radiation. The sensitive wavelengths for the detection of the investigated elements were: 330.3 nm for Na^{+}, 404.4 nm for K^{+}, 422.6 nm for Ca^{2+}, 202.6 nm for Mg^{2+}, 248.3 nm for Fe^{3+}, 279.5 nm for Mn^{2+}, 213.9 nm for Zn^{2+}, and 324.7 nm for Cu^{2+}.

2.4. *Total Phenolic Content and Antioxidant Capacity*

The sample (5 g) was poured with 12.5 mL of ethanol:water mixture (80:20, *v/v*) and extracted for 15 min in an ultrasonic bath at 40 °C, centrifuged (3000 rpm/10 min) (Centrifuga Tehtnica, Železniki, Slovenia), and the supernatant was evaporated to dryness in a stream of nitrogen at 40 °C on the Reacti-Therm I device (Thermo Fisher Scientific, Waltham, MA, USA). The dry residue was stored at −18 °C until analysis of the fraction of free polyphenolic compounds (Free PPs). The separated solid phase (precipitate) was used for determining the fraction of bound polyphenolic compounds (Bound PPs).

Bound PPs were released using alkaline hydrolysis with reflux (20 min). The precipitate was transferred to a flat-bottom flask and hydrolyzed using 50 mL of methanol, 5 mL of potassium hydroxide:water (1:1, *w/w*), and butyl-hydroxytoluene (BHT). The cooled hydrolysates were filtered through a Büchner funnel through qualitative filter paper (Whatman, Grade 4 Chr, Maidstone, Great Britain), and the filtrate was transferred to a separation funnel. The hydrolysate was initially neutralized with an HCl concentration of 6 mol/L, and protein precipitation was carried out with NaCl. The liquid–liquid extraction procedure was performed using a separatory funnel with 50 mL of the diethylether:ethylacetate (1:1, *v/v*). The water–methanol layer was extracted two more times, and the collected fractions were then evaporated on a rotary vacuum evaporator. The evaporated residue was dissolved using an ethanol:water mixture (80:20, *v/v*), after which the polyphenol content was determined on a Multiscan GO microtiter plate reader (Thermo Fisher Scientific Inc., Waltham, MA, USA).

2.4.1. Free and Bound Polyphenols

Singleton and Rossi [18] described a procedure for determining the content of polyphenols based on their reaction with the Folin–Ciocalteu reagent, which was modified for 96-well microtiter plates [19].

2.4.2. Antioxidant Capacity

The antioxidant capacity, expressed as mg Trolox equivalent (TE) per g of dry sample (mg TE/g), was determined by the 2,2-diphenyl-1-picrylhydrazyl method (DPPH) described by Gironés-Vilaplana et al. [20], 2,2′-azino-bis-3-ethylbenzothiazoline-6-sulphonic acid method (ABTS) adapted for microtiter plate [21], and reducing capacity (RC) method detailed by the Oyaizu [22]. All of the tests were carried out in triplicate, both for the free (DPPHF, ABTSF, RCF) and bound (DPPHB, ABTSB, RCB) polyphenolic fractions.

2.5. *Artificial Neural Network*

Three multi-layer perceptron (MLP) models, which have three layers (input, hidden, and output), were applied to model the ANN for the prediction of inulin, lactone, mineral, and polyphenol content and antioxidant activity as a function of moisture content, screw speed, and chicory root content. Before ANN computation, the database was normalized to improve the result accuracy of the ANN model. The normalization of the data was performed according to the min–max normalization criteria [23].

The Broyden–Fletcher–Goldfarb–Shanno (BFGS) algorithm was applied to solve unconstrained nonlinear problems during the optimization of the ANN models.

The exploratory data for ANN were randomly separated into training, cross-validation, and testing data (with 60%, 20%, and 20% of the experimental database, accordingly). A sequence of diverse topologies was employed, in which the number of hidden neurons ranged from 3 to 10, and the training process of the network was run in 100,000 repetitions using random weights and biased initial values. The optimization process was accomplished based on error minimization during the ANN validation cycle. The successful training was completed when the learning and the cross-validation curves approached zero [23–26].

Coefficients related to the hidden layer (weights and biases) were introduced into matrices W_1 and B_1. Similarly, coefficients related to the output layer were described in matrices W_2 and B_2. The neural network model (Y) can be represented using a matrix notation [23]:

$$Y = f_1(W_2 \cdot f_2(W_1 \cdot X + B_1) + B_2) \quad (1)$$

where, f_1 and f_2 are transfer functions in the hidden and output layers, respectively, and X is the matrix of input variables.

ANN models and global sensitivity analysis of the obtained results were completed using Statistica 10.0® software (StatSoft, Tulsa, OK, USA).

Local Sensitivity Analysis

Yoon's local sensitivity formula for the developed ANN model was used to evaluate the relative influence of the input parameters on output variables based on the weight coefficients of the developed ANN models [27].

$$RI_{ij}(\%) = \frac{\sum\limits_{k=0}^{n} (w_{ik} \cdot w_{kj})}{\sum\limits_{i=0}^{m} \left| \sum\limits_{k=0}^{n} (w_{ik} \cdot w_{kj}) \right|} \cdot 100\% \quad (2)$$

where: w—weight coefficient in ANN models; i—input variable; j—output variable; k—hidden neuron; n—number of hidden neurons; and m—number of inputs.

2.6. Statistical Analysis

The central composite design generated 20 snack formulations with six repetitions in the central point (samples 2, 4, 9, 14, 19, and 20). All of the experiments were performed in appropriate replications, and the results are presented as mean ± SD. The determination of the differences between the means and the samples was conducted using one-way ANOVA and Tukey's multiple range test ($p < 0.05$). Statistical data processing was carried out using the statistical package XLSTAT 2020.5.1 (Addinsoft, New York, NY, USA).

3. Results and Discussion

3.1. Inulin Content

Up to now, the recommended daily intake of inulin has not been defined, and no toxic effect caused by high doses of inulin has been reported. Inulin can be used as a food ingredient and is most often labeled as "dietary fiber" on the packaging of the product, while mentioning the "bifidogenic effect" is also legal in several countries [28].

The inulin content in the analyzed samples varied from 3.29% to 10.10% (Table 1), while inulin was not detected in the control sample made from rice flour. Similarly, Radovanovic et al. reported an inulin content of 5.23–18.23% after the addition of Jerusalem artichoke (30–80%) into buckwheat extrudates [29].

Table 1. Inulin and sesquiterpene lactones (SLs) contents in extrudates and control sample (CS).

CCD Design					SLs				
S.No.	M, %	V, rpm	P, %	Inulin, %	Lactucin, μg/g	Lactucopicrin, μg/g	Dihydrolactucin, μg/g	Dihydrolactucopicrin, μg/g
1	21.2	820	35.9	5.52 ± 0.26^{c}	7.35 ± 0.89^{e}	4.15 ± 0.49^{gh}	4.15 ± 0.49^{gh}	9.79 ± 0.97^{h}
2	19.4	700	30.0	7.26 ± 0.16^{e}	4.29 ± 0.56^{cd}	2.74 ± 0.34^{ef}	2.64 ± 0.34^{de}	5.09 ± 0.57^{de}
3	17.6	820	35.9	7.15 ± 0.06^{e}	9.13 ± 0.93^{f}	4.61 ± 0.41^{h}	4.61 ± 0.41^{h}	9.47 ± 0.86^{h}
4	19.4	700	30.0	7.55 ± 0.41^{e}	4.35 ± 0.47^{cd}	2.72 ± 0.29^{ef}	2.62 ± 0.29^{de}	4.97 ± 0.43^{de}
5	21.2	820	24.1	4.28 ± 0.42^{b}	1.92 ± 0.21^{b}	1.13 ± 0.13^{ab}	1.13 ± 0.13^{ab}	2.45 ± 0.25^{b}
6	22.5	700	30.0	9.17 ± 0.18^{g}	4.97 ± 0.52^{d}	2.89 ± 0.12^{ef}	2.89 ± 0.12^{ef}	5.68 ± 0.46^{ef}
7	19.4	900	30.0	8.39 ± 0.14^{f}	4.77 ± 0.61^{d}	2.53 ± 0.23^{de}	2.53 ± 0.23^{de}	4.76 ± 0.51^{cde}
8	17.6	820	24.1	5.64 ± 1.05^{c}	3.97 ± 0.38^{cd}	1.64 ± 0.16^{bc}	1.64 ± 0.16^{bc}	3.68 ± 0.33^{bcd}
9	19.4	700	30.0	7.77 ± 0.55^{e}	4.53 ± 0.50^{cd}	2.50 ± 0.36^{de}	2.50 ± 0.36^{de}	5.05 ± 0.48^{de}
10	21.2	580	24.1	4.18 ± 0.49^{b}	3.12 ± 0.43^{bc}	1.94 ± 0.20^{cd}	1.94 ± 0.20^{cd}	4.16 ± 0.53^{cde}
11	16.3	700	30.0	5.36 ± 0.10^{c}	5.05 ± 0.57^{d}	2.22 ± 0.41^{cde}	2.22 ± 0.41^{cde}	4.54 ± 0.40^{cde}
12	19.4	700	40.0	9.49 ± 0.69^{g}	7.28 ± 0.75^{e}	4.18 ± 0.52^{h}	4.18 ± 0.52^{h}	8.61 ± 0.76^{gh}
13	19.4	500	30.0	7.61 ± 0.64^{e}	0.51 ± 0.12^{a}	0.56 ± 0.10^{a}	0.56 ± 0.10^{a}	0.49 ± 0.09^{a}
14	19.4	700	30.0	7.66 ± 0.66^{e}	4.42 ± 0.34^{cd}	2.62 ± 0.44^{de}	2.62 ± 0.44^{de}	5.14 ± 0.55^{de}
15	17.6	580	24.1	6.58 ± 0.14^{d}	2.16 ± 0.23^{b}	1.52 ± 0.17^{bc}	1.52 ± 0.17^{bc}	3.28 ± 0.39^{bc}
16	21.2	580	35.9	4.23 ± 0.03^{b}	8.72 ± 0.88^{f}	5.52 ± 0.64^{i}	5.52 ± 0.64^{i}	12.39 ± 0.99^{i}
17	17.6	580	35.9	10.10 ± 0.27^{h}	4.64 ± 0.63^{d}	3.38 ± 0.39^{fg}	3.38 ± 0.39^{fg}	5.38 ± 0.67^{e}
18	19.4	700	20.0	3.29 ± 0.02^{a}	6.40 ± 0.72^{e}	2.86 ± 0.38^{ef}	2.86 ± 0.38^{ef}	7.21 ± 0.78^{fg}
19	19.4	700	30.0	7.42 ± 0.64^{e}	4.46 ± 0.54^{cd}	2.63 ± 0.40^{de}	2.63 ± 0.40^{de}	5.22 ± 0.72^{de}
20	19.4	700	30.0	7.59 ± 1.31^{e}	4.45 ± 0.47^{cd}	2.60 ± 0.48^{de}	2.60 ± 0.48^{de}	4.97 ± 0.68^{de}
CS	18.0	800	00.0	n.d.	n.d.	n.d.	n.d.	n.d.
		CV		2.87	1.93	3.31	1.98	1.93

CCD—central composite design; S.no.—Sample number; M—moisture content; SS—screw speed; P—chicory root flour content; SLs—sesquiterpene lactones; CS—control sample; n.d.—not detected; CV—coefficient of variation for six central points (samples 2, 4, 9, 14, 19, and 20). Values in the same column marked with different letters were statistically significantly ($p < 0.05$) different (Tukey HSD test).

The intake of 8–10 g of inulin per day may reduce the level of triglycerides, cholesterol, and LDL cholesterol [30]. Furthermore, consuming 5–15 g of inulin per day for several weeks has been shown to have prebiotic activity [31]. Considering these facts, it is notable from Table 1 that most of the gluten-free chicory-enriched extrudates designed within this study might cause some health effects.

The influence of M, V, and P were calculated using Yoon's local sensitivity formula (Equation (1)), while the relative influence of these variables on the inulin content is shown in Figure 1.

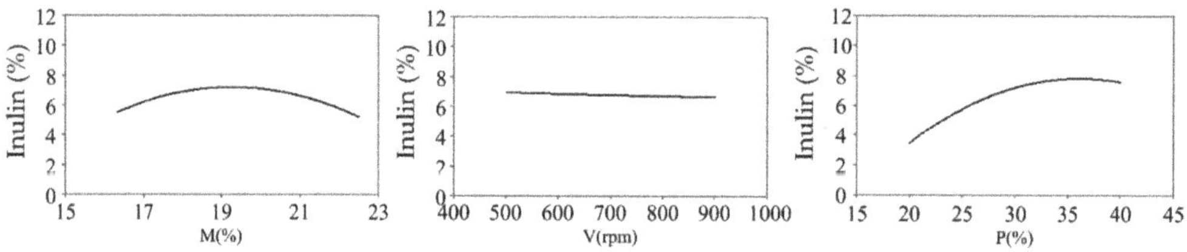

Figure 1. Influence of extrusion conditions (M—moisture content; V—screw speed; P—chicory root content) on inulin content.

The content of inulin decreases at higher moisture values (M > 19%, Figure 1), similar to the research of Ferreira et al. [32]. Namely, higher moisture content can encourage the dissolution of inulin, which makes it more exposed to shear forces and degradation during extrusion. Therefore, there is a noticeable decrease in the content of inulin when the moisture content is higher than 19% (Figure 1). Similar to obtained results, Sharma and Gujral [33] point out that a moisture content of up to 17% does not harm the content of dietary fibers, such as inulin.

The increase in screw speed showed a slightly decreased tendency for the inulin content (Figure 1), probably due to the intensified shear forces responsible for the degradation of inulin [34]. Similar conclusions were drawn by Tsokolar-Tsikopoulos et al. in a study examining the properties of expanded products with the added inulin [35].

The inulin content increased with the addition of chicory root (Figure 1), confirmed with a high positive correlation (r = 0.72, $p < 0.05$). This was expected, considering that chicory root is one of the richest sources of inulin [5,36]. Similarly, a significant increase in the inulin content ($p < 0.05$) was recorded in buckwheat-based extrudates with the addition of Jerusalem artichoke root rich in inulin [29].

3.2. *Sesquiterpene Lactones (SLs)*

Chicory sesquiterpene lactones have been reported to possess considerable biological activities and have been used in traditional medicines for centuries. Moreover, foods rich in SLs might be considered to be part of a healthy, balanced diet [37]. There is a noticeable lack of scientific studies that analyzed the content of SLs in food products, as well as the influence of processing conditions on the content of these bioactive compounds. Kulkarni et al. examined the influence of extrusion cooking on the bioavailability of the sesquiterpene lactone artemisinin, concluding that adjusting the pH (slightly acidic environment) can contribute to the preservation of this bioactive compound [38].

The presences of SLs in the extrudates were within the following ranges: 0.51–9.13 μg/g for lactucin, 0.56–5.52 μg/g for lactucopyrin, 0.49–12.39 μg/g for dihydrolactucin, and 0.00–4.07 μg/g for dihydorolactucopicrin (Table 1). Sample 16 contained the highest concentrations of all of the examined SLs (Table 1), which was expected considering that this sample had the highest proportion of chicory root (40%). The control sample did not contain any of the analyzed SLs.

The effect of the M, V, and P variables was evaluated using Yoon's local sensitivity formula, and the relative influence of these variables on the sesquiterpene lactone content is shown in Figure 2.

Increased moisture content had a positive effect on the content of all of the examined SLs (Figure 2), probably due to the facilitated release of SLs from the complexes. Similarly, moisture affected the solubility of artemisinin, a bioactive compound containing a sesquiterpene ring, thereby increasing its bioavailability [38]. Aberham et al. [39] examined the stability of sesquiterpene lactones (such as absinthe) in an aqueous environment, showing that the aqueous solutions of the lactone were stable for up to 6 months.

A screw speed up to 700 rpm caused an increase in SLs, while a further increase in V (above 700 rpm) negatively affected the content of the lactones (Figure 2). It is assumed that the higher screw speeds cause stronger shearing, frictional forces, and heating, whereby the partial or complete degradation of the analyzed lactones can occur. Screw speeds of 700 rpm and lower probably cause the release of lactones from the complexes, causing an increase in their bioavailability.

The addition of chicory root caused an increase in the content of all SLs, followed by high positive correlations (r = 0.65 to 0.75, $p < 0.05$). These results were expected due to the fact that chicory root is rich in SLs [40].

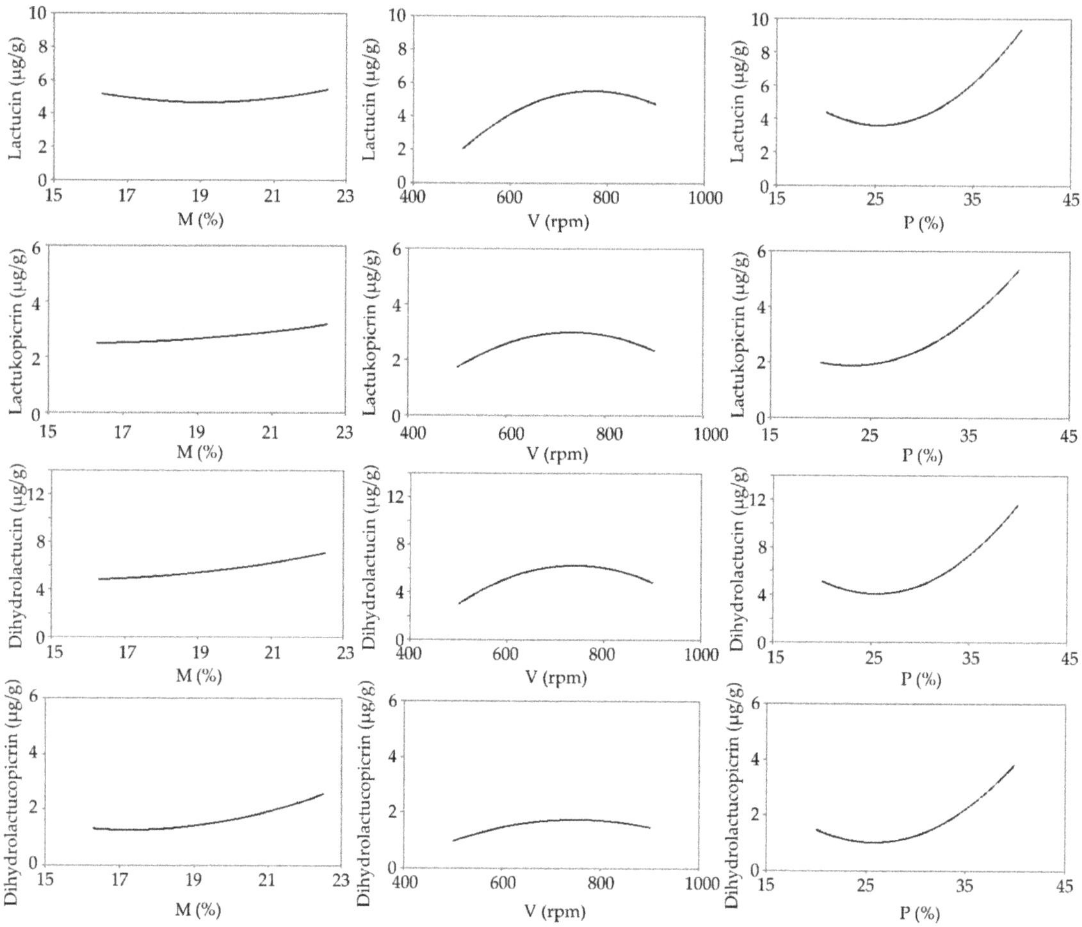

Figure 2. Influence of extrusion conditions (M—moisture content; V—screw speed; P—chicory root content) on sesquiterpene lactones content.

3.3. Total Phenolic Content and Antioxidant Capacity

Although selective and sensitive high-performance liquid chromatography provides reliable information about individual polyphenols, for more numerous samples predicted by the experimental design, the application of rapid and low-cost alternative methods was recommended [41]. The content of free and bound polyphenolic compounds ranged from 16.83–36.87 mg GAE/g d.m. and 4.12–9.06 mg GAE/g d.m., respectively (Table 2). Similar free and bound polyphenols value ranges were also noted in the purple potato and dry pea flour extrudates investigated by Nayak et al. [42]. Moreover, the extrusion of different sorghum genotypes also showed higher content of free polyphenols compared to bounded polyphenolic compounds [43]. As anticipated, the control sample recorded lower values of Free PPs and Bound PPs compared to extrudates supplemented with chicory root (8.39 mg GAE/g d.m. of free polyphenols and 2.45 mg GAE/g d.m. of bound polyphenolic compounds) (Table 2). This trend can be justified by the fact that chicory root is a rich source of polyphenols. According to Nwafor et al. [44], the most abundant polyphenols in chicory roots are coumaric, caffeic, chlorogenic, and protocatechuic acids.

Table 2. Contents of free (Free PPs) and bound (Bound PPs) (poly) phenolics (PPs), and the antioxidant capacity tests for both free and bound (polyphenolics) by the DPPH and ABTS methods. Total reducing capacity (RC) test for free (RCF) and bound phenolics (RCB) for extrudates and the control (CS).

	CCD Design										
S.No.	M, %	V, rpm	P, %	Free PPs, mg GAE/g d.w.	Bound PPs, mg GAE/g d.w.	DDPHF, mmol TE/g d.w.	DDPHB, mmol TE/g d.w.	ABTSF, mmol TE/g d.w.	ABTSB, mmol TE/g d.w.	RCF, mmol TE/g d.w.	RCB, mmol TE/g d.w.
1	21.2	820	35.9	33.26 ± 0.64 ^{h}	8.13 ± 0.06 ^{fg}	0.79 ± 0.00 ^{efg}	0.05 ± 0.00 ^{f}	400.17 ± 28.52 ^{efg}	80.41 ± 3.26 ^{bc}	91.99 ± 7.06 ^{efg}	4.54 ± 0.00 ^{h}
2	19.4	700	30.0	27.12 ± 0.13 ^{efg}	6.96 ± 0.11 ^{def}	0.61 ± 0.07 ^{cde}	0.03 ± 0.00 ^{d}	356.28 ± 12.64 ^{cdef}	71.28 ± 3.69 ^{bc}	79.54 ± 4.16 ^{cdefg}	3.97 ± 0.00 ^{f}
3	17.6	820	35.9	36.87 ± 0.98 ^{i}	9.06 ± 0.32 ^{g}	1.03 ± 0.00 ^{h}	0.06 ± 0.00 ^{g}	473.69 ± 12.97 ^{gh}	96.47 ± 15.02 ^{c}	108.97 ± 2.71 ^{g}	5.14 ± 0.01 ^{j}
4	19.4	700	30.0	27.01 ± 0.02 ^{efg}	6.57 ± 0.24 ^{def}	0.62 ± 0.07 ^{cde}	0.03 ± 0.00 ^{d}	347.23 ± 11.54 ^{cdef}	70.79 ± 1.29 ^{bc}	79.1 ± 3.15 ^{cdefg}	3.76 ± 0.00 ^{f}
5	21.2	820	24.1	23.11 ± 0.54 ^{cd}	5.51 ± 0.12 ^{bcd}	0.54 ± 0.09 ^{bcd}	0.02 ± 0.00 ^{c}	313.69 ± 2.69 ^{bcde}	65.97 ± 13.06 ^{bc}	60.12 ± 7.56 ^{bcdef}	3.11 ± 0.00 ^{d}
6	22.5	700	30.0	20.47 ± 0.06 ^{c}	4.25 ± 0.57 ^{bc}	0.48 ± 0.01 ^{bc}	0.01 ± 0.00 ^{b}	289.36 ± 24.56 ^{bcd}	61.17 ± 4.15 ^{bc}	51.89 ± 2.61 ^{bc}	2.83 ± 0.00 ^{c}
7	19.4	900	30.0	24.55 ± 0.69 ^{de}	6.05 ± 0.09 ^{de}	0.55 ± 0.01 ^{bcd}	0.02 ± 0.00 ^{c}	316.16 ± 5.25 ^{bcde}	67.69 ± 5.50 ^{bc}	62.06 ± 2.97 ^{bcde}	3.09 ± 0.00 ^{d}
8	17.6	820	24.1	26.98 ± 0.39 ^{efg}	6.90 ± 0.25 ^{def}	0.60 ± 0.00 ^{de}	0.03 ± 0.00 ^{d}	340.17 ± 23.69 ^{bcde}	89.33 ± 16.58 ^{bc}	77.14 ± 9.88 ^{bcdefg}	3.47 ± 0.01 ^{e}
9	19.4	700	30.0	26.68 ± 0.21 ^{efg}	6.85 ± 0.04 ^{def}	0.63 ± 0.00 ^{cde}	0.03 ± 0.00 ^{d}	353.69 ± 16.87 ^{cdef}	69.79 ± 2.58 ^{bc}	77.47 ± 1.65 ^{cdefg}	3.86 ± 0.02 ^{f}
10	21.2	580	24.1	20.81 ± 0.03 ^{c}	4.41 ± 0.39 ^{bc}	0.48 ± 0.07 ^{bc}	0.01 ± 0.00 ^{b}	279.36 ± 21.49 ^{bc}	62.09 ± 7.28 ^{bc}	53.71 ± 8.63 ^{bcd}	2.91 ± 0.00 ^{c}
11	16.3	700	30.0	28.54 ± 0.03 ^{g}	7.01 ± 0.03 ^{def}	0.66 ± 0.02 ^{cde}	0.04 ± 0.00 ^{e}	374.63 ± 13.64 ^{def}	75.87 ± 2.14 ^{bc}	84.65 ± 9.23 ^{defg}	3.94 ± 0.00 ^{f}
12	19.4	700	40.0	36.14 ± 0.87 ^{i}	8.94 ± 0.01 ^{g}	0.91 ± 0.01 ^{gh}	0.06 ± 0.00 ^{g}	498.21 ± 14.36 ^{h}	91.63 ± 2.41 ^{bc}	103.89 ± 1.47 ^{g}	4.69 ± 0.02 ^{i}
13	19.4	500	30.0	25.77 ± 0.36 ^{def}	6.13 ± 0.68 ^{de}	0.56 ± 0.00 ^{bcd}	0.03 ± 0.00 ^{d}	324.61 ± 2.57 ^{bcde}	67.48 ± 2.11 ^{bc}	71.62 ± 9.46 ^{bcdef}	3.51 ± 0.00 ^{e}
14	19.4	700	30.0	27.36 ± 1.06 ^{efg}	7.16 ± 0.01 ^{def}	0.61 ± 0.00 ^{cde}	0.03 ± 0.00 ^{d}	351.98 ± 2.65 ^{cdef}	72.24 ± 5.79 ^{bc}	78.73 ± 2.65 ^{cdefg}	3.92 ± 0.00 ^{f}
15	17.6	580	24.1	23.55 ± 0.28 ^{d}	5.81 ± 0.08 ^{cde}	0.53 ± 0.03 ^{bcd}	0.02 ± 0.00 ^{c}	301.21 ± 14.65 ^{bcd}	65.26 ± 1.36 ^{bc}	62.35 ± 7.45 ^{bcde}	3.15 ± 0.00 ^{d}
16	21.2	580	35.9	32.05 ± 0.43 ^{h}	8.01 ± 0.09 ^{fg}	0.71 ± 0.03 ^{def}	0.05 ± 0.00 ^{f}	395.27 ± 14.57 ^{efg}	78.25 ± 11.69 ^{bc}	91.43 ± 6.93 ^{efg}	4.15 ± 0.03 ^{g}
17	17.6	580	35.9	34.56 ± 0.91 ^{hi}	8.32 ± 0.11 ^{fg}	0.84 ± 0.01 ^{fgh}	0.04 ± 0.00 ^{e}	436.11 ± 23.14 ^{fgh}	81.24 ± 6.54 ^{bc}	96.36 ± 5.87 ^{fg}	4.69 ± 0.00 ^{i}
18	19.4	700	20.0	16.83 ± 0.07 ^{b}	4.12 ± 0.08 ^{b}	0.39 ± 0.00 ^{b}	0.01 ± 0.00 ^{b}	254.96 ± 6.43 ^{b}	57.37 ± 0.12 ^{b}	45.62 ± 2.32 ^{b}	1.87 ± 0.03 ^{b}
19	19.4	700	30.0	26.96 ± 0.13 ^{efg}	6.87 ± 0.21 ^{def}	0.61 ± 0.06 ^{cde}	0.03 ± 0.00 ^{d}	352.21 ± 17.61 ^{cdef}	70.06 ± 5.69 ^{bc}	78.14 ± 0.39 ^{cdefg}	3.69 ± 0.00 ^{f}
20	19.4	700	30.0	27.08 ± 0.09 ^{efg}	6.99 ± 0.56 ^{def}	0.62 ± 0.00 ^{cde}	0.03 ± 0.00 ^{d}	353.28 ± 29.64 ^{cdef}	70.31 ± 6.91 ^{bc}	80.11 ± 5.28 ^{cdefg}	3.76 ± 0.04 ^{f}
CS	18.0	800	00.0	8.39 ± 0.11 ^{a}	2.45 ± 0.35 ^{a}	0.11 ± 0.01 ^{a}	0.00 ± 0.00 ^{a}	85.73 ± 0.97 ^{a}	35.36 ± 1.05 ^{a}	14.85 ± 0.41 ^{a}	0.76 ± 0.01 ^{a}
		CV		0.82	2.84	1.68	0.00	0.85	1.28	1.21	2.81

CCD—central composite design; S.no.—Sample number; M—moisture content; V—screw speed; P—chicory root flour content; PPs—polyphenols; DDPHF—DPPH antioxidant activity of free polyphenolic fraction; DDPHB_DPPH antioxidant activity of bound polyphenolic fraction; ABTSF—ABTS antioxidant activity of free polyphenolic fraction; ABTSB—ABTS antioxidant activity of bound polyphenolic fraction; RCF_reducing capacity of free polyphenolic fraction; RCB—reducing capacity of bound polyphenolic fraction; GAE—gallic acid equivalent; TE—trolox equivalent; d.w.—dry weight; CS—control sample; CV— coefficient of variation for six central points (samples 2, 4, 9, 14, 19, and 20). Values in the same column marked with different letters were statistically significantly ($p < 0.05$) different (Tukey HSD test).

The measurement of the antioxidant activity of the biological samples largely depends upon the free radical or the oxidants used in the assays and the degree and type of antioxidants. Hence, it is important to use different antioxidant assays instead of relying on a single assay to assess and compare the antioxidant activity of the extrudates. Synergistic effects and concentration may also change the results that are not observed when the individual constituents are tested.

The antioxidant activity against DPPH• radicals in the fraction of free polyphenolic compounds was in the range of 0.39 mmol TE/g d.m. to 1.03 mmol TE/g d.m., while in the fraction of bound polyphenolic compounds, the activity was recorded in the range of 0.01 mmol TE/g d.m. to 0.06 mmol TE/g d.m. (Table 2). The control sample noted 0.11 mmol TE/g d.m. for DPPHF and 0.00 mmol TE/g d.m. for DPPHB. The highest antioxidant activity was determined in sample 3, containing 35.9% chicory root (1.03 and 0.06 mmol TE/g d.m for DPPHF and DPPHB, respectively).

The antioxidant activity against ABTS+• radicals was in the range of 254.96 mmol TE/g d.m. up to 498.21 mmol TE/g d.m. for the ABTSF, i.e., of 57.37 mmol TE/g d.m. up to 96.47 mmol TE/g d.m for the ABTSB. The control sample recorded 85.73 mmol TE/g d.m. and 35.36 mmol TE/g d.m. for ABTSF and ABTSB, respectively.

The reducing capacity was in ranges of 45.62–108.97 mmol TE/g d.m. for RCF, i.e., 1.87–5.14 mmol TE/g d.m. for RCB (Table 2), while the control showed 14.85 mmol TE/g d.m. for RCF, i.e., 0.76 mmol TE/g d.m. for RCB (Table 2).

The gluten-free chicory-enriched snacks were reported to have higher antioxidant capacity when compared to the control sample. Chicory root addition positively contributed to the increased antioxidative capacity, which is confirmed by the high positive correlations between the chicory root content and antioxidative capacity (R^2 = 0.89; 0.80; 0.94; 0.82; 0.90, and 0.92 for DPPHF, DPPHB, ABTSF, ABTSB, RCF, and RCB, respectively). The addition of polyphenolic-rich materials also improved the extrudate's antioxidant capacity in the studies of Vallée et al. [45] and Igual et al. [46], when compared to the control samples.

The impact of the M, V, and P variables was computed using Yoon's local sensitivity formula, and the relative influence of these variables on the total phenolic content and antioxidant capacity is shown in Figure 3.

Increased moisture negatively affected the content of polyphenolic compounds (Figure 3). High moisture content possibly favors the decarboxylation of phenolic acids [46]. Moreover, the polymerization of polyphenolic compounds can occur during extrusion, reducing its extractability and, consequently, the antioxidant activity [47]. Similar observations were noted by Yağci and Göğüş [48], where the content of polyphenolic compounds decreased with the increasing moisture content of corn-based extrudates enriched with broccoli flour.

The increase in the screw speed raised the content of PPs (Figure 3), probably due to the release of polyphenolic compounds from the matrixes [49]. Furthermore, high screw speeds (400 rpm) may accelerate the formation of products of non-enzymatic browning of the phenolic structure [50]. Moreover, the increase in screw speed could shorten the time of exposure of polyphenolic compounds to thermal destruction, due to which they remain largely preserved, as noted by Natabirwa et al. [51].

Chicory root supplementation contributed to an increase in polyphenolic compounds (Figure 3) due to the richness of chicory root in polyphenolic compounds [5]. High positive correlations were determined between chicory root and polyphenolic contents (r = 0.92 for Free PPs and r = 0.86 for Bound PPs, $p < 0.05$). The control sample made from pure rice flour was indigent in polyphenolic content due to the loss of these compounds during rice milling [52].

The increase in the screw speed and chicory root content had a positive effect on the antioxidant and reduction capacity of the extruded products (Figure 3). Such extrusion conditions increased the polyphenolic content of the extrudates, which are directly related to the antioxidative activity and reducing capacity, while moisture showed the opposite effect [53].

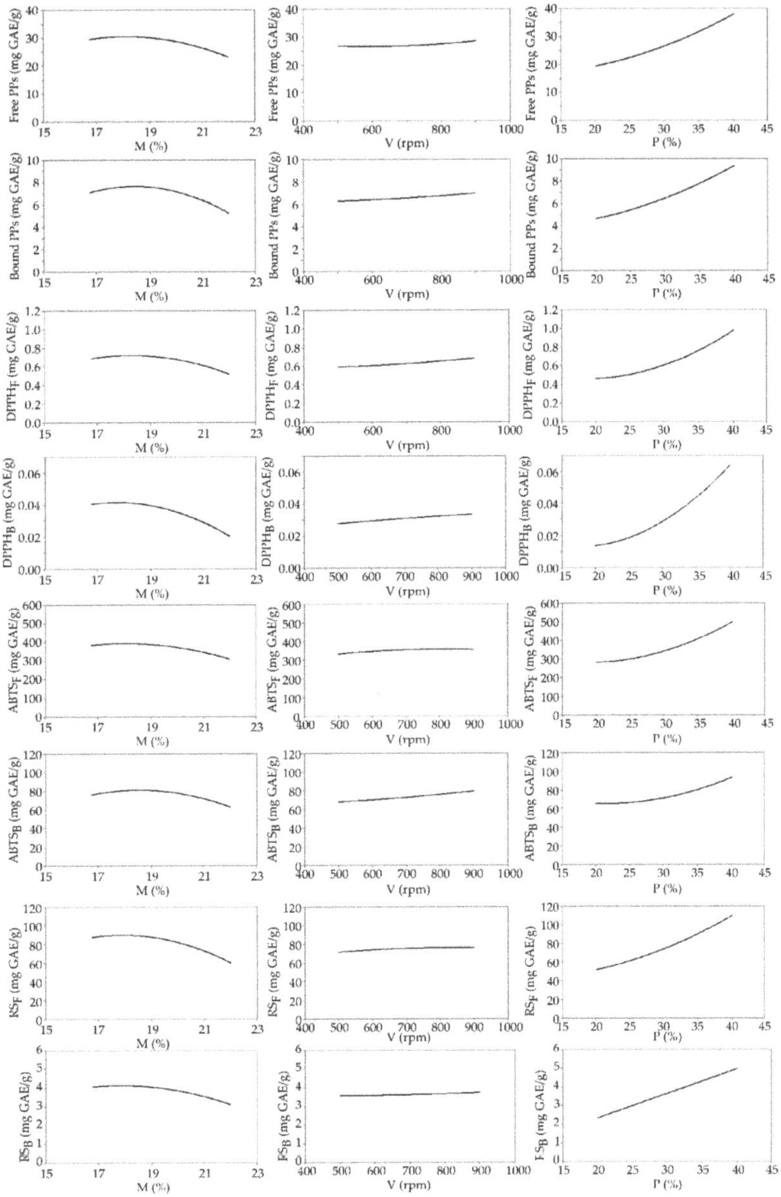

Figure 3. Influence of extrusion conditions (M—moisture content; V—screw speed; P—chicory root content) on polyphenols (PPs) content and antioxidative activity. Index F represent fraction of free polyphenols, and index B represents fraction of bound polyphenols.

3.4. Mineral Elements

Chicory root might be a relevant ingredient in foods due to the notable presence of minerals. These elements have been considered important traits for a balanced diet [5]. Chicory root proved to be rich in K, Na, Ca, and Mg, while significant amounts of Fe, Zn, Mg, and Cu were also noted (Table 3). These results are in agreement with those reported by Nwafor et al. [44] and Zarroug et al. [54].

Table 3. Mineral contents in extrudates and control sample (CS).

S.No.	CCD Design M, %	CCD Design V, rpm	CCD Design P, %	Ca	K	Mg	Na	Fe	Mn	Zn	Cu
1	21.2	820	35.9	639.29 ± 1.32	2242.72 ± 4.32 ^{f}	501.48 ± 0.98 ^{de}	2076.30 ± 3.03 ^{j}	54.13 ± 0.26 ^{i}	7.36 ± 0.06 ^{de}	15.75 ± 0.22 ^{i}	3.75 ± 0.09 ^{f}
2	19.4	700	30.0	534.07 ± 1.04	2292.48 ± 3.13 ^{h}	512.78 ± 0.83 ^{fg}	1562.49 ± 2.76 ^{e}	33.63 ± 0.13 ^{e}	6.91 ± 0.11 ^{bcde}	14.77 ± 0.37 ^{ghi}	2.83 ± 0.03 ^{bcde}
3	17.6	820	35.9	427.90 ± 1.12	2159.61 ± 2.59 ^{d}	514.25 ± 1.01 ^{fg}	1358.71 ± 2.02 ^{d}	28.41 ± 0.11 ^{c}	6.40 ± 0.09 ^{bc}	15.59 ± 0.45 ^{i}	2.27 ± 0.11 ^{bc}
4	19.4	700	30.0	535.51 ± 2.01	2290.68 ± 3.24 ^{h}	539.55 ± 0.76 ^{fg}	1561.92 ± 2.14 ^{e}	35.25 ± 0.27 ^{e}	6.93 ± 0.09 ^{bcde}	14.76 ± 0.51 ^{ghi}	2.95 ± 0.09 ^{bcde}
5	21.2	820	24.1	483.27 ± 1.00	2214.95 ± 2.11 ^{e}	534.14 ± 1.12 ^{ij}	1398.47 ± 1.98 ^{d}	31.04 ± 0.31 ^{d}	6.90 ± 0.07 ^{bcde}	11.21 ± 0.20 ^{bc}	2.26 ± 0.21 ^{bc}
6	22.5	700	30.0	556.71 ± 0.98	2544.48 ± 2.27 ^{j}	545.94 ± 1.22 ^{k}	1700.58 ± 2.04 ^{fg}	34.47 ± 0.44 ^{e}	6.53 ± 0.10 ^{bcd}	15.60 ± 0.11 ^{i}	2.78 ± 0.17 ^{bcde}
7	19.4	900	30.0	533.90 ± 1.03	2565.08 ± 3.01 ^{k}	533.04 ± 0.99 ^{jk}	1702.68 ± 2.15 ^{fg}	35.18 ± 0.34 ^{e}	6.84 ± 0.12 ^{bcde}	15.48 ± 0.27 ^{hi}	3.15 ± 0.23 ^{def}
8	17.6	820	24.1	417.43 ± 1.15	2070.44 ± 1.09 ^{c}	487.20 ± 0.73 ^{c}	1242.16 ± 1.72 ^{c}	26.05 ± 0.25 ^{bc}	6.07 ± 0.07 ^{b}	14.50 ± 0.35 ^{gh}	2.29 ± 0.31 ^{b}
9	19.4	700	30.0	531.03 ± 1.02	2293.79 ± 1.56 ^{h}	508.63 ± 1.01 ^{fg}	1560.18 ± 1.84 ^{e}	36.34 ± 0.61 ^{e}	6.91 ± 0.09 ^{bcde}	15.01 ± 0.19 ^{ghi}	2.88 ± 0.14 ^{bcde}
10	21.2	580	24.1	482.28 ± 0.89	2240.59 ± 1.48 ^{f}	510.00 ± 0.94 ^{ef}	1729.49 ± 2.01 ^{g}	26.25 ± 0.23 ^{bc}	6.43 ± 0.10 ^{bc}	10.72 ± 0.29 ^{b}	3.07 ± 0.19 ^{def}
11	16.3	700	30.0	563.83 ± 1.17	2526.92 ± 3.22 ^{i}	534.87 ± 1.04 ^{ij}	1693.07 ± 1.87 ^{fg}	36.02 ± 0.56 ^{e}	6.44 ± 0.11 ^{bc}	11.32 ± 0.35 ^{bc}	2.63 ± 0.09 ^{bcd}
12	19.4	700	40.0	713.51 ± 2.09	3018.60 ± 4.02 ^{n}	526.52 ± 0.73 ^{hi}	2074.80 ± 2.90 ^{j}	51.11 ± 0.72 ^{h}	7.32 ± 0.07 ^{e}	13.30 ± 0.27 ^{ef}	3.75 ± 0.07 ^{f}
13	19.4	500	30.0	555.55 ± 1.23	2525.85 ± 3.12 ^{i}	564.65 ± 1.12 ^{l}	1668.48 ± 1.78 ^{f}	36.62 ± 0.45 ^{e}	6.59 ± 0.12 ^{bcde}	12.48 ± 0.41 ^{de}	2.96 ± 0.08 ^{cdef}
14	19.4	700	30.0	532.15 ± 1.11	2285.53 ± 2.12 ^{h}	552.90 ± 1.01 ^{fg}	1560.16 ± 1.09 ^{e}	36.77 ± 0.37 ^{e}	6.90 ± 0.15 ^{bcde}	14.91 ± 0.61 ^{ghi}	2.86 ± 0.11 ^{bcde}
15	17.6	580	24.1	472.51 ± 0.99	2260.06 ± 1.87 ^{g}	522.93 ± 0.95 ^{gh}	1381.76 ± 1.23 ^{d}	24.21 ± 0.22 ^{b}	6.87 ± 0.07 ^{bcde}	12.16 ± 0.37 ^{cd}	2.43 ± 0.12 ^{bcd}
16	21.2	580	35.9	644.01 ± 2.01	2777.22 ± 2.35 ^{l}	493.60 ± 0.84 ^{cd}	1899.85 ± 2.34 ^{h}	46.03 ± 0.48 ^{g}	7.09 ± 0.09 ^{cde}	14.12 ± 0.59 ^{fg}	3.76 ± 0.21 ^{f}
17	17.6	580	35.9	643.34 ± 1.92	2824.91 ± 2.47 ^{m}	507.18 ± 1.01 ^{ef}	2027.11 ± 2.76 ^{i}	43.99 ± 0.58 ^{g}	7.29 ± 0.11 ^{de}	14.20 ± 0.63 ^{fg}	3.49 ± 0.31 ^{ef}
18	19.4	700	20.0	385.94 ± 1.35	1835.95 ± 1.03 ^{b}	427.24 ± 0.78 ^{b}	1145.93 ± 1.03 ^{b}	41.78 ± 0.33 ^{f}	6.81 ± 0.08 ^{bcde}	12.60 ± 0.43 ^{de}	3.08 ± 0.19 ^{def}
19	19.4	700	30.0	532.69 ± 1.22	2295.36 ± 2.25 ^{h}	506.46 ± 1.17 ^{fg}	1560.14 ± 1.54 ^{e}	35.45 ± 0.28 ^{e}	6.92 ± 0.07 ^{bcde}	14.68 ± 0.52 ^{ghi}	2.82 ± 0.11 ^{bcde}
20	19.4	700	30.0	533.35 ± 1.24	2293.32 ± 2.11 ^{h}	528.88 ± 0.72 ^{fg}	1562.28 ± 1.86 ^{e}	35.14 ± 0.31 ^{e}	6.93 ± 0.08 ^{bcde}	14.47 ± 0.47 ^{ghi}	2.98 ± 0.10 ^{bcde}
CS	18.0	800	00.0	43.82 ± 0.26 ^{a}	934.98 ± 0.89 ^{a}	380.18 ± 0.45 ^{a}	56.10 ± 0.37 ^{a}	3.42 ± 0.41 ^{a}	4.79 ± 0.07 ^{a}	9.91 ± 0.06 ^{a}	1.08 ± 0.05 ^{a}
CV				0.29	0.15	3.58	0.07	3.09	0.18	1.27	2.25

CCD—central composite design; S.no.—Sample number; M—moisture content; SS—screw speed; P—chicory root flour content; CS—control sample; CV—coefficient of variation for six central points (samples 2, 4, 9, 14, 19, and 20). Values in the same column marked with different letters were statistically significantly ($p < 0.05$) different (Tukey HSD test).

The mineral contents in the obtained extrudates ranged within the following limits: 385.94–713.51 mg/kg for Ca, 1835.95–3018.60 mg/kg for K, 427.24–564.56 mg/kg for Mg, 1145.93–2076.30 mg/kg for Na, 24.21–54.13 mg/kg for Fe, 6.07–7.36 mg/kg for Mn, 12.16–15.75 mg/ kg for Zn, and 2.26–3.76 mg/kg for Cu. The achieved values of minerals are close to those corresponding to the recommended daily allowance (RDA) of 1000 mg for Ca, 3500 mg for K, 350 mg for Mg, 2400 mg for Na, 15 mg for Fe, 5 mg for Mn, 15 mg for Zn, and 2 mg for Cu [55].

The influence of the M, V, and P variables was calculated utilizing Yoon's local sensitivity formula, and the relative influence of these variables on the mineral content is shown in Figure 4.

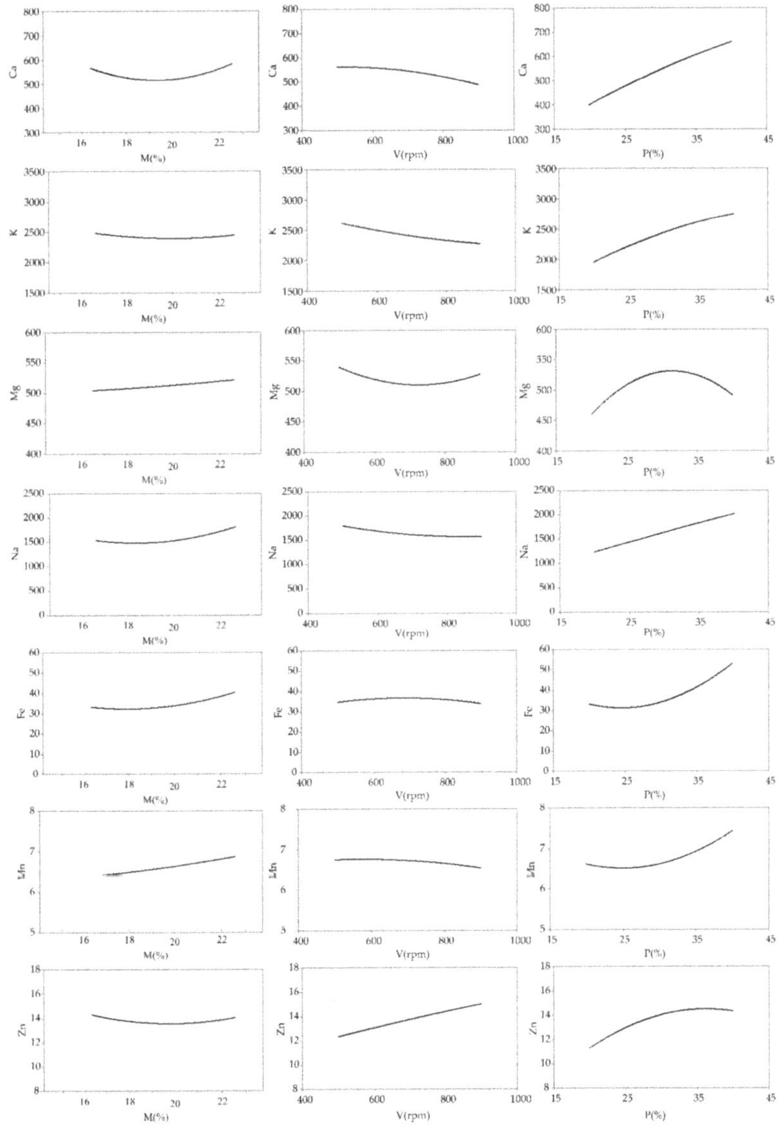

Figure 4. Influence of extrusion conditions (M—moisture content; V—screw speed; P—chicory root content) on mineral content.

Moisture increase had a positive effect on the mineral content (Figure 4). Similar observations were made by Danbaba et al. [56], who investigated the mineral content of rice extrudates supplemented with peas. The increase in mineral content in the extrudates might be due to their accumulation in the water [57].

Increasing the screw speed negatively affected the mineral content, except Zn (Figure 4). Increased screw speed can promote the binding of zinc ions with fibers (such as inulin) more efficiently, which might preserve this element during extrusion [58]. Screw speeds above 700 rpm resulted in increased Mg (Figure 4), which is potentially a consequence of the destruction of anti-nutritive compounds (such as tannins and phytates that build insoluble complexes with minerals), which favorably affects the availability of minerals [59].

Chicory root improved the mineral content of the extrudates when compared to the control sample (Figure 4). This observation might be related to the richness of chicory roots in minerals [5]. Positive correlations were noted between chicory root addition and all of the investigated minerals (r = 0.64 to 0.92, $p < 0.05$). The decreased content of Mg after the addition of more than 30% chicory root may be related to fibers (inulin), which might act as chelating agents, interfering with the extraction of Mg [60].

3.5. *Artificial Neural Network*

The optimal numbers of neurons in the hidden layers were chosen according to ANN performance (network MLP) with the aim of reaching high values of R^2. The quality of the ANN models was tested through the coefficient of determination (R^2), which should be close to 1 [61].

The optimal number of neurons in the hidden layer for inulin content prediction was four (MLP 3-4-1 network), according to the highest value of R2 (0.97). The optimal network was constructed using the BFGS 1000 algorithm, while the activation functions in hidden and output layers were exponential and identical functions [23–26]. A similar R2 value (up to 0.94) of the ANN model was noted in the study of Pandiselvam et al. [62] during the analysis of the extrusion conditions (screw speed, barrel temperature, and formulation) on different product characteristics.

The optimal number of neurons in the hidden layer for antioxidant activity prediction was three (MLP 3-3-8 network), according to the high values of R2 (0.87, 0.81, 0.85, 0.82, 0.91, 0.66, 0.83, and 0.78, for the prediction of Free PPs, Bound PPs, DDPHF, DDPHB, ABTSF, ABTSB, RCF, and RCB, respectively). The optimal network was constructed using the BFGS 57 algorithm, while the activation functions in hidden and output layers were hyperbolic tangents and identical functions [23–26]. The R2 values obtained within this study follow the results of Pandey et al., who investigated the antioxidant activity and the total phenolic contents of raw banana and defatted soy composite extrudates (R2 = 0.99 and R2 = 0.96, respectively) [63].

The optimal number of neurons in the hidden layer for sesquiterpene lactones prediction was nine (MLP 3-9-4 network), according to the high values of R2 (1.00, 0.99, 0.99, and 0.99, for prediction of lactucin, lactucopyrin, dihydrolactucin, and dihydorolactucopyrin, respectively). The optimal network was constructed using the BFGS 771 algorithm, while the activation functions in hidden and output layers were hyperbolic tangents and identical functions [23–26]. This study represents one of the first investigations in the ANN prediction of the SLs content in extruded food, while high R2 values indicate the excellent predictive performance of the ANN model.

The optimal number of neurons in the hidden layer for mineral content prediction was nine (MLP 3-9-8 network), according to the high values of R2 (0.99, 0.99, 0.92, 0.99, 0.99, 0.95, 0.99, and 0.99, for the prediction of Ca, K, Mg, Na, Fe, Mn, Zn, and Cu content, respectively). The optimal network was constructed using the BFGS 1000 algorithm, while the activation functions in hidden and output layers were hyperbolic tangents and identical functions [23–26]. Kothakota et al. noted R2 up to 0.99 during the ANN prediction of mineral content (Ca, P, and Fe) in enzymatic milled rice [64].

4. Conclusions

The extrusion process conditions (moisture content, screw speed, and chicory root content) affected the bioactive profile and antioxidant activity of rice-based chicory-enriched snacks.

Increased moisture content contributed to an increase in SLs and minerals, while high screw speed enhanced the polyphenol content. Chicory root addition contributed to the improved bioactive features of the obtained snacks, compared to the control sample in terms of inulin, SLs, and polyphenols, as well as antioxidant activities. Mineral content was also positively influenced by chicory root addition.

The achieved results presented the important impact of extrusion on the bioactive profile of the obtained snacks and promoted chicory root as an attractive food ingredient in terms of functionality. Furthermore, the consumers' acceptability of novel gluten-free chicory snacks will be the goal of future investigation.

Author Contributions: Conceptualization, J.B., J.K. (Jovana Kojić); methodology, V.T.Š., J.B., J.K. (Jovana Kojić) and J.K. (Jelena Krulj); formal analysis, J.B., V.T. and D.A.M.; investigation, J.B., V.B.; data curation, L.P.; writing—original draft, J.B., V.T., L.P.; writing—review and editing, J.K. (Jovana Kojić), J.K. (Jelena Krulj), L.P., V.T., V.B., D.A.M.; visualization, V.T.Š., L.P., M.B.-S.; All authors have read and agreed to the published version of the manuscript.

Funding: The study was funded by the Ministry of Education, Science and Technological Development of the Republic of Serbia (451-03-68/2022-14/200222).

Data Availability Statement: Authors can confirm that all relevant data are included in the article and/or its Supplementary Material files. The datasets generated during and/or analyzed during the current study are available from the corresponding author upon reasonable request.

Acknowledgments: The authors are grateful to Strahinja Vidosavljević and Viktor Stojkov (Institute of Food Technology in Novi Sad, Novi Sad, Serbia) for technical assistance.

Conflicts of Interest: The authors declare no conflict of interest.

References

1. Výrostková, J.; Regecová, I.; Zigo, F.; Marcinčák, S.; Kožárová, I.; Kováčová, M.; Bertová, D. Detection of Gluten in Gluten-Free Foods of Plant Origin. *Foods* **2022**, *11*, 2011. [CrossRef]
2. Taylor, J.R.N.; Taylor, J.; Campanella, O.H.; Hamaker, B.R. Functionality of the storage proteins in gluten-free cereals and pseudocereals in dough systems. *J. Cereal Sci.* **2016**, *67*, 22–34. [CrossRef]
3. Torbica, A.; Hadnadev, M.; Dapčević, T. Rheological, textural and sensory properties of gluten-free bread formulations based on rice and buckwheat flour. *Food Hydrocoll.* **2010**, *24*, 626–632. [CrossRef]
4. Arribas, C.; Cabellos, B.; Cuadrado, C.; Guillamón, E.; Pedrosa, M.M. Bioactive compounds, antioxidant activity, and sensory analysis of rice-based extruded snacks-like fortified with bean and carob fruit flours. *Foods* **2019**, *8*, 381. [CrossRef] [PubMed]
5. Perović, J.; Tumbas Šaponjac, V.; Kojić, J.; Krulj, J.; Moreno, D.A.; García-Viguera, C.; Bodroža-Solarov, M.; Ilić, N. Chicory (*Cichorium intybus* L.) as a food ingredient—Nutritional composition, bioactivity, safety, and health claims: A review. *Food Chem.* **2021**, *336*, 127676. [CrossRef] [PubMed]
6. Pouille, C.L.; Jegou, D.; Dugardin, C.; Cudennec, B.; Ravallec, R.; Hance, P.; Rambaud, C.; Hilbert, J.L.; Lucau-Danila, A. Chicory root flour—A functional food with potential multiple health benefits evaluated in a mice model. *J. Funct. Foods* **2020**, *74*, 104174. [CrossRef]
7. Nishimura, M.; Ohkawara, T.; Kanayama, T.; Kitagawa, K.; Nishimura, H.; Nishihira, J. Effects of the extract from roasted chicory (*Cichorium intybus* L.) root containing inulin-type fructans on blood glucose, lipid metabolism, and fecal properties. *J. Tradit. Complement. Med.* **2015**, *5*, 161–167. [CrossRef]
8. Ripoll, C.; Schmidt, M.B.; Ilic, N.; Poulev, A.; Dey, M.; Kurmukov, G.A.; Raskin, I. Anty-inflammatory effects of a sesquiterpene lactone extract from chicory (*Cichorium intybus* L) roots. *Nat. Prod. Commun.* **2007**, *2*, 717–722.
9. Mihaylova, D.; Vrancheva, R.; Petkova, N.; Ognyanov, M.; Desseva, I.; Ivanov, I.; Popova, M.; Popova, A. Carotenoids, tocopherols, organic acids, charbohydrate and mineral content in different medicinal plant extracts. *Z. Naturforsch.-C J. Biosci.* **2018**, *73*, 439–448. [CrossRef]
10. Harrington, K.C.; Thatcher, A.; Kemp, P.D. Mineral composition and nutritive value of some common pasture weeds. *N. Z. Plant Prot.* **2006**, *59*, 261–265. [CrossRef]
11. Patil, S.S.; Kaur, C. Current trends in extrusion: Development of functional foods and novel ingredients. *Food Sci. Technol. Res.* **2018**, *24*, 23–34. [CrossRef]

12. Morales, P.; Berrios, J.D.J.; Varela, A.; Burbano, C.; Cuadrado, C.; Muzquiz, M.; Pedrosa, M.M. Novel fiber-rich lentil flours as snack-type functional foods: An extrusion cooking effect on bioactive compounds. *Food Funct.* **2015**, *6*, 3135–3143. [CrossRef] [PubMed]
13. Dilrukshi, H.N.N.; Torrico, D.D.; Brennan, M.A.; Brennan, C.S. Effects of extrusion processing on the bioactive constituents, in vitro digestibility, amino acid composition, and antioxidant potential of novel gluten-free extruded snacks fortified with cowpea and whey protein concentrate. *Food Chem.* **2022**, *389*, 133107. [CrossRef] [PubMed]
14. Bokić, J.; Kojić, J.; Krulj, J.; Pezo, L.; Banjac, V.; Škrobot, D.; Tumbas Šaponjac, V.; Vidosavljević, S.; Stojkov, V.; Ilić, N.; et al. Development of a Novel Rice-Based Snack Enriched with Chicory Root: Physicochemical and Sensory Properties. *Foods* **2022**, *11*, 2393. [CrossRef]
15. Perović, J.; Kojić, J.; Krulj, J.; Pezo, L.; Tumbas Šaponjac, V.; Ilić, N.; Bodroža-Solarov, M. Inulin determination by an improved HPLC-ELSD method. *Food Anal. Methods* **2022**, *15*, 1001–1010. [CrossRef]
16. Schmidt, B.M.; Ilic, N.; Poulev, A.; Raskin, I. Toxicological evaluation of a chicory root extract. *Food Chem. Toxicol.* **2007**, *45*, 1131–1139. [CrossRef]
17. *SRPS EN ISO 6869/2008*; Animal feeding stuffs—Determination of the Contents of Calcium, Copper, Iron, Magnesium, Manganese, Potassium, Sodium and Zinc—Method Using Atomic Absorption Spectrometry. Institute for Standardization of Serbia: Belgrade, Serbia, 2008.
18. Singleton, L.V.; Rossi, A.J. Colorimetry of total phenolics with phospho¬molybdic-phosphotungstic acid reagents. *Am. J. Enol. Vitic.* **1965**, *16*, 144–158.
19. Gaonkar, A.G.; Vasisht, N.; Khare, A.R.; Sobel, R. *Microencapsulation in the Food Industry: A Practical Implementation Guide*; Elsevier: Amsterdam, The Netherlands, 2014; pp. 267–282.
20. Gironés-Vilaplana, A.; Mena, P.; Moreno, D.A.; García-Viguera, C. Evaluation of sensorial, phytochemical and biological properties of new isotonic beverages enriched with lemon and berries during shelf life. *J. Sci. Food Agric.* **2014**, *94*, 1090–1100. [CrossRef]
21. Mena, P.; Garcia-Viguera, C.; Navarro-Rico, J.; Moreno, A.D.; Bartual, J.; Saura, D.; Marti, N. Phytochemical characterisation for industrial use of pomegranate (*Punica granatum* L.) cultivars grown in Spain. *J. Sci. Food Agric.* **2011**, *91*, 1893–1906. [CrossRef]
22. Oyaizu, M. Studies on products of browning reactions: Antioxidative activities of product of browning reaction prepared from glucosamine. *Jpn. J. Nutr.* **1986**, *44*, 307–315. [CrossRef]
23. Kollo, T.; von Rosen, D. *Advanced Multivariate Statistics with Matrices*; Springer: Dordrecht, The Netherlands, 2005.
24. Trelea, I.C.; Raoult-Wack, A.L.; Trystram, G. Application of neural network modelling for the control of dewatering and impregnation soaking process (osmotic dehydration). *Food Sci. Technol. Int.* **1997**, *3*, 459–465. [CrossRef]
25. Mohieddin, J.; Yinyin, W.; Ali, A.; Jing, T. Unsupervised Learning and Multipartite Network Models: A Promising Approach for Understanding Traditional Medicine. *Front. Pharmacol.* **2020**, *11*, 1319.
26. Basheer, I.A.; Hajmeer, M. Artificial neural networks: Fundamentals, computing, design, and application. *J. Microbiol. Methods* **2000**, *43*, 3–31. [CrossRef]
27. Yoon, Y.; Swales, G.; Margavio, T.M. A Comparison of Discriminant Analysis versus Artificial Neural Networks. *J. Oper. Res. Soc.* **2017**, *44*, 51–60. [CrossRef]
28. Coussement, P.A.A. Inulin and oligofructose: Safe intakes and legal status. *J. Nutr.* **1999**, *129* (Suppl. 7), 1412–1417. [CrossRef]
29. Radovanovic, A.; Stojceska, V.; Plunkett, A.; Jankovic, S.; Milovanovic, D.; Cupara, S. The use of dry Jerusalem artichoke as a functional nutrient in developing extruded food with low glycaemic index. *Food Chem.* **2015**, *177*, 81–88. [CrossRef] [PubMed]
30. Chaito, C.; Judprasong, K.; Puwastien, P. Inulin content of fortified food products in Thailand. *Food Chem.* **2016**, *193*, 102–105. [CrossRef]
31. Khuenpet, K.; Jittanit, W.; Sirisansaneeyakul, S.; Srichamnong, W. Inulin powder production from Jerusalem artichoke (Helianthus tuberosus L.) tuber powder and its application to commercial food products. *J. Food Process. Preserv.* **2017**, *41*, e13097. [CrossRef]
32. Ferreira, S.M.; Capriles, V.D.; Conti-Silva, A.C. Breakfast cereals with inulin obtained through thermoplastic extrusion: Chemical characteristics and physical and technological properties. *LWT* **2020**, *137*, 110390. [CrossRef]
33. Sharma, P.; Gujral, H.S. Extrusion of hulled barley affecting β-glucan and properties of extrudates. *Food Bioproc. Technol.* **2013**, *6*, 1374–1389. [CrossRef]
34. Katsavou, I.D.; Tsokolar-Tsikopoulos, K.C.; Eleni, P.N.; Krokida, M.K. Sensorial, functional, optical and thermal properties of inulin enriched expanded products. *Int. Food Res. J.* **2019**, *26*, 671–687.
35. Tsokolar-Tsikopoulos, K.C.; Katsavou, I.D.; Krokida, M.K. The effect of inulin addition on structural and textural properties of extruded products under several extrusion conditions. *J. Food Sci. Technol.* **2015**, *52*, 6170–6181. [CrossRef] [PubMed]
36. Singh, R.S.; Singh, T.; Larroche, C. Biotechnological applications of inulin-rich feedstocks. *Bioresour. Technol.* **2019**, *273*, 641–653. [CrossRef]
37. Weng, H.; He, L.; Zheng, J.; Li, Q.; Liu, X.; Wang, D. Low oral bioavailability and partial gut microbiotic and phase ii metabolism of brussels/witloof chicory sesquiterpene lactones in healthy humans. *Nutrients* **2020**, *12*, 3675. [CrossRef] [PubMed]
38. Kulkarni, C.; Kelly, A.L.; Gough, T.; Jadhav, V.; Singh, K.K.; Paradkar, A. Application of hot melt extrusion for improving bioavailability of artemisinin a thermolabile drug. *Drug Dev. Ind. Pharm.* **2017**, *44*, 206–214. [CrossRef]
39. Aberham, A.; Cicek, S.S.; Schneider, P.; Stuppner, H. Analysis of sesquiterpene lactones, lignans, and flavonoids in wormwood (*Artemisia absinthium* L.) using high-performance liquid chromatography (HPLC)-mass spectrometry, reversed phase HPLC, and HPLC-Solid phase extraction-nuclear magnetic resonance. *J. Agric. Food Chem.* **2010**, *58*, 10817–10823. [CrossRef]

40. Willeman, H.; Hance, P.; Fertin, A.; Voedts, N.; Duhal, N.; Goossens, J.; Hilbert, J. A method for the simultaneous determination of chlorogenic acid and sesquiterpene lactone content in industrial chicory root foodstuffs. *Sci. World J.* **2014**, *2014*, 583180. [CrossRef] [PubMed]
41. Jandrić, Z.; Cannavan, A. An investigative study on differentiation of citrus fruit/fruit juices by UPLC-QToF MS and chemometrics. *Food Control* **2017**, *72*, 173–180. [CrossRef]
42. Nayak, B.; Liu, R.H.; Berrios, J.D.J.; Tang, J.; Derito, C. Bioactivity of Antioxidants in Extruded Products Prepared from Purple Potato and Dry Pea Flours. *J. Agric. Food Chem.* **2011**, *59*, 8233–8243. [CrossRef]
43. dos Santos D'Almeida, C.T.; Mameri, H.; dos Santos Menezes, N.; de Carvalho, C.W.P.; Queiroz, V.A.V.; Cameron, L.C.; Morel, M.H.; Takeiti, C.Y.; Ferreira, M.S.L. Effect of extrusion and turmeric addition on phenolic compounds and kafirin properties in tannin and tannin-free sorghum. *Food Res. Int.* **2021**, *149*, 110663. [CrossRef]
44. Nwafor, I.C.; Shale, K.; Achilonu, M.C. Chemical composition and nutritive benefits of chicory (*Cichorium intybus*) as an ideal complementary and / or alternative livestock feed supplement. *Sci. World J.* **2017**, *2017*, 7343928. [CrossRef] [PubMed]
45. Vallée, M.; Lu, X.; Narciso, J.O.; Li, W.; Qin, Y.; Brennan, M.A.; Brennan, C.S. Physical, predictive glycaemic response and antioxidative properties of black ear mashroom (*Auricularia auricula*) extrudates. *Plant Foods Hum. Nutr.* **2017**, *72*, 301–307. [CrossRef] [PubMed]
46. Igual, M.; Chiş, M.S.; Socaci, S.A.; Vodnar, D.C.; Ranga, F.; Martínez-Monzó, J.; García-Segovia, P. Effect of Medicago sativa Addition on Physicochemical, Nutritional and Functional Characteristics of Corn Extrudates. *Foods* **2021**, *10*, 928. [CrossRef]
47. Brennan, C.; Brennan, M.; Derbyshire, E.; Tiwari, B.K. Effects of extrusion on the polyphenols, vitamins and antioxidant activity of foods. *Trends Food Sci. Technol.* **2011**, *22*, 570–575. [CrossRef]
48. Yağci, S.; Göğüş, F. Response surface methodology for evaluation of physical and functional properties of extruded snack foods developed from food-by-products. *J. Food Eng.* **2008**, *86*, 122–132. [CrossRef]
49. Leonard, W.; Zhang, P.; Ying, D.; Fang, Z. Application of extrusion technology in plant food processing byproducts: An overview. *Compr. Rev. Food Sci. Food Saf.* **2020**, *19*, 218–246. [CrossRef]
50. Chalermchaiwat, P.; Jangchud, K.; Jangchud, A.; Charunuch, C.; Prinyawiwatkul, W. Antioxidant activity, free gamma-aminobutyric acid content, selected physical properties and consumer acceptance of germinated brown rice extrudates as affected by extrusion process. *LWT-Food Sci. Technol.* **2015**, *64*, 490–496. [CrossRef]
51. Natabirwa, H.; Nakimbugwe, D.; Lungáho, M.; Muyonga, J.H. Optimization of Roba1 extrusion conditions and bean extrudate properties using response surface methodology and multi-response desirability function. *LWT* **2018**, *96*, 411–418. [CrossRef]
52. Boue, S.M.; Daigle, K.; Beaulieu, J.C.; Heiman, M. Rice flour and bran enriched with blueberry polyphenols increases storage stability and decreases arsenic content in bran. *Foods* **2019**, *8*, 276. [CrossRef]
53. Olszowy, M. What is responsible for antioxidant properties of polyphenolic compounds from plants? *Plant Physiol. Biochem.* **2019**, *144*, 135–143. [CrossRef] [PubMed]
54. Zarroug, Y.; Abdelkarim, A.; Dorra, S.T.; Hamdaoui, G.; Felah, M.E.L.; Hassouna, M. Biochemical characterization of Tunisian *Cichorium Intybus* L. roots and optimization of ultrasonic inulin extraction. *Mediter. J. Chem.* **2016**, *6*, 674–685. [CrossRef]
55. Lenntech, B.V. *Recommended Daily Intake of Vitamins and Minerals*; Lenntech: Rotterdamseweg, The Netherlands, 2013.
56. Danbaba, N.; Nkama, I.; Badau, M.H. Application of response surface methodology (RSM) and central composite design (CCD) to optimize minerals composition of rice-cowpea composite blends during extrusion cooking. *Int. J. Food Sci. Nut. Eng.* **2015**, *5*, 40–52. [CrossRef]
57. Camire, M.E.; Camire, A.; Krumhar, K. Chemical and nutritional changes in foods during extrusion. *Crit. Rev. Food Sci. Nutr.* **1990**, *29*, 35–57. [CrossRef] [PubMed]
58. Bergman, C.J.; Gualberto, D.G.; Weber, C.W. Mineral binding capacity of dephytinized insoluble fiber from extruded wheat, oat and rice brans. *Plant Foods Hum. Nutr.* **1997**, *51*, 295–310. [CrossRef]
59. Singh, S.; Gamlath, S.; Wakeling, L. Nutritional aspects of food extrusion: A review. *Int. J. Food Sci. Technol.* **2007**, *42*, 916–929. [CrossRef]
60. Irungu, F.G.; Mutungi, C.M.; Faraj, A.K.; Affognon, H.; Tanga, C.; Ekesi, S.; Nakimbugwe, D.; Fiaboe, K.K.M. Minerals content of extruded fish feeds containing cricket (*Acheta domesticus*) and black soldier fly larvae (*Hermetia illucens*) fractions. *Int. Aquat. Res.* **2018**, *10*, 101–113. [CrossRef]
61. Vakula, A.; Pavlić, B.; Pezo, L.; Tepić Horecki, A.; Daničić, T.; Raičević, L.; Ljubojević, M.; Šumić, Z. Vacuum drying of sweet cherry: Artificial neural networks approach in process optimization. *J. Food Process. Preserv.* **2020**, *44*, e14863. [CrossRef]
62. Pandiselvam, R.; Manikantan, M.R.; Sunoj, S.; Sreejith, S.; Beegum, S. Modeling of coconut milk residue incorporated rice-corn extrudates properties using multiple linear regression and artificial neural network. *J. Food Process Eng.* **2019**, *42*, e12981. [CrossRef]
63. Pandey, S.; Kumar, A.; Rao, P.S. Optimization, modeling, and characterization study for the physicochemical properties of raw banana and defatted soy composite extrudates. *Food Chem.* **2021**, *339*, 127865. [CrossRef]
64. Kothakota, A.; Pandiselvam, R.; Siliveru, K.; Pandey, J.P.; Sagarika, N.; Srinivas, C.H.S.; Kumar, A.; Singh, A.; Prakash, S.D. Modeling and optimization of process parameters for nutritional enhancement in enzymatic milled rice by multiple linear regression (MLR) and artificial neural network (ANN). *Foods* **2021**, *10*, 2975. [CrossRef]

foods

MDPI

Article

Development of High-Fibre and Low-FODMAP Crackers

Kristina Radoš, Nikolina Čukelj Mustač *, Katarina Varga, Saša Drakula, Bojana Voučko, Duška Ćurić and Dubravka Novotni

Faculty of Food Technology and Biotechnology, University of Zagreb, Pierottijeva 6, 10000 Zagreb, Croatia
* Correspondence: nikolina.cukelj@pbf.unizg.hr

Abstract: Since there are no products in the European market labelled as low-FODMAP (low in fermentable oligosaccharides, disaccharides, monosaccharides, and polyols), patients with irritable bowel syndrome and non-celiac wheat sensitivity often consume gluten-free products. These naturally contain little FODMAP, but have poorer sensory properties and lower nutritional value. This study aimed to develop sensory attractive crackers with high-fibre and low-FODMAP content. Various gluten-free flours (wholemeal buckwheat and millet, white maize), pumpkin seed meal, chia seeds, flax seeds, rice protein, sweet potato, sourdough, and spices were used to develop nine formulations. Using a nine-point hedonic scale and ranking test, four best-scored products were selected for which descriptive sensory analysis was performed and nutritional value and fructan content were determined. Crackers made from maize and millet flour mixtures (ratio 1:2.5) with sourdough and with chia or flax seed addition were rated highest for overall impression (8.2 and 7.0, respectively). Generally, high-fibre content, hardness, chewiness, dark colour, and bitterness lower the acceptability of crackers, but the addition of spices and sourdough can improve their acceptability and marketability. The crackers could be labelled as "gluten-free", "low-FODMAP" (<0.12 g/100 g), "naturally high-fibre" (7–10 g/100 g of which 17–23% are soluble), and "high in protein" (24–26 g/100 g).

Keywords: wholemeal formulations; low fructan; sensory analysis; ready-to-eat; oilseeds; spices; sourdough; gluten-free

Citation: Radoš, K.; Čukelj Mustač, N.; Varga, K.; Drakula, S.; Voučko, B.; Ćurić, D.; Novotni, D. Development of High-Fibre and Low-FODMAP Crackers. *Foods* **2022**, *11*, 2577. https://doi.org/10.3390/foods11172577

Academic Editors: Theodoros Varzakas and Maria Papageorgiou

Received: 25 July 2022
Accepted: 22 August 2022
Published: 25 August 2022

1. Introduction

Non-celiac wheat sensitivity (NCWS) and irritable bowel syndrome (IBS) are the growing metabolic disorders for which an adapted and individualised diet is usually recommended as an integral part of the treatment. Low fermentable oligosaccharide, disaccharide, monosaccharide, and polyol (FODMAP) diets show the best effect in alleviating the symptoms of IBS and NCWS. FODMAPs are important natural prebiotics and their intake affects the structure and function of the gut microbiota. However, they are also characterised by poor absorption in the small intestine, fermentability, and high osmotic activity, which can cause digestive problems in sensitive individuals. In fact, reducing FODMAPs is cited as more likely to improve symptoms in NCWS than eliminating gluten-containing products [1].

An important principle of the low-FODMAP diet is the replacement of intolerable high-FODMAP foods with nutritionally equivalent low-FODMAP foods. This is a major challenge as avoiding FODMAP can easily lead to low intake of dietary fibre and micronutrients, which, if not compensated by suitable alternative sources of fibre, can lead to undesirable changes in the gut microbiota [2]. Therefore, more attention should be paid to the development of tasty functional ready-to-eat products with balanced nutrient composition but with low-FODMAP content [3]. Ready-to-eat products, such as snacks, have become an important part of the diet in the fast-paced modern world. Commonly, cereal snacks are high in saturated fats, simple sugars, or sodium [4], while low in fibre. However, for people suffering from NCWS, adequate intake of fibre (21 to 38 g per day) is crucial as chronic constipation is a common symptom [5]. The use of wholemeal cereals, oilseeds, and by-products from oil processing holds great potential for the development of functional snacks. Pumpkin, flax, and chia seeds are characterised by high biological and nutritional value due to their high protein, fibre, and unsaturated fatty acid content, and are also a rich source of minerals, vitamins, carotenoids, and bioactive compounds [6]. Pumpkin seed meal is the main by-product in the production of pumpkin oil and is a source of proteins and unsaturated fatty acids, but it is also a rich treasury of carotenoids, bioactive compounds, vitamins, and minerals [7,8]. Wholemeal cereals contain a high amount of dietary fibre but can contain FODMAPs such as fructans including fructooligosaccharides (FOS) and galactooligosaccharides (GOS), which are mainly concentrated in the bran [9].

Thus, there are several strategies to produce low-FODMAP products. They can be produced either by selecting low-FODMAP ingredients in the product formulations or by using biotechnological tools to degrade FODMAP during the production process (e.g., sourdough, yeast fermentation, and the application of enzymes) [9,10]. Wholemeal flours from naturally gluten-free grains offer a high content of fibre while a limited amount of FODMAP [3]. The main benefits of using sourdough in wholemeal bakery products are flavour development together with extended shelf life, reduced glycaemic index, and increased bioavailability of mineral and bioactive compounds [11]. Moreover, Loponen and Gänzle [10] showed that the use of sourdough in the production of wholemeal bread significantly reduced the content of FODMAP in bread without affecting the content of slowly fermented and well-tolerated dietary fibre. Recently, Habuš et al. [9] showed that sourdough fermentation of amaranth and wheat bran decreased fructan content by up to 93%. Regardless of the method used, low-FODMAP alternatives should have the same or better nutritional and sensory quality than conventional food products. Thus, there is need to investigate these approaches in the development of a low-FODMAP snack.

In the development of a new product, consumer acceptance continues to be the most important factor, while sensory analysis is the gold standard in assessing product acceptance. Consumers are attracted to good sensory attributes, and therefore, sensory methods are an important and integral tool that should be used in the new product development process [12]. Sensory evaluation of a product makes it possible to obtain more information about the product being analysed, its quality, and to review the factors that influence its acceptance by consumers, which facilitates work on improving the quality of the product or its reformulation [13].

Currently, very few FODMAP-labelled products are available on the European market, due to the lack of definitions and EU legislation for FODMAPs. Low-FODMAP foods can be found among gluten-free products, but these are rarely sensory attractive and have low nutritional value [3]. A good example of FODMAP awareness can be found in Australia, where products are certified as low-FODMAP by Monash University and/or can go through the FODMAP Friendly Certification Programme (FODMAP everyday. Available online: https://www.fodmapeveryday.com/ accessed on 14 August 2022). The criteria that must be met for certification are not publicly available, and each product is evaluated individually. However, in general, products or formulations must not contain added FODMAPs, including fructo-oligosaccharides, inulin, and polyols (except sorbitol and mannitol), maltitol, xylitol, erythritol, lactitol, and isomalt. Among the certified products, there are only two cracker products labelled as "Low-FODMAP".

The aim of this study was to investigate the possibility of using different gluten-free wholemeal cereals, oilseeds, plant protein, sweet potato, sourdough, and spices in the development of nutritionally enhanced and sensory attractive crackers for consumers with IBS and NCWS. The first phase was the development of nine cracker formulations with high fibre but low fructan content; and the second phase involved selecting the four best rated crackers and evaluating them for their sensory description, nutritional value, and fructan content. The chemical composition of the developed crackers was compared with the chemical composition of two similar products available on the market.

2. Materials and Methods

2.1. *Raw Materials*

The cracker formulations contained combinations of different gluten-free raw materials, as these usually contain small amounts of FODMAP. The raw materials were white maize flour, wholemeal buckwheat, and millet flour (all three from Pukanić Mill Ltd., Velika Gorica, Croatia) (median particle size D(50) ~300 µm), sweet potato puree (dm-drogerie markt GmbH + Co. KG, Karlsruhe, Germany), flax seeds (SME Osijek, Osijek, Croatia), chia seeds (Nutrigold, EU), pumpkin seed meal (D(50) ~500 µm, PoljoPosavec Ltd., Dunjkovec, Croatia), and rice protein (Nutrigold, EU). The formulations also contained fat spread consisting of 59% vegetable oil (Omegol, Zvijezda Ltd., Zagreb, Croatia), olive oil (Olitalia, Forli, Italy), LIVENDO^{TM} LV1 starter (Lessafre Inc., Paris, France), spices (dill, pepper, tumeric, chives, wild garlic, Mediterranean seasoning mix, garlic, rosemary, chili, thyme, basil, and parsley), dry yeast (Lesaffre Adriatic Inc., Prigorje Brdovečko, Croatia), salt, and tap water. Raw materials were used without milling except chia and flax seeds which were ground in a coffee grinder until a particle size of <500 µm was achieved, and sieved using a laboratory rotary vibration sieve. Median particle size was determined by laser diffraction on a Mastersizer 2000 device (Malvern Panalytical, Malvern, UK) in combination with a dry dispersion unit (Scirocco 2000) at a pressure of p = 1 bar and a cell supply rate of 60%, in three replicates. The proximate composition taken from producers' labels of main ingredients is shown in Table 1. The proximate composition of our crackers was compared with commercially available certified low-FODMAP crackers (San-J Tamari Black Sesame and San-J Tamari Brown Sesame) (FODMAP everyday. Available online: https://www.fodmapeveryday.com/product-category/collections/low-fodmap-certified-brands/ (accessed on 14 August 2022)). The ingredients in the one of the commercial crackers were black sesame, potato starch, brown rice, soy sauce, water, dextrin and salt, while the ingredients in the other commercial cracker were sesame, brown rice, rice flour and sweet potato starch and soy sauce.

Table 1. Proximate composition of raw materials used in preparation of crackers (g/100 g).

	Chia Seeds	Pumpkin Seed Meal	Flax Seeds	White Maize Flour	Millet Flour	Buckwheat Flour	Rice Protein	Sweet Potato Puree
Proteins	20.0	59.4	23.8	6.3	10.0	10.8	83.0	1.0
Fats	31.0	15.8	26.6	0.9	3.3	2.9	4.5	1.9
Carbohydrates	6.3	nd	9.0	78.0	73.0	70.1	2.9	7.6
Salt	0.05	na	0.1	0.3	0.2	0.1	<0.05	0.03

nd—not detected; na—not available

2.2. *Methods*

2.2.1. Determination of Moisture, Ash, Carbohydrates, Fats, Proteins, and Fibre Contents

The chemical composition of the four best-rated crackers was determined by standard AACC methods [14] in two replicates. The moisture content was determined according to the AACC method 62-05; the ash content was determined by incineration at 550 °C according to the AACC method 08-01. The amount of protein was determined using the Kjeldahl method, according to the AACC method 46-12. The fat content was determined using the Soxhlet method according to the AACC method 30-25.01, and fatty acids were determined by gas chromatography of their methyl esters as previously described by Balbino et al. [15]. The share of total carbohydrates was determined by subtracting the mass of water, fat, protein, and ash from the tested amount of food (100 g). Dietary fibre (insoluble and soluble) content was determined using an enzymatic kit (Integrated Total Dietary Fibre Assay kit, Megazyme, Wicklow, Ireland). Since the manufacturer's labels did not provide information on the amount of soluble fibre or fibre at all, as well as information on ash and fructan content (important data in the development of products for IBS and NCWS patients), these were also determined in flours as well as in rice protein, chia seeds, flax seeds, and pumpkin seed meal. The energy values of crackers were calculated by multiplying the factor values, i.e., 1 g of crude protein or carbohydrate provides 4 kcal/16 kJ of energy, 1 g of crude fat provides 9 kcal/36 kJ of energy, and 1 g of fibre provides 2 kcal/8 kJ.

2.2.2. Determination of the FODMAP Content

Determination of fructan and GOS content in white maize flour, millet, buckwheat, rice protein, chia seeds, flax seeds, pumpkin seed meal, and four selected crackers was performed according to the AOAC method 999.03, as previously described by Habuš et al. [9], without adding enzyme α-galactosidase. A Fructan Assay Megazyme kit (Megazyme, Wicklow, Ireland) and Mega-CalcTM calculator were used to calculate the fructan content of the samples based on the measured absorbances at 410 nm (Analytik Jena, SPECORD 50 PLUS, Jena, Germany).

2.2.3. Development of Cracker Formulations

In the first phase, nine formulations of crackers (Table 2) were developed based on the proximate composition of raw materials. Maize, proso millet, and buckwheat flour were chosen as the main raw materials. We wanted to avoid using rice flour as a well-known and commonly used ingredient in gluten-free products, as can be seen in the example of commercially available low-FODMAP crackers. Based on our previous successful use of millet, corn, and buckwheat flours in bread development (to be published), the above flours were selected for the cracker formulations. The proportions of flours were determined by preliminary experiments, while other ingredients were dosed to meet nutritional recommendations. The formulations were developed to be high in fibre which, according to EU regulation 1169/2011/EU, meant that the product contained at least 6 g of fibre per 100 g, but also satisfied the needs of patients with IBS and NCWS, since the latest recommendations are to include 25–35 g fibre per day or ≥14.6 g/1000 kcal per day for adults [16].

Table 2. Nine gluten-free formulations as the percentage of raw materials.

		F1	F2	F3	F4	F5	F6	F7	F8	F9
Main flours (% of total flour)	Maize	70	70	70	70	22	/	/	/	/
	Proso millet	30	30	30	30	45	50	20	/	85
	Buckwheat	/	/	/	/	33	50	80	100	15
Additional ingredients (% of flour weight)	Water	100	100	73	73	100	80	80	100	75
	Fat spread	20	24	60	60	42	17	17	/	15
	Olive oil	/	/	/	/	/	/	/	25	/
	Rice protein	30	30	36	36	20	30	30	30	10
	Flax seeds	/	/	/	18		12		15	5
	Chia seeds	10	10	18	/	14	/	10	/	
	Pumpkin seed meal	/	/	/	/	20			/	40
	Sweet potato puree	/	/	35	35	/	20	15	30	30
	Salt	1.25	2	2.4	2.4	1.4	0.8	1	1	1
	Spices	/	/	Med. mix 1.2	Chives 2.4	Pepper; wild garlic 0.7; 0.7	Med. mix 0.8	Wild garlic; shallot 0.5; 0.5	Tumeric; dill 0.5; 2	/
	Yeast	/	2	/	/	/	/			/
Soudough (% of dough weight)		20	-	20	20	20	20	20	20	20

Med.mix—Mediterranean seasoning mix

2.2.4. Preparation of Crackers

Eight of the nine types of crackers included the sourdough fermentation. To make sourdough, 20 g of flour mix (depending on the recipe), 35 mL of room temperature tap water, and 0.5% (flour basis) of LV1 starter were used. After fermentation in a lidded jar at 30 °C for 16 h, sourdough was added to the main dough mixture considering the amount of flour and water contained in the sourdough. The cracker F2 was made without sourdough using the "all-in" method. The cracker dough was prepared using a kitchen robot (EKM4000, Electrolux, Stockholm, Sweden) according to the modified AACC 10-50D method [14]. The method was used as a guide for formulations of crackers, while the water was adjusted in preliminary tests, taking into consideration the workability and stickiness of the dough. First, fat and sugar were mixed in a kitchen blender (slowly, one minute). Then, depending on the recipe, rice proteins, soaked chia seeds (soaked in twice the amount of water for 30 min at room temperature), flax seeds, pumpkin seed meal, salt, and spices were added and mixed slowly for three minutes. The next step was to add the sourdough (if included) whose pH value had previously been measured, the remaining water, and the sweet potato puree (stirred slowly for one minute, and then quickly for one minute). Finally, the flour was added and the dough was mixed slowly for 10 min. The prepared dough rested in the refrigerator for 30 min. Then, the dough was then rolled out manually on a board with thickness spacers to a thickness of three millimetres, cut out with a 6 cm diameter cutter, and punched. The crackers were baked in a deck oven (EBO 64-320 IS 600, Wiesheu GmbH, Großbottwar, Germany) at a temperature of 180 °C for 10 min on one side and 10 min on the other side, after which the crackers were turned again and baked for another 10 min. After baking, the crackers were left to cool for 30 min at room temperature.

2.2.5. Sensory Analysis of Crackers

The sensory analysis of the crackers was carried out at the Faculty of Food Technology and Biotechnology at the University of Zagreb, Zagreb, Croatia. A panel of nine experienced assessors (aged 23 to 52 years) who specialised in cereal products was selected. The training programme for the sensory panel included the identification and description of attributes and the procedures using the response scale [17]. All of the samples were coded using three-digit numbers, and served with water to rinse the palate. First, a hedonic sensory analysis was performed, in which all of the nine samples were evaluated according to overall impression on a scale from 1 ("extremely dislike") to 9 ("extremely like"). Then, a ranking test was performed in which the subjects ranked the samples from the most preferred (score of 1) to the least preferred (score of 9). The samples with an average score of more than 5.5 were selected. In the next step, the intensity of sensory properties of the selected crackers was evaluated on a scale from 0 (not perceived) to 10 (very intense) by descriptive sensory analysis, performed according to the standards ISO 13299:2003 and 6658:2017 [18,19]. Prior to the assessment, the panel selected relevant sensory attributes (Table 3). The scoring included appearance, odour, taste, flavour, and texture parameters. This was followed by a hedonic sensory analysis of each sensory property of the four selected samples (appearance, odour, taste and flavour, texture in the mouth, overall impression).

Table 3. Selected sensory descriptors and their definition.

Sensory Attribute		Description
Appearance	Colour	Degree of brownness, ranging from light brown to dark brown
	Uniformity of surface	Uniform–non-uniform, smooth–rough
Odour	Overall	Overall intensity of odour
Taste and flavour	Bitter taste	Basic taste produced by caffeine
	Bitter aftertaste	Bitterness after chewing
	Overall	Overall intensity of taste and flavour
Texture in mouth	Hardness	Force applied by the molar teeth to compress the cracker
	Chewiness	Number of chews necessary for food to be swallowed
	Granularity	Sense of particle size and shape (larger particles–higher granularity)
	Dryness	Amount of saliva absorbed by sample crumbs during mastication
	Solubility	Chewing required until the biscuit disintegrates (more chewing–less solubility)
	Teeth adhesiveness	Ability of food to adhere to the teeth when chewed

2.3. Statistical Analysis

Microsoft Office Excel 2016 and Statistica v.10 (StatSoft Inc., Tulsa, OK, USA) were used for the statistical data processing. The results are expressed as the mean value with standard deviation. Statistica was used to detect statistically significant differences between samples for which analysis of variance (ANOVA) was performed with the Tukey post hoc test, with a value of $p < 0.05$ set as the limit of statistical significance.

3. Results and Discussion

3.1. Chemical Composition of Raw Materials and Fructan Content

Since there is a limited number of studies dealing with FODMAP content in wholemeal cereal and oilseed flours, we determined the content of fructan in the flours used (Table 4). White maize flour contained the least fructan, while fructan in buckwheat flour was below the detection limit. Knudsen et al. [20] determined the fructan content of whole maize flour to be 0.5 g/100 g sample dry matter. A low fructan content was also found in millet flour. This was consistent with the findings of Ispiryan et al. [3], who found that millet, buckwheat, and oats, which are common raw materials for gluten-free products, have fructan contents of less than 0.1 g/100 g sample dry matter, while their GOS content is low to moderate. Nevertheless, buckwheat, which is currently on the list of low-FODMAP cereals, contains the oligosaccharide phagopyritol, which is the most soluble carbohydrate. As its structure is similar to that of GOS, it potentially negatively affects the symptoms of IBS, which requires further research [3]. A low fructan content was determined in the rice protein, while the highest concentrations were found in pumpkin seed meal and flax seeds. The content of FODMAP in foods depends on the applied production process. For example, if soluble carbohydrates, which include GOS, are not removed during processing, the product will have a higher content of FODMAP [3]. Walnuts, peanuts, and pumpkin seeds, contrary to cashews and pistachios, are examples of nuts and seeds with low-FODMAP content [21]. The limit for oligosaccharides (total fructans and GOS) is 0.3 g in a standard serving of whole grain products, nuts, legumes, and seeds [22]. Although the FODMAP content in a particular food may be slightly higher, the total amount absorbed by the body is important and should not exceed the cut-off value of 0.3 g per serving (30 g snack = 1 serving). A serving of two tablespoons of chia seeds contains sufficiently low FODMAP content that it should be tolerated by most people with IBS [23]. In this research, flax seed has been found to contain about twice the fructose content of chia seeds. Thus, eating one tablespoon of flax seed is considered to be a low source of FODMAP, while larger amounts should be avoided [23].

Table 4. Fructan, fibre, and minerals content in raw materials (g/100 g of sample, $n = 2$, expressed on sample dry matter as mean ± standard deviation).

Raw Material	Fructan and Galactooligosaccharides	Total Dietary Fibre	Soluble Fibre	Insoluble Fibre	Minerals (As ash)
Buckwheat flour	nd	6.12 ± 0.53	1.55 ± 0.03	4.57 ± 0.12	3.01 ± 0.00
Millet flour	0.29 ± 0.01	14.90 ± 0.35	8.01 ± 0.14	6.89 ± 0.11	1.26 ± 0.02
White maize flour	0.03 ± 0.06	2.34 ± 0.15	1.09 ± 0.03	1.25 ± 0.03	0.97 ± 0.09
Rice protein	0.14 ± 0.00	3.78 ± 0.69	2.21 ± 0.48	1.57 ± 0.21	1.19 ± 0.07
Chia seeds	0.36 ± 0.00	37.90 ± 0.04	7.26 ± 0.00	30.64 ± 0.00	4.76 ± 0.00
Flax seeds	0.64 ± 0.22	27.88 ± 0.00	7.24 ± 0.00	20.64 ± 0.21	3.85 ± 0.01
Pumpkin seed meal	0.83 ± 0.03	14.80 ± 0.01	6.86 ± 0.02	7.94 ± 0.09	7.49 ± 0.06

nd—not detected

Total dietary fibre content in the raw materials used ranged from 2 to 38 g per 100 g of raw material (Table 4). Chia seeds, flax seeds, and pumpkin seed meal are rich sources of fibre with pumpkin seed meal having the highest proportion of soluble fibre among them (46%). Among cereal flours, millet flour was dominant in both total and soluble (50%) fibre content. The most recent clinical guidelines on the management of IBS consider soluble fibre as a reasonable first line therapy for IBS patients with symptoms. In contrast, products containing insoluble fibre, particularly wheat bran, do not appear to be useful in treating IBS symptoms [23]. Indeed, unlike insoluble, soluble fibre significantly improved the global assessment of IBS symptoms as compared with placebo.

According to Coskuner and Karababa [24], flax seeds on average contain 30–40% fat, 20–25% protein, 20–28% total fibre, 4–8% moisture content, and 3–4% ash. The chemical composition of flax seeds used in this study was similar to the above values. As expected, white maize flour was low in fibre, while rice protein was a source of fibre [25] Plant proteins are often used to increase the nutritional value of gluten-free products or to improve the rheological properties of the dough. Similarly, chia seeds improve the rheology of the gluten-free dough due to their high ability to absorb water, up to 15 times their mass, which leads to the formation of a gel [26].

The highest mineral content was found in pumpkin seed meal, followed by chia and flax seeds. Comparing the mineral contents of the flours, buckwheat flour is dominant in terms of minerals. Although the addition of buckwheat flour has a negative impact on the sensory acceptance of the product due to its characteristic taste, it contributes to the nutritional value of the product [27]. The lowest mineral content was found in white maize flour. In the gastrointestinal tract, vitamins and minerals are important for nutrient absorption, gut motility, modulation of the human gut microbiome, and other functions [28]. In research by Roth et al. [29], intake of micronutrients by IBS patients was lower than recommended, which was associated with gastrointestinal and extraintestinal symptoms, as well as with fatigue.

3.2. *Results of Sensory Analysis*

3.2.1. Acceptability and Preference of Nine Crackers

A hedonic sensory analysis and ranking test were used to determine the general acceptability and preference of crackers produced according to nine different formulations (Table 2). Only four crackers had hedonic scores higher than 5.5 (common cut-off point for product marketability [30]) and were chosen for further chemical and sensory analyses. Furthermore, crackers were targeted to contain about 15 g of dietary fibre per 1000 kcal, which was achieved for selected crackers. Crackers prepared according to F3 and F4 formulations achieved the highest average scores for overall impression (Figure 1). The cracker made according to the F3 formulation was "moderately liked", while cracker F4 was "liked very much". They were followed by crackers F5 and F8 "neither liked nor disliked". This neutral liking is probably related to the fact that crackers F5 and F8 contained buckwheat flour which has specific sensory properties. On the contrary, Sedej et al. [31] observed no significant differences in sensory quality of wholegrain buckwheat crackers as compared with wheat crackers. The overall impression of the other crackers was rated as "slightly dislike".

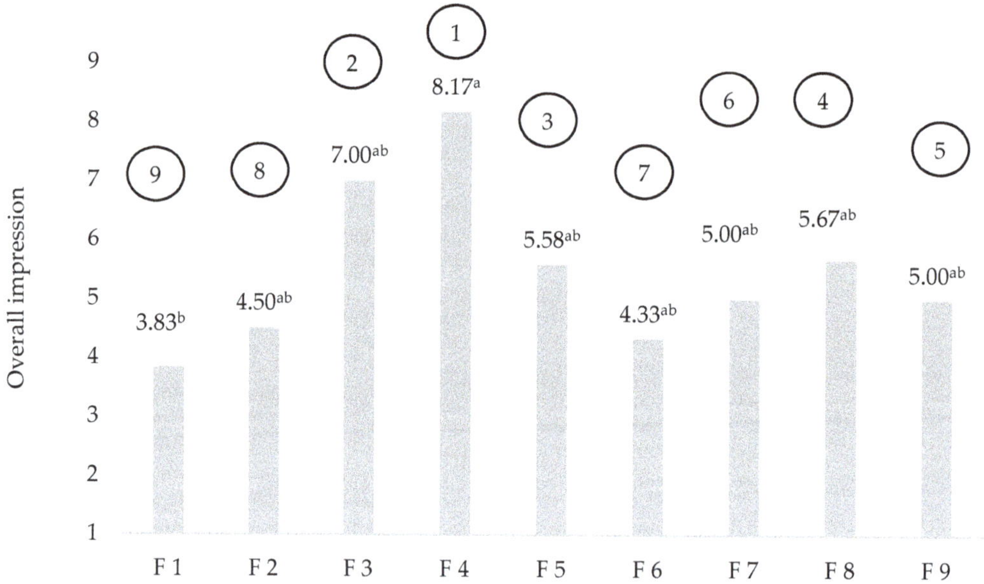

Figure 1. Results of the hedonic sensory analysis of the overall impression and the ranking test (in circles) of nine different cracker formulations (samples among the column marked with different letters are statistically significantly different ($p < 0.05$)).

The preference rank order is also presented in Figure 1 and is consistent with the results of the hedonic sensory analysis. In addition, a Friedman's ANOVA did not find statistically significant differences in the ranking of the samples (Q = 0.043, $p = 0.930$).

3.2.2. Sensory Profile of Four Best-Rated Crackers

The results of the descriptive and hedonic sensory analyses of four best-rated crackers (Figure 2) are shown in Figures 3 and 4, respectively.

Figure 2. The appearance of four best-rated crackers: (**a**) F3; (**b**) F4; (**c**) F5; (**d**) F8.

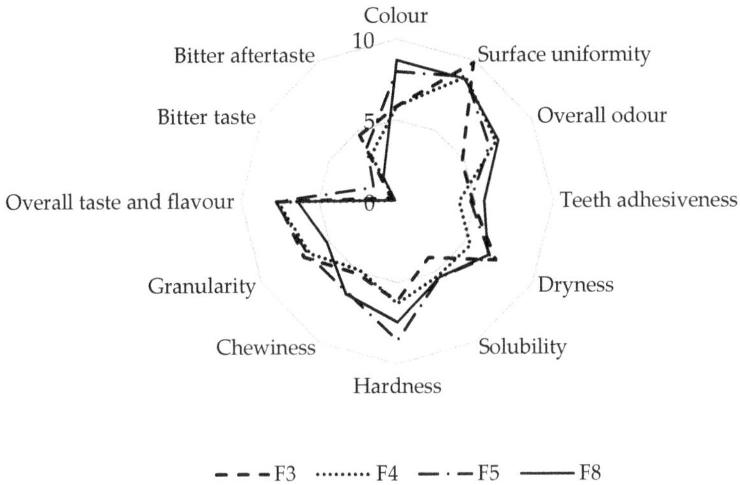

Figure 3. Intensity of sensory attributes of four best-rated crackers in the descriptive sensory analysis.

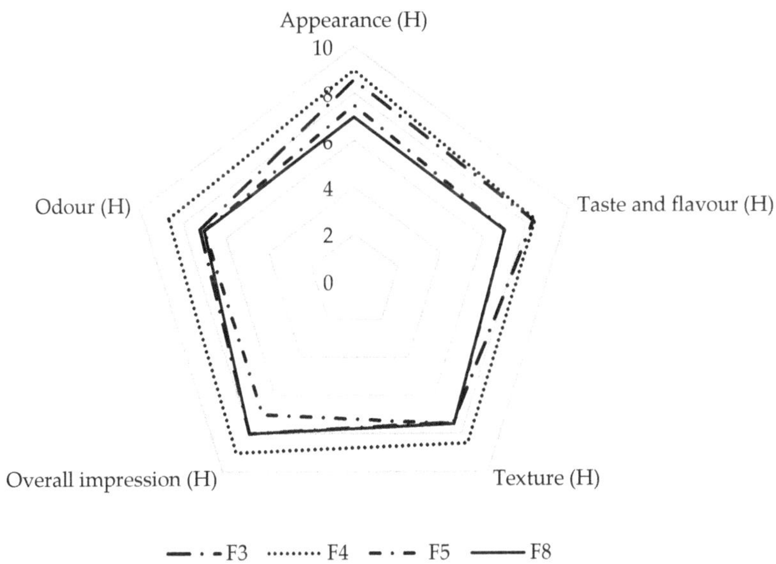

Figure 4. Acceptability of appearance, flavour, taste, odour, textural characteristics, and overall impression of four best-rated crackers in the hedonic sensory analysis.

There was a statistically significant difference ($p = 0.045$) between the colour intensity of crackers F5 and F8 with an average value $\geq$8 as compared with the lighter F3 and F4, which had an average value of 5.9 for colour (Figure 3). The surface uniformity was similarly high among crackers.

The odour intensity of the crackers was strongly influenced by the spices used. In cracker F3, Mediterranean seasoning mix (a combination of garlic, rosemary, chilli, thyme, basil, and parsley) was used which resulted in the lowest scores for odour intensity (average 4.8). Chives were used in cracker F4, pepper and wild garlic were used in cracker F5, and turmeric and dill were used in cracker F8; crackers F4, F5, and F8 were rated similarly for odour intensity (6.8–7.5) which was higher than cracker F3.

A bitter taste was not detected in the crackers during consumption, in contrast to a well detected bitter aftertaste (in the range 1.7–4.7) in all of the four crackers. No significant ($p > 0.05$) difference was found between the crackers in terms of intensity of overall taste and flavour, chewiness, granularity, solubility, dryness, and teeth adhesiveness.

Cracker F5 was rated as the hardest (mean score 8.6), significantly harder than crackers F3 and F4 ($p < 0.05$) (mean scores 6.1 and 6.3), but not significantly harder than cracker F8 (mean score 7.4) ($p > 0.05$). The fat content in cracker F5 was lower than the fat content in the other three types of crackers, which most likely caused their higher hardness (Table 5). In addition to physical and chemical properties, fat has a positive effect on product viability by delaying the absorption of water from starch [32]. Therefore, the texture stability of cracker F5 could be the lowest.

Table 5. Energy and nutritive value of the different types of crackers (g/100 g of sample, $n = 2$, expressed on sample dry matter as mean ± standard deviation).

	Cracker F3	Cracker F4	Cracker F5	Cracker F8
Energy (kJ/kcal)	1883/450	1866/446	1820/435	1871/447
Water content	4.86 ± 0.04 ^{c}	5.48 ± 0.05 ^{b}	4.31 ± 0.04 ^{d}	8.02 ± 0.05 ^{a}
Minerals (as ash)	2.81 ± 0.01 ^{c}	2.91 ± 0.01 ^{c}	5.59 ± 0.00 ^{b}	6.33 ± 0.00 ^{a}
Fats	17.72 ± 0.12 ^{a,b}	17.11 ± 0.62 ^{b,c}	14.98 ± 0.00 ^{c}	19.60 ± 0.10 ^{a}
of which saturated	1.37 ± 0.00	1.54 ± 0.00	1.66 ± 0.00	2.71 ± 0.00
Proteins	24.94 ± 0.14 ^{b,c}	25.97 ± 0.04 ^{a}	26.08 ± 0.04a ^{b}	24.02 ± 0.03 ^{c}
Carbohydrates	44.00 ± 0.28 ^{b}	43.51 ± 0.69 ^{b}	44.41 ± 2.67 ^{b}	38.83 ± 2.80 ^{a}
Dietary fibre	7.49 ± 0.21 ^{c}	7.12 ± 0.18 ^{d}	8.92 ± 0.58 ^{b}	9.53 ± 0.78 ^{a}
of which soluble	1.25 ± 0.13 ^{d}	1.65 ± 0.08 ^{c}	2.06 ± 0.78 ^{b}	2.34 ± 0.51 ^{a}
Fructan	0.10 ± 0.00 ^{a}	0.12 ± 0.00 ^{a}	0.12 ± 0.00 ^{a}	0.11 ± 0.00 ^{a}

Samples within the same row marked with different letters are statistically significantly different ($p < 0.05$)

The results of the hedonic sensory analysis of the individual properties of the four best-rated crackers are shown in Figure 4. Cracker F4 was best-liked in all examined properties (appearance, odour, taste and flavour, and texture, as well as overall), with average scores for each property above 8 ("like very much"), followed by cracker F3 with slightly lower average scores, but not significantly different from F4 ($p > 0.05$). The differences in the best-rated cracker F4 as compared with cracker F3 were the addition of flax seeds in the recipe and the use of chives, while in the cracker F3, chia seeds and Mediterranean seasoning mix was used.

Previous studies on the addition of flax seeds in biscuit production have shown that partial replacement of wheat flour with ground flax seeds (up to 12%) does not affect the physical and sensory properties of the product [33]. However, Čukelj et al. [6] achieved sensory acceptability of a multigrain flax seed biscuit similar to that of white wheat biscuits by using a suitable combination of cereal flour with ground flax seeds.

If we compare the results of the hedonic and descriptive sensory analyses, crackers F3 and F4 differed significantly only in intensity of overall odour, which was higher for cracker F4, suggesting that the better acceptance of cracker F4 was influenced by the spices used.

If we compare the results of the hedonic sensory analysis conducted on nine crackers with the sensory analysis conducted on four crackers in terms of overall impression, the best-rated crackers in both analyses were cracker F4, followed by cracker F3. However, in the second sensory analysis, crackers F5 were rated better than F8, while in first analysis it was vice versa.

Crackers F5 and F8 differed significantly from the better-rated crackers F3 and F4 ($p < 0.05$). Appearance, taste, and flavour, as well as texture in the mouth were rated better for cracker F5 than for cracker F8. Odour is the only sensory property that was rated better for cracker F8 than for cracker F5.

There are several factors that might influenced the better acceptance of cracker F5 as compared with cracker F8. First, several types of flour (maize, millet and buckwheat) were combined in cracker F5, while only buckwheat flour was used in cracker F8. In the research of Šimurina et al. [34], a wholegrain buckwheat cracker was rated better in terms of taste as compared with control wheat cracker, while in terms of odour, the control wheat cracker was rated better. In addition, chia seeds and pumpkin seed meal were included in our cracker F5, while flax seeds and sweet potato puree were used our cracker F8. We can conclude that combining buckwheat flour with other flours leads to better acceptability of crackers.

3.3. Nutritive Value, Fructan Content, and Labelling of Crackers

The moisture content in the final products was below 5 g in 100 g of sample, except for cracker F8 in which the moisture content was around 8% and could cause crackers to have a lower shelf life as compared with the other three crackers. Significantly higher content ($p < 0.001$) of minerals was found in crackers F5 and F8 as compared with crackers F3 and F4. The addition of pumpkin seed meal and buckwheat flour (Table 5) contributed to higher mineral content in crackers F5 and F8.

Crackers made according to the F8 formulation had the highest fat content, significantly differing from crackers F4 and F5 ($p = 0.048$ and 0.004, respectively), but not from cracker F3 ($p = 0.183$). In addition, all the crackers were low in saturated fat. In the F3, F4, and F5 formulations, 31–35% of the energy value originated from fats rich in unsaturated fatty acids.

Although significantly lower protein content was found in crackers F3 and F8 than in crackers F4 and F5, the differences were small. Pumpkin seed meal, which contains up to 65% protein [35], contributed to the higher protein content of cracker F5. Proteins were a source of 22–24% of the total energy value of crackers. Given that more than 20% of the energy value of crackers in four different recipes comes from protein, it could be labelled as "high in protein" [25]. Nutritional or health claims in the labelling, presentation and advertising of products while ensuring the effective functioning of the market also aim to ensure a high level of consumer protection.

In crackers F3, F4, and F5, the proportion of calories derived from carbohydrates was around 44%, while in cracker F8 it was slightly lower, around 40%. Thus, the stated calorie proportions derived from high-value proteins and carbohydrates were slightly lower than the recommended values of ~25–32% and 45–55%, respectively (Academy of Nutrition and Dietetics, Chicago, IL, USA). To achieve the target values for carbohydrates and proteins, the amount of fat should be reduced and the amount of proteins increased in our formulations. All the types of crackers were "high in fibre", since they contained more than 15 g of fibre per 1000 kcal [25]. One serving of the crackers would contribute around 10% of the recommended daily intake (25 g). Nevertheless, crackers F4, F5, and F8 had higher proportions of soluble fibre (23–24%) as compared with cracker F3 with 17% soluble fibre (Table 5). Although insoluble fibre promotes regularity in a healthy digestive system, in patients with IBS, it can increase bloating, gas, cramping, etc., whereas soluble fibre helps improve IBS symptoms and slows the digestion of food [36]. Moreover, fructan levels in all four of the analysed crackers were below the cut-off value (Table 5). Thus, our crackers could be labelled as "low-FODMAP", "naturally rich in fibre", and "gluten-free".

Considering the content of soluble fibre, crackers F4, F5, and F8 would be suitable for IBS patients, while all of the four crackers would be suitable for patients who suffer from NCWS and celiac disease. Hence, it can be observed that higher total dietary fibre content contrasts important descriptive sensory properties of crackers such as taste and flavour, while soluble fibre content contrasts undesirable descriptive texture attributes and bitter taste (Table 5 and Figure 4).

To the best of our knowledge, there are only two certified low-FODMAP crackers on the market, and they can be found in Australia (FODMAP everyday. Available online: https://www.fodmapeveryday.com/product-category/collections/low-fodmap-certified-brands/fodmap-friendly-low-fodmap-certified/?paged=8 (accessed on 14 August 2022)). The crackers contain about 21 g of fats, 55–60 g of carbohydrates, 4–7 g of fibre, 10–14 g of proteins, and approximately 450 kcal per 100 g of product. If we compare our crackers with those from the market, it is evident that they have less fat, much more proteins and fibre, and still have low-FODMAP content. The reason for this is that our crackers had more complex compositions and more nutritionally valuable ingredients such as wholemeal flours and oilseed flours. Furthermore, our crackers were prepared with sourdough which contributed to their flavour and taste, but also assured low-FODMAPs [9]. In particular, we observed about a 2.5-fold reduction in expected fructan level (almost 0.3 g/100 g) in cracker R5 containing pumpkin seed cake. In conclusion, oilseeds in a moderate amount can be safely used in the formulation of low-FODMAP crackers with sourdough.

4. Conclusions

The development of high-fibre low-FODMAP crackers requires the selection of a larger number of raw materials, as well as chemical and sensory evaluations. Wholemeal millet is a suitable raw material for such crackers because it is high in fibre, but more importantly, high in soluble fibre while low in FODMAPs. Furthermore, oilseed flours, as well as flours from by-products (pumpkin seed meal) contribute to a higher nutritional value of crackers in terms of a higher content of protein, unsaturated fatty acids, and minerals. Moreover, crackers from wholemeal flours with a milder taste (maize and millet) are better accepted than those from buckwheat flour, which is characterised by a specific taste. At the same time, sweet potato and spices have a significant influence on the acceptance of the product, as does the addition of sourdough, which has the potential to improve the taste and appearance of the crackers. Unlike total dietary fibre content that obstructs important descriptive sensory properties of crackers such as taste and flavour, soluble fibre content confines undesirable descriptive texture attributes and bitter taste. Nevertheless, soluble fibre and fat content contrast crackers' hedonic qualities. In the development of products with a high fibre content for IBS sufferers, special attention should be given to the proportion of soluble fibre in total fibre, not only for consumer acceptance, but also for health reasons. Future studies should investigate novel processing techniques in the production of snacks for IBS and NCWS patients, as well as their shelf life.

Author Contributions: Conceptualisation, N.Č.M., D.N.; methodology, K.R., K.V., S.D., and B.V.; validation, N.Č.M., S.D., D.Ć., and D.N.; formal analysis, K.R., K.V., and S.D.; investigation, K.R., N.Č.M., and D.N.; resources, D.N.; data curation, K.R. and K.V.; writing—original draft preparation, K.R. and N.Č.M.; writing—review and editing, S.D., N.Č.M., D.Ć., and D.N.; visualisation, K.R. and N.Č.M.; supervision, D.N.; project administration, B.V.; funding acquisition, D.N. All authors have read and agreed to the published version of the manuscript.

Funding: This research was funded by the Croatian Science Foundation (IP-2020-3829 and DOK-2021-02). Any opinions, findings, conclusions, or recommendations expressed in this material are those of the authors and do not necessarily reflect the views of the Croatian Science Foundation.

Institutional Review Board Statement: This study was conducted in accordance with the Declaration of Helsinki, and the protocol was approved by the Ethics Committee of the University of Zagreb, Faculty of Food Technology and Biotechnology.

Informed Consent Statement: All subjects gave their informed consent for inclusion before they participated in the study.

Data Availability Statement: The data used to support the findings of this study are included in the article.

Acknowledgments: This work is based upon the work from COST Action 18101 SOURDOMICS—Sourdough biotechnology network towards novel, healthier and sustainable food and bioprocesses (SOURDOMICS. Available online: https://sourdomics.com/; https://www.cost.eu/actions/CA18101/, accessed on 14 August 2022), where the author D.N. is a Management Committee member and the author N.Č.M. is a member of the Working Group 7. SOURDOMICS are supported by COST (European Cooperation in Science and Technology). COST is a funding agency for research and innovation networks. COST Actions help connect research initiatives across Europe and enable scientist to grow their ideas by sharing them with their peers, thus, boosting their research, career, and innovation. We thank Lesaffre Adriatic Inc. for donating Livendo™ LV1 starter and SME Osijek for donating flax seeds.

Conflicts of Interest: The authors declare no conflict of interest.

References

1. Biesiekierski, J.R.; Peters, S.L.; Newnham, E.D.; Rosella, O.; Muir, J.G.; Gibson, P.R. No effects of gluten in patients with self-reported non-celiac gluten sensitivity after dietary reduction of fermentable, poorly absorbed, short-chain carbohydrates. *Gastroenterology* **2013**, *145*, 320–328. [CrossRef] [PubMed]
2. Brouns, F.; Delzenne, N.; Gibson, G. The dietary fibres—FODMAP controversy. *Cereal Food World* **2017**, *62*, 98–103. [CrossRef]
3. Ispiryan, L.; Zannini, E.; Arendt, E.K. Characterization of the FODMAP-profile in cereal-product ingredients. *J. Cereal Sci.* **2020**, *92*, 102916. [CrossRef]
4. Kahlon, T.S.; Avena-Bustillos, R.J.; Chiu, M.-C.M. Sensory evaluation of gluten-free quinoa whole grain snacks. *Heliyon* **2016**, *2*, e00213. [CrossRef] [PubMed]
5. Bellini, M.; Tonarelli, S.; Barracca, F.; Rettura, F.; Pancetti, A.; Ceccarelli, L.; Ricchiuti, A.; Costa, F.; de Bortoli, N.; Marchi, S.; et al. Chronic constipation: Is a nutritional approach reasonable? *Nutrients* **2021**, *13*, 3386. [CrossRef]
6. Čukelj, N.; Novotni, D.; Sarajlija, H.; Drakula, S.; Voučko, B.; Ćurić, D. Flaxseed and multigrain mixtures in the development of functional biscuits. *LWT Food Sci. Technol.* **2017**, *86*, 85–92. [CrossRef]
7. Gao, D.; Helikh, A.; Duan, Z.; Liu, Y.; Shang, F. Development of pumpkin seed meal biscuits. *East. Eur. J. Enterp. Technol.* **2022**, *2*, 36–42. [CrossRef]
8. Balbino, S.; Dorić, M.; Vidaković, S.; Kraljić, K.; Škevin, D.; Drakula, S.; Voučko, B.; Čukelj, N.; Obranović, M.; Ćurić, D. Application of cryogenic grinding pretreatment to enhance extractability of bioactive molecules from pumpkin seed cake. *J. Food Process Eng.* **2019**, *42*, 1–13. [CrossRef]
9. Habuš, M.; Mykolenko, S.; Iveković, S.; Pastor, K.; Kojić, J.; Drakula, S.; Ćurić, D.; Novotni, D. Bioprocessing of wheat and amaranth bran for the reduction of fructan levels and application in 3D-printed snacks. *Foods* **2022**, *11*, 1649. [CrossRef]
10. Loponen, J.; Gänzle, M. Use of sourdough in low-FODMAP baking. *Foods* **2018**, *7*, 96. [CrossRef]
11. Siepmann, F.B.; Ripari, V.; Waszczynskyj, N.; Spier, M.R. Overview of sourdough technology: From production to marketing. *Food Bioprocess Technol.* **2018**, *11*, 242–270. [CrossRef]
12. Świąder, K.; Florowska, A.; Konisiewicz, Z.; Chen, Y.-P. Functional tea-infused set yoghurt development by evaluation of sensory quality and textural properties. *Foods* **2020**, *9*, 1848. [CrossRef]
13. Heiniö, R.L.; Noort, M.W.J.; Katina, K.; Alam, S.A.; Sozer, N.; de Kock, H.L.; Hersleth, M.; Poutanen, K. Sensory characteristics of wholegrain and bran-rich cereal foods—A review. *Trends Food Sci. Technol.* **2016**, *47*, 25–38. [CrossRef]
14. AACC International. *Approved Methods of Analysis*, 11th ed.; Cereals & Grains Association: Saint Paul, MN, USA, 2010; ISBN 978-1-891127-68-2.
15. Balbino, S.; Repajić, M.; Obranović, M.; Medved, A.M.; Tonković, P.; Dragović-Uzelac, V. Characterization of lipid fraction of Apiaceae family seed spices: Impact of species and extraction method. *J. Appl. Res. Med. Aromat. Plants* **2021**, *25*, 100326. [CrossRef]
16. Gill, S.K.; Rossi, M.; Bajka, B.; Whelan, K. Dietary fibre in gastrointestinal health and disease. *Nat. Rev. Gastroenterol. Hepatol.* **2021**, *18*, 101–116. [CrossRef]
17. *ISO Standard No. 11035*; Sensory Analysis—Identification and Selection of Descriptors for Establishing a Sensory Profile by a Multidimensional Approach. International Organization for Standardization: Geneva, Switzerland, 1994.
18. *ISO Standard No. 13299*; Sensory Analysis—Methodology—General Guidance for Establishing a Sensory Profile. International Organization for Standardization: Geneva, Switzerland, 1994.
19. *ISO Standard No. 6658*; Sensory Analysis—Methodology—General Guidance. International Organization for Standardization: Geneva, Switzerland, 1994.
20. Knudsen, K.E.B.; Nørskov, N.P.; Bolvig, A.K.; Hedemann, M.S.; Laerke, H.N. Dietary fibres and associated phytochemicals in cereals. *Mol. Nutr. Food Res.* **2017**, *61*, 1–37.
21. Varney, J. Getting Enough Fibre. Monash University. 2021. Available online: https://www.monashfodmap.com/blog/getting-enough-fibre/ (accessed on 14 August 2022).
22. Varney, J.; Barrett, J.; Scarlata, K.; Catsos, P.; Gibson, P.R.; Muir, J.G. FODMAPs: Food composition, defining cutoff values and international application. *J. Gastroenterol. Hepatol.* **2017**, *32*, 53–61. [CrossRef]
23. Currò, D. Current evidence on the therapeutic use of fiber in irritable bowel syndrome. *Expert Rev. Gastroenterol. Hepatol.* **2021**, *16*, 426–436. [CrossRef]
24. Coskuner, Y.; Karababa, E. Some physical properties of flaxseed (*Linum usitatissimum* L.). *J. Food Eng.* **2007**, *78*, 1067–1073. [CrossRef]
25. Commission Regulation (EU) No 1047/2012 of 8 November 2012 Amending Regulation (EC) No 1924/2006 with Regard to the List of Nutrition Claims Text with EEA Relevance OJ L 310, 9.11.2012; p. 36–37. Available online: https://eur-lex.europa.eu/legal-content/EN/TXT/?uri=celex%3A32012R1047 (accessed on 14 August 2022).

26. Ullah, R.; Nadeem, M.; Khalique, A.; Imran, M.; Mehmood, S.; Javid, A.; Hussain, J. Nutritional and therapeutic perspectives of Chia (*Salvia hispanica* L.): A review. *J. Food Sci. Technol.* **2016**, *53*, 1750–1758. [CrossRef]
27. Selimović, A.; Miličević, D.; Selimović, A.; Žuljević, S.O.; Jašića, A.; Vranac, A. Properties of crackers with buckwheat sourdough. *Acta Chim. Slov.* **2017**, *10*, 152–158. [CrossRef]
28. Pham, V.T.; Fehlbaum, S.; Seifert, N.; Richard, N.; Bruins, M.J.; Sybesma, W.; Rehman, A.; Steinert, R.E. Effects of colon-targeted vitamins on the composition and metabolic activity of the human gut microbiome—A pilot study. *Gut Microbes* **2021**, *13*, 1875774. [CrossRef] [PubMed]
29. Roth, B.; Larsson, E.; Ohlsson, B. Poor intake of vitamins and minerals is associated with symptoms among patients with irritable bowel syndrome. *J. Gastroen. Hepatol.* **2022**, *37*, 1253–1262. [CrossRef] [PubMed]
30. Severini, C.; Derossi, A.; Ricci, I.; Caporizzi, R.; Fiore, A. Printing a blend of fruit and vegetables. New advances on critical variables and shelf life of 3D edible objects. *J. Food Eng.* **2018**, *220*, 89–100. [CrossRef]
31. Sedej, I.; Sakač, M.; Mandic, A.; Misan, A.; Pestoric, M.; Šimurina, O. Quality assessment of gluten-free crackers based on buckwheat flour. *LWT Food Sci. Technol.* **2011**, *44*, 694–699. [CrossRef]
32. Colla, K.; Costanzo, A.; Gamlath, S. Fat replacers in baked food products. *Foods* **2018**, *7*, 192. [CrossRef]
33. Khouryieh, H.; Aramouni, F. Physical and sensory characteristics of cookies prepared with flaxseed flour. *J. Sci. Food Agric.* **2012**, *92*, 2366–2372. [CrossRef] [PubMed]
34. Šimurina, O.; Sedej, I.; Sakač, M.; Hadnađev, M.; Filipčev, B.; Pestorić, M.; Pribiš, V. Buckweat in the wholegrain cracker production. *Prog. Theor. Exp. Phys.* **2009**, *13*, 153–156.
35. Majchrzak, K.K.; Sumara, A.; Fornal, E.; Montowska, M. Oilseed proteins—Properties and application as a food ingredient. *Trends Food Sci. Technol.* **2020**, *106*, 160–170. [CrossRef]
36. Moayyedi, P.; Andrews, C.N.; MacQueen, G.; Korownyk, C.; Marsiglio, M.; Graff, L.; Kvern, B.; Lazarescu, A.; Liu, L.; Paterson, W.G.; et al. Canadian association of gastroenterology clinical practice guideline for the management of irritable bowel syndrome (IBS). *J. Can. Assoc. Gastroenterol.* **2019**, *2*, 6–29. [CrossRef]

foods

MDPI

Article

Effects of Teff-Based Sourdoughs on Dough Rheology and Gluten-Free Bread Quality

Rosen Chochkov [1], Daniela Savova-Stoyanova [1], Maria Papageorgiou [2], João Miguel Rocha [3,4], Velitchka Gotcheva [5] and Angel Angelov [5,*]

1 Technology of Cereals, Feed, Bakery and Confectionery Products Department, University of Food Technologies, 26 Maritza Blvd., 4002 Plovdiv, Bulgaria; rosen4o4kov@abv.bg (R.C.); gourmetacademy@abv.bg (D.S.-S.)
2 Department of Food Science and Technology, International Hellenic University, P.O. Box 141, 57400 Thessaloniki, Greece; mariapapage@ihu.gr
3 LEPABE—Laboratory for Process Engineering, Environment, Biotechnology and Energy, Faculty of Engineering, University of Porto, Rua Dr. Roberto Frias, 4200-465 Porto, Portugal; jmfrocha@fe.up.pt
4 ALiCE—Associate Laboratory in Chemical Engineering, Faculty of Engineering, University of Porto, Rua Dr. Roberto Frias, 4200-465 Porto, Portugal
5 Department of Biotechnology, University of Food Technologies, 26 Maritza Blvd., 4002 Plovdiv, Bulgaria; gotcheva_v@uft-bio.com
* Correspondence: angelov@uft-bio.com

Citation: Chochkov, R.; Savova-Stoyanova, D.; Papageorgiou, M.; Rocha, J.M.; Gotcheva, V.; Angelov, A. Effects of Teff-Based Sourdoughs on Dough Rheology and Gluten-Free Bread Quality. *Foods* **2022**, *11*, 1012. https://doi.org/10.3390/foods11071012

Academic Editor: Alessandra Marti

Received: 10 February 2022
Accepted: 25 March 2022
Published: 30 March 2022

Abstract: Production of gluten-free bread (GFB) with good quality characteristics represents a technological challenge. Our study aimed to obtain nongluten bread from cereals and pseudocereals with applying single cultures of *Pediococcus acidilactici*, *Pediococcus pentosaceus* and *Enteroccocus durans* as sourdoughs. The effect of sourdoughs on the quality traits of gluten-free (GF) dough and GFB was explored. The structural and baking properties of GF dough composed of teff, rice, corn, and sorghum flours were improved by adding xanthan gum (0.6%), guar gum (1.0%) and carboxymethyl cellulose (1.0%). The tested strains reached 10^8 cfu/g in teff flour and produced sourdoughs with a pleasant lactic aroma. The sourdough-fermented doughs were softer and more elastic compared to control dough and yielded reduced baking loss. Strain *Enterococcus durans* ensured the best baking characteristics of GF dough and the highest softness of the GFB during storage. Strain *Pediococcus pentosaceus* had the most pronounced positive effect on aroma, taste and aftertaste. Pan baking was found to be more appropriate to obtain stable shape and good-looking products. A careful starter culture selection is necessary for GFB development since a significant effect of strain specificity on dough rheology and baking characteristics was observed.

Keywords: teff; sorghum; sourdough; lactic acid bacteria; starter cultures; gluten-free bread; celiac disease; cereals; pseudocereals

1. Introduction

The incidence of gluten intolerance, or so-called 'celiac disease' (CD), has increased significantly over the last 50 years, affecting approximately 1% of the world's population. If untreated, it can cause a number of serious health complications. The only way to control celiac condition is to maintain a strict gluten-free diet throughout the affected person's life [1–3].

In the literature, various strategies for improving the quality of nongluten dough are suggested. However, they are mostly focused on the composition of flour mixes and food additives that aim to reproduce the viscoelastic properties of the gluten network and hence increase the development of bread loaf volume [4–7]. Dairy, soy and egg ingredients have also been used to improve the quality of gluten-free bread (GFB), but with limited success [8–11]. The need to improve the nutritional profile, appearance, taste and aroma of gluten-free breads still represents a challenge for researchers [12–14]. The latest innovative technological approaches include the application of high hydrostatic pressure, new baking methods (Ohmic heating) and sourdough technology [15,16].

The use of sourdough is the oldest biotechnological process to leaven bakery goods and to improve bread texture, aroma, and shelf life [17–19]. A number of studies have shown that sourdough can be used successfully to improve the processing characteristics of gluten-free doughs [20–23]. In some cases, this effect may be attributed to the potential of lactic acid bacteria to secrete extracellular polysaccharides, which could be a beneficial alternative to conventional thickeners used to modify the viscosity, structure and stability of a wide range of gluten-free foods [24]. These extracellular polysaccharides may act as a substitute for hydrocolloids used as food additives and, therefore, the application of lactic acid bacteria could represent a cost-efficient approach to improving the rheology of gluten-free doughs. In addition, these extracellular polysaccharides have a beneficial effect on the intestinal microbiome by selectively stimulating the growth of bifidobacteria and other beneficial microorganisms [25–32].

Cereals such as oats, sorghum and teff and various pseudocereals (buckwheat, amaranth, quinoa), tubers (cassava, potatoes), legumes (chickpea, soy, pea, lupin), nuts (acorns) and oil seeds (rapeseed, sunflower, canola) are used for the production of non-traditional flour types [6,10,11,33–39]. In most cases, they are superior to traditional cereal-based flours in terms of protein, essential amino acids, dietary fiber and vitamins.

One of the most widely used gluten-free cereals is rice (*Oryza sativa*). It is rich in carbohydrates (75–80%) and minerals, but protein content is only 6–7%. Potassium, magnesium, phosphorus, calcium, manganese and zinc are commonly found in rice flour. However, it is poor in trace elements due to the specific process of grain milling. Compared to corn, wheat and potatoes, rice is a relatively good source of thiamine (pantothenic acid, folic acid and vitamin E [40]. Corn flour (*Zea mays*) is combined in most cases with other types of flour to make gluten-free bread. In bakery products, it contributes to obtaining a dense and moist bread crumb. Corn flour is rich in essential nutrients, such as choline (21.6 mg/100 g), folate (48 µg/100 g), vitamins B1 (0.25 mg/100 g), B5 (0.66 mg/100 g), B6 (0.37 mg/100 g) and especially in minerals—potassium (315 mg/100 g), phosphorus (272 mg/100 g) and magnesium (93 mg/100 g) [41–45].

Teff flour (*Eragrostis tef*) is found to have a number of valuable nutritional characteristics compared to flours from more common crops. Its starch is slowly digestible and, therefore, has a low glycemic index (GI). It has a favorable amino acid composition and does not contain gluten [46,47]. It is a good source of unsaturated fatty acids (1.66 g/100 g) and is high in minerals, especially iron (7.63 mg/100 g) and calcium (180 mg/100 g) [48]. In addition, the high content of protein (13.3 g/100 g) and dietary fiber (8 g/100 g) makes teff flour a desirable raw material for various functional foods, some of which may bear nutritional claims such as "rich in protein" and "source of fiber" [49,50]. Teff starch has a slow retrogradation tendency, which could have a potentially positive impact on the shelf-life of baked products [51]. In addition, some authors report that the addition of teff to cereal-based sourdoughs (rice and buckwheat) modified the aroma profile of the breads, increasing the fruity, toasty and cereal notes [52–57], which contributes to developing more diverse baking products with positive consumer acceptance.

Sorghum (*Sorghum bicolor* L. Moench) is the fifth largest crop in the world. Sorghum grain has a high protein content varying from 4.4 to 21.1% with an average value of 11.4% [33,58]. Sorghum grain is gluten free, high in resistant starch, a rich source of nutrients, and most importantly, contains a diverse range of bioactive phenolic compounds [59,60]. It has been proven safe for people with celiac disease, but literature on its use in gluten-free foods is relatively scarce compared to that for corn and rice.

Since none of these gluten-free crops has satisfactory bakery performance when used alone, appropriate combinations of raw materials and processing techniques should be sought to develop successful gluten-free bakery products. Therefore, the aim of the present study was to explore the application of teff-based sourdoughs in the development of gluten-free bread with improved sensory and quality characteristics.

2. Materials and Methods

2.1. Raw Materials

The flours used in this study were teff flour (Adan Village Ltd., Sungurlare, Bulgaria), rice flour (Bodpie Food Ltd., Varna, Bulgaria), corn flour (Lubeks Ltd., Asenovgrad, Bulgaria) and sorghum flour obtained by laboratory grain milling at the University of Food Technologies, Bulgaria. Compressed yeast was supplied by Lesaffre Ltd. (Sofia, Bulgaria). Guar gum (E412) and xanthan gum (E415) were supplied by Cargill (Minneapolis, MN, USA), and carboxymethyl cellulose (E466) was supplied by Dow Wolff Cellulosics GmbH (Bomlitz, Germany).

2.2. Flour Characterization

Crude fat, protein, moisture, fiber and starch contents in the flours were determined according to the American Association of Cereal Chemists (AACC) methods 30–10, 46–12, 44–15, 32–05 and 76–13, respectively [61]. Protein content was calculated with a protein factor of 6.25.

2.3. Lactic Acid Bacteria (LAB)

Three lactic acid bacteria (LAB) strains were used in this study: *Pediococcus acidilactici* 02P108 (PA), *Pediococcus pentosaceus* 12R2187 (PP) and *Enteroccocus durans* 09B374 (ED) from the culture collection of the department of biotechnology, University of Food Technologies, Plovdiv, Bulgaria. The strains originate from typical Bulgarian sourdoughs [62,63]. Starter cultures for the sourdoughs were prepared from stock cultures of each strain stored in Microbank™ (Pro Lab Diagnostics Inc., Richmond Hill, ON, CA) by cultivation in MRS broth (de Man–Rogosa–Sharpe, Merck KGaA, Darmstadt, Germany) at 37 °C for 48 h.

2.4. Sourdough Preparation

Teff flour was used to prepare a separate sourdough with each LAB strain. Equal weights of flour and sterile water were mixed to obtain a dough yield (DY) of 280. Each starter culture was added to a batch of teff dough at an inoculum amount of 5 log colony-forming units (CFU)/g of dough. The sourdough variants were then fermented in an incubator at 37 °C for 24 h.

2.5. Analyses of Sourdoughs

Active acidity (pH) was determined by a pH meter (FiveEasy FE20, Mettler-Toledo GmbH, Greifensee, Switzerland). Total titratable acidity (TTA) was determined by titration with 0.1 N NaOH to pH 8.4 and expressed as mL of NaOH/10 g of sourdough, as described in AACC 02-31 [61]. The measurements were performed in triplicate. LAB viable cell counts were determined on MRS agar plates at the beginning and the end of the sourdough fermentation. The identity of the starter cultures in each sourdough batch was confirmed by colony morphology and microscopic observations.

2.6. Bread Preparation

In order to select the most appropriate formulation of a nongluten dough, gluten-free flours were mixed with an equal amount of baker's yeast and different combinations and ratios of xanthan, guar gum and carboxymethyl cellulose. The formulations of the three tested dough variants (A, B and C) are presented in Table 1.

After selecting the most appropriate dough variant, test gluten-free breads with sourdoughs and control bread with baker's yeast were prepared. Their formulations are presented in Table 2.

Table 1. Composition of gluten-free bread formulations.

Ingredients		Variants A	Variants B	Variants C
		Quantity, %		
Teff flour	Gluten-free flour base	40	40	40
Rice flour		40	40	40
Sorghum flour		10	10	10
Corn flour		10	10	10
Other ingredients, g/100 g Gluten-free flour base				
Water		65	65	65
Baker's yeast		3.0	3.0	3.0
Salt		1.5	1.5	1.5
Xanthan gum		1.0	0.6	0.6
Guar gum		1.0	1.0	1.0
Carboxymethyl cellulose (CMC)			1.0	3.0

Table 2. Composition of gluten-free breads with sourdoughs and control bread.

Ingredients		Control Sample	Sample PP	Sample PA	Sample ED
		Quantity, %			
Teff flour	Gluten-free flour base	40		32.8	
Rice flour		40		40.0	
Sorghum flour		10		10.0	
Corn flour		10		10.0	
Other ingredients, g/100 g Gluten-free flour base					
Water		65		52.4	
Baker's yeast		3.0		-	
Salt		1.5		1.5	
Xanthan gum		0.6		0.6	
Guar gum		1.0		1.0	
Carboxymethyl cellulose (CMC)		1.0		1.0	
Sourdough (Teff flour + Water + LAB)				21.5	

Kneading was performed by a single-phase process of dough preparation to obtain a dough with a homogeneous mass and an initial temperature of 25–26 °C. The dough was left to rest for 20 min and was then divided into pieces of 230 g—for floor bread and 440 g—for pan bread. After shaping, the dough was subjected to a final fermentation at 33 °C for 60 min in a fermenting chamber (Tecnopast CRN 45-12, Novacel Rovimpex Novaledo, Italy). The dough was then baked in an electric floor oven Salva E-25 (Salva Industrial S.L.U., Lezo, Spain), preheated to a temperature of 220–230 °C, for 17–18 min for floor bread (plain bread, baked by placing the loaf directly on the floor of the oven) and 22–25 min for pan bread (the loaf is baked in a pan). After baking, the breads were allowed to cool for 3 h at room temperature [64].

2.7. Degree of Immersion of the Doughs

The degree of immersion of the prepared doughs was measured by an automatic penetrometer AP-4/2 (Steinmeyer Mechatronik GmbH, Dresden, Germany). Each dough was divided into pieces of 13 g and placed in the sleeve of the penetrometer, which was then placed in a thermostat at 35 °C for 60 min. The immersion of the calibrated body in the dough for 5 s was automatically measured and expressed in penetrometer units (PU).

2.8. Rheological Properties of the Doughs

The following dough characteristics were determined by a farinograph (Brabender GmbH&Co. KG, Duisburg, Germany): water absorption (%), development time (min), stability (min), degree of softening (farinograph units (FU)) and consistency (FU), with modification of AACC Method 54-21.02 [61]. Farinographic analysis is only applied for wheat flour doughs where the reference value is 500 FU. In the present study it was used for nongluten flour combinations, with which the highest values achieved for FU were much lower (Table 3). Flour samples of 300 g were analyzed using the ICC standard method 115/1 [65]. Water absorption (WA) of the formulations was first adjusted to 65%, and then the other parameters of the doughs were measured.

Table 3. Effect of sourdoughs on the rheological characteristics of nongluten dough.

Samples	Rheological Characteristics				
	Water Absorption, %	Consistency, FU	DDT, min	Stability, min	Degree of Softening, FU
Control sample	65 ± 3.56 ^{a}	350 ± 1.41 ^{a}	6.0 ± 0.82 ^{a}	10.0 ± 0.82 ^{a}	10 ± 0.82 ^{a}
Sample PP	65 ± 5.65 ^{a}	290 ± 1.83 ^{b}	2.0 ± 0.82 ^{b}	7.5 ± 0.08 ^{b}	30 ± 0.82 ^{b}
Sample PA	65 ± 0.82 ^{a}	290 ± 2.58 ^{b}	1.5 ± 0.22 ^{b}	7.5 ± 0.08 ^{b}	30 ± 0.82 ^{b}
Sample ED	65 ± 5.72 ^{a}	290 ± 0.82 ^{b}	1.5 ± 0.08 ^{b}	6.5 ± 0.29 ^{c}	30 ± 0.82 ^{b}

Mean values with different letter in superscript within the same column differ significantly ($p < 0.05$). Note: *Pediococcus acidilactici* 02P108 (PA); *Pediococcus pentosaceus* 12R2187 (PP); *Enteroccocus durans* 09B374 (ED). DDT, dough development time.

2.9. Bread Quality

2.9.1. Physical Properties

The quality of the prepared breads was assessed by the following characteristics. Bread loaf volume was determined after baking and cooling the breads for 3 h at room temperature by a rapeseed displacement method [61]. The specific volume was calculated by the ratio between volume (cm^3) and mass (g) of each sample. Bread height and diameter were measured by a caliper, and the shape stability (height/diameter) was calculated [66]. Bake loss (%) was determined following weighing each loaf before and after baking [67]. The bread loaves were wrapped in plastic bags and stored at room temperature (27 ± 2 °C) to determine the storage time (in days) until mold growth became visible [68,69].

2.9.2. Crumb Elasticity

Total and plastic deformation were measured by an automatic penetrometer and expressed as penetrometric units (PU) [69]. A 4-mm thick crumb sample was cut from the bread and placed on the flat surface of the lifting table, which was raised until the upper surface of the sample lightly touched the lower end of the immersion body. The value of penetration of the immersion body in the sample after 5 s represented total deformation. The steel disk was removed and the immersion system unloaded. Then the measurement was repeated for 10 s, and the recorded value in PU represented plastic deformation. Elastic deformation was calculated as the difference between total and plastic deformation.

2.10. Sensory Analysis

Sensory analyses of the obtained breads were performed by a descriptive panel consisting of 25 panelists (52% women and 48% men) aged 22–60 years, who were familiar with sensory analysis of foods but not specifically trained in the evaluation of sourdough breads. The analyses were carried out according to ISO 6658:2017 [70]. The panelists were asked to score eight parameters, namely shape, crust color, crumb color, porosity, aroma, chewability, taste and aftertaste. They expressed the intensity of each attribute on a 9-point hedonic scale (9—extremely good; 1—extremely bad).

2.11. Statistical Analyses

All analyses were conducted in triplicate. The obtained data were subject to one-way ANOVA using XLSTAT version 2019.1.2 (AddinSoft Inc., New York, NY, USA). Comparison among least-squares means were performed by Tukey's test; differences were considered significant at p-value < 0.05 [71].

3. Results and Discussion

In recent years, there has been an increased interest toward using nonconventional raw materials for sourdough preparation, including both fermenting matrices and starter cultures [72–77]. In this line, the benefits of nongluten flours and lactic acid bacteria as well as applied sourdough technology were studied with the goal of developing gluten-free breads with improved nutritional and quality characteristics.

3.1. Flour Composition

Four types of nongluten flours were selected as raw materials for gluten-free bread preparation: teff, rice, sorghum and corn. The raw materials selection was based on literature data for the cultivars' chemical composition. Table 4 shows the detailed chemical composition (content of moisture, protein, fiber, starch and lipids, in dry-matter (DM) basis) of the nongluten flours used in this study.

Table 4. Chemical composition of the nongluten flours.

Flour	Moisture Content, %	Protein Content, % (d.m.)	Fiber Content, % (d.m.)	Crude Fat, % (d.m.)	Starch Content, % (d.m.)
Teff	10.18 ± 0.16^{a}	10.20 ± 0.13^{a}	12.39 ± 0.13^{a}	3.09 ± 0.16^{a}	74.5 ± 0.80^{a}
Rice	10.00 ± 0.14^{a}	6.99 ± 0.10^{b}	1.47 ± 0.16^{b}	0.59 ± 0.09^{b}	84.7 ± 1.07^{b}
Sorghum	11.04 ± 0.24^{b}	12.49 ± 0.46^{c}	9.56 ± 0.08^{c}	4.54 ± 0.12^{c}	72.2 ± 0.92^{c}
Corn	9.05 ± 0.16^{c}	4.20 ± 0.09^{d}	3.74 ± 0.12^{d}	3.73 ± 0.08^{d}	73.4 ± 0.77^{d}

Mean values with different letter in superscript within the same column differ significantly ($p < 0.05$).

The obtained data shows that protein and fiber content of teff and sorghum flour are much higher compared to the values of these parameters for rice and corn. Protein content of sorghum was 12.49%, and in teff it was 10.20%. Furthermore, fiber content was 12.39 and 9.56%, while the values for corn and rice flour were very low—3.74 and 1.47%, respectively. These results confirm that teff and sorghum are excellent sources of protein and dietary fiber as reported by other authors [48,78–81], and could, therefore, be preferred raw materials for the development of nongluten breads with additional functionalities (rich in protein and fiber). Other authors reported approximately 11–13% of protein, approximately 80% of complex carbohydrates, and 2.4–3.0% of fat for teff [34,82]. Sorghum and teff are also generally characterized by good technological properties [83]. Apart from fiber, teff is also an excellent source of iron and contains far more calcium, potassium and other essential minerals than other grains [34,84–89]. Except for rice flour (0.59% fat), the other three studied flours contained relatively high amounts of crude fat (3.09–4.54%). It is important to point out that teff contains fat that is not easily oxidized—which results in a longer shelf-life of teff flour compared to other nongluten flours [72]. Our analyses showed that starch content was of similar levels for teff, sorghum and corn—viz., 74.5, 72.2 and 73.4%, respectively, while starch in rice flour amounted to 84.7%, which was more than 10% higher compared to the other nongluten flours. Other authors report similar results for starch content of the analyzed raw materials [72,73,81,90,91].

3.2. Formulation of the Control Gluten-Free Dough Matrix

To develop a gluten-free bread with good nutritional and quality characteristics, the initial step of the study was to formulate a flour blend based on the functional chemical composition as well as the technological characteristics of nongluten cereals and pseudocereals. Based on the available literature data and preliminary trials, four types of flour were selected to formulate the bread matrix—teff, rice, sorghum and corn flour.

A technological challenge with developing nongluten bakery products is that the doughs do not have an adequate structure consistency (elasticity) to retain the gas formed during fermentation. Therefore, it is necessary to use other ingredients/food additives such as proteins, starches, gums and hydrocolloids with water-binding and structure-building properties, which are able to compensate for the lack of gluten in the dough [92].

The appropriate combination of structure-forming additives and their amounts depends on the type of the nongluten flour or the specific flour combination, since these have a significant effect on the quality characteristics of the baked products. Ćuric et al. (2007) [93] supplemented a nongluten flour mix (rice flour, corn starch) with 1, 2 and 3% of xanthan, guar gum, pectin or cellulose as stabilizing additives toward the improvement of the structure of the obtained gluten-free bread and found that 3% guar gum had the best structure-forming effect.

In the present study, different combinations of xanthan, guar gum and carboxymethyl cellulose at different ratios to the nongluten flour base were tested aiming to find the most efficient gluten-replacement strategy (Materials and Methods, Section 2.6). Dough and baking characteristics of the three formulations are presented in Tables 5 and 6.

Table 5. Fermentation characteristics of nongluten dough formulations.

Dough Variant	Preliminary Degree of Immersion, PU	Degree of Immersion After 20 min, PU
A	70 ± 1.63 [a]	117 ± 3.74 [a]
B	195 ± 4.55 [b]	366 ± 0.82 [b]
C	118 ± 1.63 [c]	190 ± 3.46 [c]

Variant A—xanthan 1.0% and guar gum 1.0%; Variant B—xanthan 0.6%, guar gum 1.0% and CMC 1.0%; Variant C—xanthan 0.6%, guar gum 1.0% and CMC 3.0%. Mean values with different letter in superscript within the same column differ significantly ($p < 0.05$).

Table 6. Baking characteristics of nongluten dough formulations.

Dough Variant	Height/Diameter	Specific Volume, cm^3/g		Bake Loss, %	
	Floor Bread	Floor Bread	Pan Bread	Floor Bread	Pan Bread
A	0.45 ± 0.03 [a]	0.90 ± 0.08 [a]	1.12 ± 0.09 [a]	22.61 ± 0.07 [a]	20.68 ± 0.09 [a]
B	0.45 ± 0.02 [a]	1.40 ± 0.06 [b]	1.85 ± 0.09 [b]	20.87 ± 0.11 [b]	17.72 ± 0.05 [b]
C	0.36 ± 0.01 [b]	0.95 ± 0.05 [a]	0.88 ± 0.09 [c]	21.30 ± 0.64 [b]	16.14 ± 0.67 [c]

Variant A—xanthan 1.0% and guar gum 1.0%; Variant B—xanthan 0.6%, guar gum 1.0% and CMC 1.0%; Variant C—xanthan 0.6%, guar gum 1.0% and CMC 3.0%. Mean values with different letter in superscript within the same column differ significantly ($p < 0.05$).

The obtained results show that Variant B yielded a significantly higher specific volume of both floor bread (1.40 cm^3/g) and pan bread (1.85 cm^3/g) compared to Variants A and C (Table 6). These values were positively correlated with the highest values of the degree of immersion before and after fermentation of Variant B (Table 6) and with low bake loss—that is, 20.87% for floor bread and 17.72% for pan bread of dough variant B (Table 6). Variant C had a significantly lower shape stability (height/diameter = 0.36) compared to Variants A and B, which did not differ in this parameter. These results clearly show that the higher hydration of the nongluten dough enables a more active fermentation, but the appropriate combination of structure-forming additives is of key importance for the baking characteristics of the nongluten breads. Based on the obtained results, Variant B was selected as the control nongluten dough formulation for the next steps of the study. In terms of baking method, pan breads showed better baking characteristics, and to commercialize such nongluten formulations, pan baking would be more appropriate to obtain good-looking products.

3.3. Sourdough Fermentation

The sourdoughs applied for leavening of the main gluten-free flour mix were prepared only from teff flour with the addition of the three different cultures of lactic acid bacteria. The kinetics of sourdough fermentation was monitored by the changes in pH, TTA and the total viable counts of the respective starter culture. Results from these analyses are presented in Table 7.

Table 7. Fermentation kinetics of teff sourdoughs.

Strain/Time	*Pediococcus acidilactici* 02P108 (PA)			*Pediococcus pentosaceus* 12R2187 (PP)			*Enteroccocus durans* 09B374 (ED)		
	pH	TTA	log CFU/g	pH	TTA	log CFU/g	pH	TTA	log CFU/g
0 h	5.92 ± 0.06	4.2 ± 0,25	3.00 ± 0.11	5.92 ± 0.18	4.2 ± 0.29	2.95 ± 0.19	5.92 ± 0.18	4.2 ± 0.40	2.90 ± 0.57
4 h	5.48 ± 0.10	4.6 ± 0.26	3.08 ± 0.21	5.40 ± 0.23	4.7 ± 0.31	3.04 ± 0.18	5.26 ± 0.14	4.6 ± 0.31	2.95 ± 0.38
8 h	5.26 ± 0.10	5.8 ± 0.13	4.63 ± 0.64	5.32 ± 0.18	5.6 ± 0.26	4.62 ± 0.31	5.12 ± 0.36	5.9 ± 0.26	4.71 ± 0.59
12 h	4.88 ± 0.14	9.6 ± 0.31	6.26 ± 0.35	5.02 ± 0.28	8.8 ± 0.25	6.75 ± 0.21	4.65 ± 0.26	10.0 ± 0.26	6.86 ± 0.67
16 h	4.32 ± 0.15	13.2 ± 0.25	6.79 ± 0.49	4.64 ± 0.19	13.4 ± 0.38	6.95 ± 0.47	4.28 ± 0.35	15.3 ± 0.29	7.00 ± 0.71
20 h	4.12 ± 0.10	14.9 ± 0.45	8.15 ± 0.24	4.04 ± 0.38	14.2 ± 0.28	8.28 ± 0.26	4.15 ± 0.12	17.1 ± 0.31	8.53 ± 0.49
24 h	4.02 ± 0.10	15.2 ± 0.26	8.72 ± 0.59	3.88 ± 0.27	14.6 ± 0.33	8.81 ± 0.51	4.10 ± 0.21	18.4 ± 0.34	8.83 ± 0.27
	a	a	a	a	a	a	a	b	a

Mean values of the same parameters with different letter within the same row differ significantly ($p < 0.05$).

The obtained results clearly indicate a good capacity of the three tested strains to ferment teff flour, while increasing their biomass and producing organic acids, which resulted in sourdoughs with a pleasant lactic acid aroma. Indeed, the genera *Pediococcus* and *Enterococcus* are homofermentative or homolactic bacteria, thus releasing solely lactic acid from fermentation—and known as hexose fermentation pathway or Embden–Meyerholf–Parnas (EMP) pathway. Data analysis showed no statistical differences in the yielded biomass concentration at the end of the fermentations among the three tested LAB strains (8.72–8.83 log CFU/g). The lowest pH value (3.88) was reached in the sourdough with strain *Pediococcus pentosaceus* 12R2187, but it was not significantly different compared to the other two sourdoughs. One of the strains used to prepare sourdough was of *Enteroccocus durans* species, which is not commonly found in sourdoughs. It was interesting to observe that this strain yielded a final TTA value of 18.4, which was significantly higher compared to the other two LAB strains and indicated the highest capacity of strain *Enteroccocus durans* 09B374 to produce organic acids.

3.4. Development of Gluten-Free Bread with Sourdough

3.4.1. Degree of Immersion in the Nongluten Sourdoughs

Three nongluten doughs were prepared by using as a matrix the preselected dough variant B (Section 3.2) without yeast and by adding equal amounts of teff sourdoughs prepared with the three selected lactic acid bacteria strains—*Pediococcus acidilactici* 02P108, *Pediococcus pentosaceus* 12R2187 and *Enteroccocus durans* 09B374. Dough variant B with yeast was used as the control. All nongluten doughs were kneaded according to the established methodology (Section 2.6). The results from analyzing the degree of immersion in the nongluten doughs are presented in Figure 1.

The initial degree of immersion of the control dough was significantly higher (195 PU) ($p < 0.01$) compared to each of the three samples with added sourdoughs (160–164 PU). However, the difference between these three samples was not significant. After 20 min of fermentation, the degree of immersion of the control sample increased by 87.7%, and for all sourdough-fermented doughs, the values more than doubled (335–345 PU). However, for all three sourdough-fermented samples, the degree of immersion was significantly lower than that of the control sample (366 PU). These results indicate that the sourdough-fermented doughs are more resistant and less elastic compared to the control dough fermented by baker's yeast (*Saccharomyces cerevisiae*). These observations differ from the results reported by Wolter et al. (2014) [94], where sourdough addition led to decreased dough strength resulting in softer dough. Other authors also confirm that sourdough fermentation increased the elasticity and reduced the stiffness of doughs [95]. However, these teams studied wheat-based doughs, while with nongluten doughs this effect is the opposite, as observed in our study. The reduced elasticity could be attributed to the nongluten flour composition and the specific fermentation capacity of the tested strains. Other studies on the rheological properties of gluten-free sourdoughs prepared with lactic acid bacteria also found that the addition of sourdough reduced the elasticity of the dough and improved the dough strength [69,96,97].

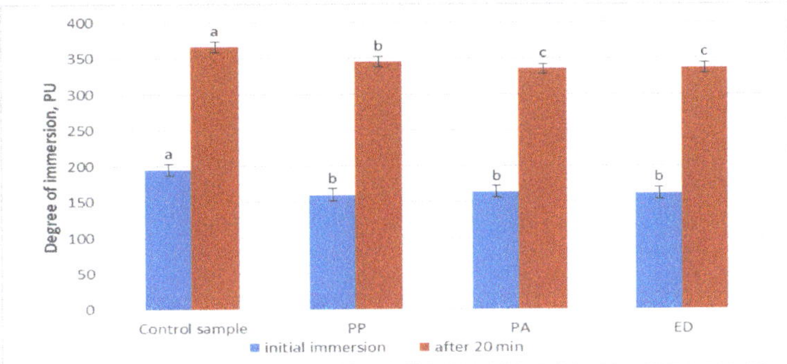

Figure 1. Degree of immersion in nongluten doughs prepared with the addition of sourdoughs with single cultures lactic acid bacteria. Note: PU—penetrometric units; PA—sample with *Pediococcus acidilactici* 02P108; PP—sample with *Pediococcus pentosaceus* 12R2187; ED—sample with *Enteroccocus durans* 09B374. Different small letters for each time measurement indicate significant difference between mean values ($p < 0.01$).

In our study, after 20 min of fermentation, strain PP produced a significantly softer dough compared to the other two strains. Such findings confirmed that the reduced elasticity was attributed to the LAB fermentation and that strain specificity has an effect on nongluten dough stability.

3.4.2. Effect of Sourdoughs on the Rheological Characteristics of Nongluten Dough

The results from the rheological analysis of the nongluten doughs obtained with the addition of sourdoughs with different lactic acid bacteria are depicted in Table 3.

A farinograph is a useful tool for the determination of the optimal water content for dough preparation. In addition, it provides information about dough stability and dough development time [98]. In general, the water-absorbing capacity of flour depends on the protein content, the particle size, the amount of starchy grains with impaired integrity and some other factors. In the present study, the addition of hydrocolloids also contributed to the water absorption and to the improved and rheological properties. Our preliminary experiment with different WA values (55%, 65% and 70%) of nongluten flour basis showed that 65% provided best dough consistency as well as best specific volume and porosity of the bread (data not shown). Based on these findings, the WA of the formulations in the current study was first adjusted to 65%, and then the other parameters of the doughs were measured.

The consistency of the control sample was 20.69% higher than that of the experimental samples. No significant differences were observed in the consistency (290 FU) of the samples with added sourdoughs. Significant differences between the control and the sourdough-leavened doughs were also observed for the other tested parameters: dough development time (DDT), stability and relaxation of the dough. The DDT of the sourdough-added samples was reduced by 3 times (2 min) for sample PP and by 4 times (1.5 min) for samples PA and ED. These results are in contrast to the observations of Tafti et al. (2013) [99] who found no effect of spray-dried sourdough addition on wheat dough development time. Again, the difference might be attributed to the different kind of dough matrix.

In our study, the control sample had significantly higher stability (10 min) than the sourdough-fermented samples. However, the values of 6.5–7.5 min also indicated good stability of the doughs obtained with sourdough addition and with dough samples PP and PA showing significantly better effect compared to the sample ED. The degree of dough softening of the control sample was 10 FU, while the sourdough-fermented doughs showed a significantly higher value (30 FU), which is still not too high for this parameter and was the same for the three test samples. Tafti et al. (2013) [99] also observed that the degree of softening significantly increased with an increase in the sourdough level, whereas dough stability was significantly reduced. The results obtained in our study indicate that the use of sourdoughs requires some additional matrix optimization to achieve the same dough rheology as the yeast-fermented dough.

3.4.3. Quality Assessment of Gluten-Free Breads Leavened with Sourdough

To assess the baking characteristics of the gluten-free breads with added sourdough, two types of bread—floor and pan bread—were prepared from each sourdough-leavened bread variety. Fermentation and baking of all dough samples were carried out under equal conditions, according to the adopted technology. Results from the quality assessment of their baking characteristics are presented in Table 8.

Table 8. Baking characteristics of nongluten breads with sourdough.

Dough Samples	Specific Volume, cm^3/g		Height/ Diameter	Baking Loss, %	
	Floor Bread	Pan Bread	Floor Bread	Floor Bread	Pan Bread
Control sample	1.49 ± 0.01^{a}	1.55 ± 0.01^{a}	0.36 ± 0.01^{a}	20.86 ± 0.06^{a}	17.72 ± 0.05^{a}
Sample PP	1.55 ± 0.06^{a}	1.24 ± 0.03^{b}	0.30 ± 0.04^{a}	18.90 ± 0.06^{b}	13.02 ± 0.03^{b}
Sample PA	1.59 ± 0.01^{b}	1.46 ± 0.01^{c}	0.36 ± 0.03^{a}	16.70 ± 0.01^{c}	12.25 ± 0.17^{c}
Sample ED	1.70 ± 0.03^{c}	1.56 ± 0.02^{a}	0.48 ± 0.04^{b}	15.65 ± 0.04^{d}	12.59 ± 0.07^{d}

Mean values with different letter in superscript within the same column differ significantly ($p < 0.05$). Note: *Pediococcus acidilactici* 02P108 (PA); *Pediococcus pentosaceus* 12R2187 (PP); *Enteroccocus durans* 09B374 (ED).

The specific volumes of the floor bread samples prepared with strains PA and ED were significantly higher compared to the control and sample PP, with strain *Enteroccocus durans* yielding the highest value (1.70 cm^3/g). Comparison of pan bread samples showed that samples PP and PA had significantly lower (20% for sample PP) specific volumes than the control sample and sample ED. These results indicate that strain specificity is important in terms of the generated specific volume of nongluten breads, and in our study, strain *Enteroccocus durans* gave the best performance in leavening the nongluten dough.

Literature data on the effect of LAB on the specific volume of gluten-free breads are diverse. According to some authors, the addition of sourdough to GF breads does not have a significant influence on the specific volume [100,101]. Cappa et al. (2016) [17] reported that sourdoughs have been effective in improving bread volume and softness, which confirms the positive effects observed in our experiments. Other studies also showed that sourdough gluten-free bread had a higher specific volume and was less firm than GF bread fermented with baker's yeast alone [102,103].

In terms of shape stability (height/diameter ratio, H/D), strain ED gave a significantly higher value of 0.48 compared to the other three tested doughs. Similar to this result, Falade et al. (2014) [104] reported that the addition of lactic acid bacteria increased the bread height after baking.

It is interesting to note that all three LAB-leavened samples had significantly lower baking loss compared to the yeast-leavened control sample, with the lowest value (15.65%) among floor breads observed for sample ED, and the lowest among pan breads (12.25%) observed for sample PA. This positive effect could be attributed to the organic acids produced by the bacteria, which strengthen the structure of the gels in the gluten-free doughs. Therefore, the gas retention in the bread is greater. Our observations confirm the findings of Wolter et al. (2014) [94] and indicate that the application of sourdoughs does affect the baking characteristics of the gluten-free bread, mostly by a significantly pronounced reduction in baking loss. A strain-specific effect on the analyzed parameters was also observed, and this effect differed with respect to the type of bread—i.e., floor or pan.

3.4.4. Sensory Profile of Nongluten Breads with Sourdough

In general, the use of LAB-inoculated sourdough can improve the quality of bread with regards to various characteristics such as taste, staleness, odor, chewability, softness, moisture content, pH, acidity and texture [105]. In the present study, the gluten-free breads leavened with sourdoughs differed in appearance from the control mainly by a better color of the crust (Figure 2), especially of sample ED. The crust of the three samples was thin, smooth and soft.

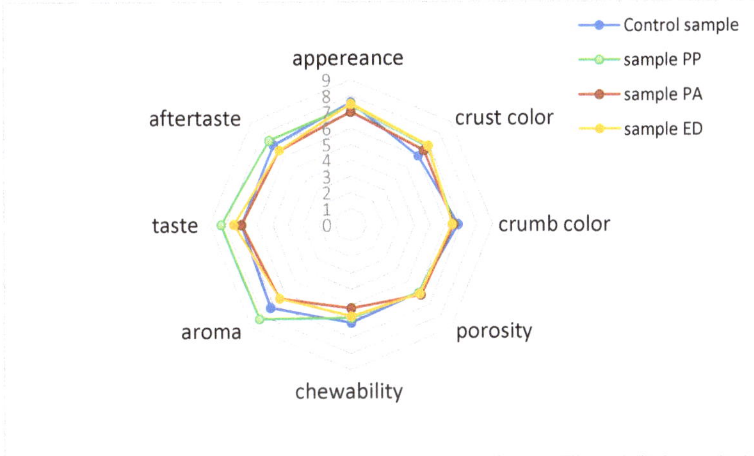

Figure 2. Sensory profile of nongluten pan breads with sourdough. Note: *Pediococcus acidilactici* 02P108 (PA); *Pediococcus pentosaceus* 12R2187 (PP); *Enteroccocus durans* 09B374 (ED).

Generally, all three strains reached biomass content at the level of 10^8 CFU/g, but the differences in the fermenting capacity showed variations in acid production as well as the composition of the produced organic acids. These differences resulted in variations in the sensory characteristics of the obtained nongluten breads (Section 3.4.4).

The ratio between lactic and acetic acids s an important factor affecting the aroma of bread [106], and its influenced by the fermenting microorganisms, the fermentation temperature and the type of flour or flours [107]. In the present study, the aroma, taste and aftertaste were found pleasant for all samples prepared from the nongluten flour mix. However, these characteristics were most pronounced in the sample with strain *Pediococcus pentosaccus* 12R2187 (PP), with a significant difference compared to the other samples.

According to Moore et al. (2008) [100], acidification during sourdough fermentation increases polysaccharide swelling that can partially replace gluten and improve the structure of gluten-free bread. In our study, the development of bread porosity in the sourdough-fermented bread samples did not differ from the control sample. The middle of all samples was soft and slightly moist, and slight sticking to the teeth was observed while chewing. Indeed, only sample PA gave a significant difference compared to control regarding this parameter. Data from the sensory analysis showed positive acceptance of the prepared nongluten breads. Some sensory characteristics were significantly influenced by the LAB strain specificity, with the most pronounced positive effect shown by strain *Pediococcus pentosaceus* 12R2187 (PP).

3.4.5. Shelf-Life Estimation of Nongluten Bread with Sourdough

Bourne (1978) [108] described the use of instrumental texture profile analysis extensively, using force, deformation, and work measurement to determine the texture parameters for hardness, fracturability, cohesiveness, adhesiveness, springiness, gumminess and chewiness. In the present research, estimation of the shelf life of the prepared gluten-free pan breads with sourdoughs was based on the time of occurrence of mold growth, and the analysis of the deformation characteristics—viz., total, elastic and plastic deformation—were measured by an automatic penetrometer of the bread crumb. Pan breads were selected to explore deformation characteristics since this method of baking showed better baking characteristics of the nongluten formulations. The total deformation of the breads leavened with lactic acid bacteria (33–35 PU) was more than two times higher than that of the yeast-leavened control sample (15 PU) (Figure 3). This trend was generally preserved until the end of the experiment (72 h), while total deformation gradually decreased for all samples.

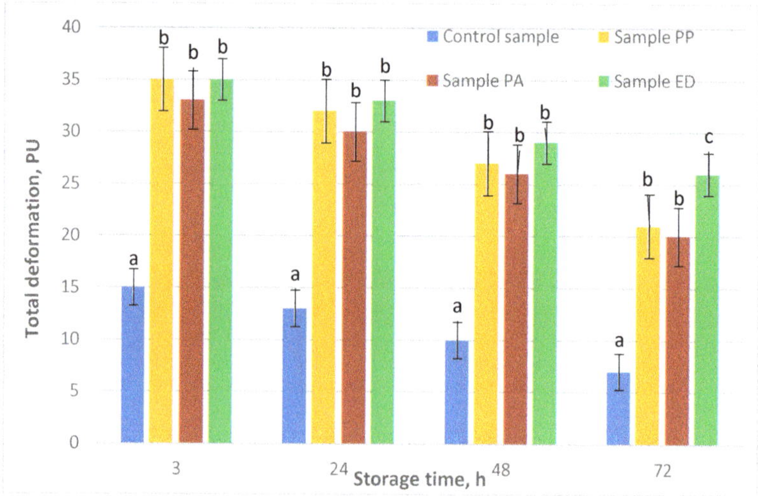

Figure 3. Effect of the different LAB strains on the total deformation of nongluten bread. Note: PU—penetrometric units; PA—sample with *Pediococcus acidilactici* 02P108; PP—sample with *Pediococcus pentosaceus* 12R2187; ED—sample with *Enteroccocus durans* 09B374. Different small letters for each time measurement indicate significant difference between mean values ($p < 0.01$).

It is interesting to note that the total deformation differences between the three samples with sourdoughs were not significant during the 48 h of fermentation, and only at the end of the process did samples PP and PA have a significantly lower total deformation (21 and 20 PU, respectively) compared to sample ED (26 PU).

The total deformation of the control sample was reduced by 53% for 72 h, while the average reduction for the breads with sourdough addition was 25%, with the largest decrease observed for sample PP (40%), and the lowest for sample ED (26%). These results clearly show that the application of LAB improves the softness of bread throughout storage time, with the strongest effect observed for strain *Enteroccocus durans* 09B374 (ED). In a study on the application of sourdough *Lactobacillus* strains to obtain gluten-free bread, Di Cagno et al. (2008) [103] also found that the addition of lactic acid bacteria resulted in lower hardness of gluten-free bread crumb during storage.

Results from analyzing the plastic deformation of sourdough-leavened nongluten breads are displayed in Figure 4.

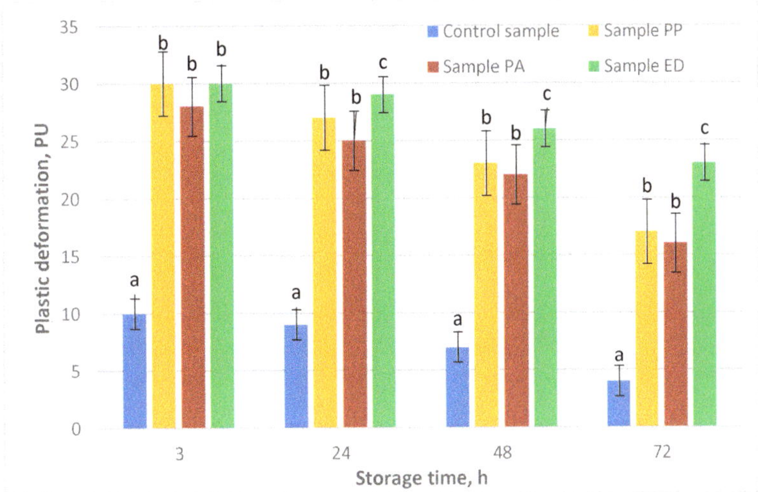

Figure 4. Effect of lactic acid bacteria on the plastic deformation of nongluten bread. Note: PU—penetrometric units; PA—sample with *Pediococcus acidilactici* 02P108; PP—sample with *Pediococcus pentosaceus* 12R2187; ED—sample with *Enteroccocus durans* 09B374. Different small letters for each time measurement indicate significant difference between mean values ($p < 0.01$).

The plastic deformation of the sourdough-leavened breads was considerably greater (around threefold more) compared to the control, and it gradually decreased until the end of the experiment. The crumb of sample ED retained its plasticity to the greatest extent, which was most pronounced at the end of the 72 h test. Plasticity of this sample decreased by 25% between the third and the 72nd hour compared to a 60% reduction observed in the control. For the other two LAB-leavened samples, plasticity decreased by an average of 44%. After 24 h storage, the plastic deformation of sample ED became significantly higher compared to sample PA, and this trend was maintained until the end of the experiment, when plastic deformation of sample ED was significantly higher than sample PP as well. The difference between samples PP and PA was not significant during the course of the entire duration of the experiment. These results indicate that the microbial strain specificity affects the plastic deformation of nongluten breads during storage, and in this study strain *Enteroccocus durans* 09B374 (ED) provided the highest softness of the nongluten bread in terms of total and plastic deformation.

The third analyzed shelf-life parameter for the nongluten breads prepared with the additions of sourdoughs was elastic deformation (Figure 5).

Elastic deformation measurement at 3 h of storage showed the same average values (5 PU) for all tested samples. The relaxation capacity of the gluten-free bread samples PP and PA did not change after 24 h and remained at 4 PU at 48 and 72 h, while sample ED behaved similar to the control sample and also did not change between 48 and 72 h. Results for samples PP and PA were significantly different ($p < 0.01$) compared to control and sample ED, which indicates the effect of strain specificity on the elastic properties during storage.

Many other studies also demonstrated that sourdough gluten-free bread was less firm than gluten-free bread leavened with baker's yeast alone [101,103,109,110]. Shelf-life rheological tests made by Moore et al. (2008) [101], however, showed that the addition of sourdough to a gluten-free mix led to increased firmness and elasticity overtime, which indicated that a LAB strain could be used to produce gluten-free bread with increased quality and shelf life. A study of Moroni et al. (2009) [24] also showed that the use of sourdough in nongluten bread development had positive effects on the crumb structure. Staling was delayed and longer shelf-life was achieved. These effects are mostly associated with the production of lactic and acetic acids, as well as exopolysaccharides during fermentation with lactic acid bacteria. In addition, the use of LAB-inoculated sourdoughs in bread preparation may contribute antifungal properties, thereby increasing the shelf-life of the bread even at a reduced salt content [111].

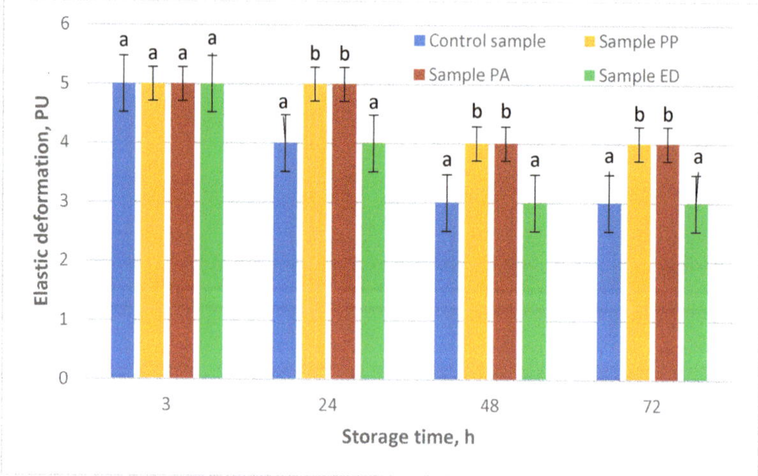

Figure 5. Effect of lactic acid bacteria on the elastic deformation of nongluten bread. Note: PU—penetrometric units; PA—sample with *Pediococcus acidilactici* 02P108; PP—sample with *Pediococcus pentosaceus* 12R2187; ED—sample with *Enteroccocus durans* 09B374. Different small letters for each time measurement indicate significant difference between mean values ($p < 0.01$).

4. Conclusions

The selected combination of nongluten flours and additives was adequate for obtaining nongluten bread with good quality characteristics. The three tested LAB strains demonstrated good capacity to ferment teff flour into sourdough, reaching different levels of acidification. The application of teff-based sourdoughs had a positive effect on various technological and sensory characteristics of the nongluten doughs and breads. The sourdough-fermented doughs were softer and more elastic compared to the control dough fermented by baker's yeast. The application of sourdoughs resulted in a significantly pronounced reduction of baking loss. A strain-specific effect on the analyzed quality parameters was observed, and this effect also differed with respect to the bread type—floor or pan bread. Pan baking resulted in better bread characteristics and proved to be more appropriate for commercialization of nongluten bread formulations.

The obtained nongluten breads had positive sensory acceptance. Strain specificity had a significant effect on some sensory characteristics of the products and on bread softness during storage.

The study demonstrated that the application of sourdoughs in the nongluten flour matrix is a successful approach for gluten-free bread development. Strain specificity is significant for dough rheology and the baking characteristics, and it is, therefore, important to perform a careful starter culture selection.

Author Contributions: Conceptualization, A.A., V.G., R.C.; methodology, A.A., R.C., M.P.; experimental work, R.C., D.S.-S.; data analysis, A.A., V.G.; resources, A.A., M.P.; writing, R.C., V.G., A.A., J.M.R., M.P.; supervision, A.A. All authors have read and agreed to the published version of the manuscript.

Funding: This research was funded by the Bulgarian National Science Fund, grant number KP-06-PN-36/11 and by the Bulgarian Ministry of Education and Science under the National Research

Programme "Healthy Foods for a Strong Bio-Economy and Quality of Life" approved by DCM # 577/17.08.2018.

Institutional Review Board Statement: Not applicable.

Informed Consent Statement: Not applicable.

Data Availability Statement: Not applicable.

Acknowledgments: The present study is part of the work of COST Action 18101 SOURDOMICS—Sourdough biotechnology network towards novel, healthier and sustainable food and bioprocesses (https://sourdomics.com/ (accessed on 10 March 2022); https://www.cost.eu/actions/CA18101/ (accessed on 10 March 2022) supported by COST (European Cooperation in Science and Technology) (https://www.cost.eu/ (accessed on 10 March 2022). COST is a funding agency for research and innovation networks. The author J.M.R is the Chair and Grant Holder Scientific Representative, the author A.A. is a Management Committee Member and member of Working Groups "Screening and characterization of cereal flours and sourdough microbiota" and "Design and development of sourdough starter cultures for breadmaking and other agri-food products", the author M.P. is the vice-leader and leader of the working group "Recovery, characterization and selection of autochthonous conventional and nonconventional (pseudo)cereal seeds", and the author V.G. is a member of Working Group "Screening and characterization of cereal flours and sourdough microbiota".

Conflicts of Interest: The authors declare no conflict of interest.

Abbreviations

AACC, American Association of Cereal Chemists; CD, celiac disease; CFU, colony-forming units; C-GFB, control gluten-free bread; CMC, carboxymethyl cellulose; DDT, dough development time; DM, dry-matter; DY, dough yield; ED, *Enteroccocus durans* 09B374; EMP, Embden-Meyerhof-Parnas; FU, farinograph units; GFB, gluten-free bread; GI, glycemic index; H/D, Height/Diameter ratio; LAB, lactic acid bacteria; MRS, de Man-Rogosa-Sharpe; PA, *Pediococcus acidilactici* 02P108; PP, *Pediococcus pentosaceus* 12R2187; PU, penetrometer units; TTA, total titratable acidity.

References

1. Cappelli, A.; Oliva, N.; Cini, E. A systematic review of gluten-free dough and bread: Dough rheology, bread characteristics, and improvement strategies. *Appl. Sci.* **2020**, *10*, 6559. [CrossRef]
2. Novotni, D.; Gänzle, M.; Rocha, J.M. Composition and Activity of Microbiota in Sourdough and Their Effect on Bread Quality and Safety. In *Trends in Wheat and Bread Making*; Galanakis, C.M., Ed.; Elsevier-Academic Press: Cambridge, MA, USA, 2020; p. 469.
3. Rocha, J.M. Microbiological and Lipid Profiles of Broa: Contributions for the Characterization of a Traditional Portuguese Bread. Ph.D. Thesis, Instituto Superior de Agronomia, Universidade de Lisboa, Lisbon, Portugal, 2011; p. 705.
4. Galle, S.; Schwab, C.; dal Bello, F.; Coffey, A.; Ganzle, M.G.; Arendt, E.K.K. Influence of in situ synthesized exopolysaccharides on the quality of gluten-free sorghum sourdough bread. *Int. J. Food Microbiol.* **2012**, *155*, 105–112. [CrossRef] [PubMed]
5. Padalino, L.; Conte, A.; del Nobile, A.M. Overview on the general approaches to improve gluten-free pasta and bread. *Foods* **2016**, *5*, 87. [CrossRef] [PubMed]
6. Skendi, A.; Papageorgiou, M.; Varzakas, T. High protein substitutes for gluten in gluten-free bread. *Foods* **2021**, *10*, 1997. [CrossRef] [PubMed]
7. Culetu, A.; Duta, D.E.; Papageorgiou, M.; Varzakas, T. The role of hydrocolloids in gluten-free bread and pasta; rheology, characteristics, staling and glycemic index. *Foods* **2021**, *10*, 3121. [CrossRef] [PubMed]
8. Guimaraes, R.M.; Pimentel, T.C.; de Rezende, T.A.M.; Silva, J.S.; Falcao, H.G.; Ida, E.I.; Egea, M.B. Gluten-free bread: Effect of soy and corn co-products on the quality parameters. *Eur. Food Res. Technol.* **2019**, *245*, 1365–1376. [CrossRef]
9. Han, A.; Romero, H.M.; Nishijima, N.; Ichimura, T.; Handa, A.; Xu, C.; Zhang, Y. Effect of egg white solids on the rheological properties and bread making performance of gluten-free batter. *Food Hydrocoll.* **2019**, *87*, 287–296. [CrossRef]
10. Pico, J.; Reguilón, M.P.; Bernal, J.; Manuel Gómez, M. Effect of rice, pea, egg white and whey proteins on crust quality of rice flour-corn starch based gluten-free breads. *J. Cereal Sci.* **2019**, *86*, 92–101. [CrossRef]
11. Taghdir, M.; Mazloomi, S.M.; Honar, N.; Sepandi, M.; Ashourpour, M.; Salehi, M. Effect of soy flour on nutritional, physicochemical, and sensory characteristics of gluten-free bread. *Food Sci. Nutr.* **2017**, *5*, 439–445. [CrossRef]
12. Păcularu-Burada, B.; Turturică, M.; Rocha, J.M.; Bahrim, G.-E. Statistical approach to potentially enhance the postbiotication of gluten-free sourdough. *Appl. Sci.* **2021**, *11*, 5306. [CrossRef]

13. Păcularu-Burada, B.; Georgescu, L.A.; Vasile, M.A.; Rocha, J.M.; Bahrim, G.-E. Selection of wild lactic acid bacteria strains as promoters of postbiotics in gluten-free sourdoughs. *Microorganisms* **2020**, *8*, 643. [CrossRef]
14. Bartkiene, E.; Lele, V.; Ruzauskas, M.; Mayrhofer, S.; Domig, K.; Starkute, V.; Zavistanaviciute, P.; Bartkevics, V.; Pugajeva, I.; Klupsaite, D.; et al. Lactic acid bacteria isolation from spontaneous sourdough and their characterization including antimicrobial and antifungal properties evaluation. *Microorganisms* **2020**, *8*, 64. [CrossRef]
15. Vallons, K.J.R.; Ryan, L.A.M.; Arendt, E.K. Promoting structure formation by high pressure in gluten-free flours. *LWT Food Sci. Technol.* **2011**, *44*, 1672–1680. [CrossRef]
16. Bender, D.; Gratz, M.; Vogt, S.; Fauster, T.; Wicki, B.; Pichler, S.; Kinner, M.; Jäger, H.; Schoenlechner, R. Ohmic heating—A novel approach for gluten-free bread baking. *Food Bioprocess Technol.* **2019**, *12*, 1603–1613. [CrossRef]
17. Cappa, C.; Lucisano, M.; Raineri, A.; Fongaro, L.; Foschino, R.; Mariotti, M. Gluten-free bread: Influence of sourdough and compressed yeast on proofing and baking properties. *Foods* **2016**, *5*, 69. [CrossRef]
18. Sun, L.; Li, X.; Zhang, Y.; Yang, W.; Ma, G.; Ma, N.; Hu, Q.; Pei, F. A novel lactic acid bacterium for improving the quality and shelf life of whole wheat bread. *Food Control* **2020**, *109*, 106914. [CrossRef]
19. Teleky, B.E.; Martau, G.A.; Vodnar, D.C. Physicochemical effects of *Lactobacillus plantarum* and *Lactobacillus casei* cocultures on soy-wheat flour dough fermentation. *Foods* **2020**, *9*, 1894. [CrossRef]
20. Moore, M.M.; Heinbockel, M.; Dockery, P.; Ulmer, H.M.; Arendt, E.K. Network formation in gluten-free bread with application of transglutaminase. *Cereal Chem.* **2006**, *83*, 28–36. [CrossRef]
21. Moore, M.M.; Juga, B.; Schober, T.J.; Arendt, E.K. Effect of lactic acid bacteria on properties of gluten-free sourdoughs, batters, and quality and ultrastructure of gluten-free bread. *Cereal Chem.* **2007**, *84*, 357–364. [CrossRef]
22. Ngemakwe, P.H.; le Roes-Hill, M.; Jideani, V.A. Advances in gluten-free bread technology. *Food Sci. Technol. Int.* **2015**, *21*, 256–276. [CrossRef]
23. Matos, M.E.; Rosell, C.M. Understanding gluten-free dough for reaching breads with physical quality and nutritional balance. *J. Sci. Food Agric.* **2015**, *95*, 653–661. [CrossRef]
24. Moroni, A.V.; dal Bello, F.; Arendt, E.K. Sourdough in gluten-free bread-making: An ancient technology to solve a novel issue? *Food Microbiol.* **2009**, *26*, 676–684. [CrossRef]
25. Di Cagno, R.; de Angelis, M.; Lavermicocca, P.; de Vincenzi, M.; Giovannini, C.; Faccia, M.; Gobbetti, M. Proteolysis by sourdough lactic acid bacteria: Effects on wheat flour protein fractions and gliadin peptides involved in human cereal intolerance. *Appl. Environ. Microbiol.* **2002**, *68*, 623–633. [CrossRef]
26. Deora, N.S.; Deswal, A.; Mishra, H.N. Alternative approaches towards gluten-free dough development: Recent trends. *Food Eng. Rev.* **2014**, *6*, 89–104. [CrossRef]
27. Ren, Y.; Linter, B.R.; Linforth, R.; Foster, T.J. A comprehensive investigation of gluten free bread dough rheology, proving and baking performance and bread qualities by response surface design and principal component analysis. *Food Funct.* **2020**, *11*, 5333–5345. [CrossRef]
28. Bartkiene, E.; Özogul, F.; Rocha, J.M. Bread sourdough lactic acid bacteria—Technological, antimicrobial, toxin-degrading, immune system- and faecal microbiota-modelling biological agents for the preparation of food, nutraceuticals and feed. *Foods* **2020**, *11*, 452. [CrossRef]
29. Ağagündüz, D.; Yılmaz, B.; Şahin, T.O.; Güneşliol, B.E.; Ayten, S.; Russo, P.; Spano, G.; Rocha, J.M.; Bartkiene, E.; Özogul, F. Dairy lactic acid bacteria and their potential function in dietetics: The food-gut-health axis. *Foods* **2021**, *10*, 3099. [CrossRef]
30. Sharma, H.; Özogul, F.; Bartkiéne, E.; Rocha, J.M. Impact of lactic acid bacteria and their metabolites on the techno-functional properties and health benefits of fermented dairy products. *Crit. Rev. Food Sci. Nutr.* **2021**, *30*, 1–23. [CrossRef]
31. Skendi, A.; Zinoviadou, K.G.; Papageorgiou, M.; Rocha, J.M. Advances on the valorisation and functionalization of by-products and wastes from cereal-based processing industry. *Foods* **2020**, *9*, 1243. [CrossRef]
32. Zokaityte, E.; Cernauskas, D.; Klupsaite, D.; Lele, V.; Starkute, V.; Zavistanaviciute, P.; Ruzauskas, M.; Gruzauskas, R.; Juodeikiene, G.; Rocha, J.M.; et al. Bioconversion of milk permeate with selected lactic acid bacteria strains and apple by-products into beverages with antimicrobial properties and enriched with galactooligosaccharides. *Microorganisms* **2020**, *8*, 1182. [CrossRef]
33. Olojede, A.O.; Sanni, A.I.; Banwo, K. Effect of legume addition on the physiochemical and sensorial attributes of sorghum-based sourdough bread. *LWT Food Sci. Technol.* **2020**, *118*, 108769. [CrossRef]
34. Zhu, F. Chemical composition and food uses of teff (*Eragrostis tef*). *Food Chem.* **2018**, *239*, 402–415. [CrossRef] [PubMed]
35. Alvarez-Jubete, L.; Arendt, E.K.; Gallagher, E. Nutritive value and chemical composition of pseudocereals as gluten-free ingredients. *Int. J. Food Sci. Nutr.* **2009**, *60*, 240–257. [CrossRef] [PubMed]
36. Arendt, E.K.; Zannini, E. Amaranth. In *Cereal Grains for the Food and Beverage Industries*; Woodhead Publishing Series in Food Science, Technology and Nutrition; Woodhead Publishing: Cambridge, UK, 2013; pp. 439–473.
37. Lu, H.; Guo, L.; Zhang, L.; Xie, C.; Li, W.; Gu, B.; Li, K. Study on quality characteristics of cassava flour and cassava flour short biscuits. *Food Sci. Nutr.* **2020**, *8*, 521–533. [CrossRef]
38. Skendi, A.; Mouselemidou, P.; Papageorgiou, M.; Papastergiadis, E. Effect of acorn meal-water combinations on technological properties and fine structure of gluten-free bread. *Food Chem.* **2018**, *253*, 119–126. [CrossRef]
39. Angelov, A.; Yaneva-Marinova, T.; Gotcheva, V. Oats as a matrix of choice for developing fermented functional beverages. *J. Food Sci. Technol.* **2018**, *55*, 2351–2360. [CrossRef]

40. Verma, D.K.; Srivastav, P.P. Proximate composition, mineral content and fatty acids analyses of aromatic and non-aromatic Indian rice. *Rice Sci.* **2017**, *24*, 21–31. [CrossRef]
41. Gwirtz, J.A.; Garcia-Casal, M.N. Processing maize flour and corn meal food products. *Ann. N. Y. Acad. Sci.* **2013**, *1312*, 66–75. [CrossRef]
42. Rocha, J.M.; Malcata, F.X. Microbial ecology dynamics in Portuguese broa sourdough. *J. Food Qual.* **2016**, *39*, 634–648. [CrossRef]
43. Rocha, J.M.; Malcata, F.X. Behavior of the complex micro-ecology in maize and rye flour and mother-dough for broa throughout storage. *J. Food Qual.* **2016**, *39*, 218–233. [CrossRef]
44. Rocha, J.M.; Malcata, F.X. Microbiological profile of maize and rye flours, and sourdough used for the manufacture of traditional Portuguese bread. *Food Microbiol.* **2012**, *31*, 72–88. [CrossRef]
45. Rocha, J.M.; Malcata, F.X. On the microbiological profile of traditional Portuguese sourdough. *J. Food Prot.* **1999**, *62*, 1416–1429. [CrossRef]
46. Dereje, N.; Bekele, G.; Nigatu, Y.; Worku, Y.; Holland, R.P. Glycemic index and load of selected Ethiopian foods: An experimental study. *J. Diabetes Res.* **2019**, *5*, 8564879. [CrossRef]
47. Gebru, Y.A.; Hyun-II, J.; Young-Soo, K.; Myung-Kon, K.; Kwang-Pyo, K. Variations in amino acid and protein profiles in white versus brown teff (*Eragrostis tef*) seeds, and effect of extraction methods on protein yields. *Foods* **2019**, *8*, 202. [CrossRef]
48. Fliedel, G.; Marti, A.; Thiebaut, S. Caracterisation et valorization du sorgho. *Bibliogr. CIRAD* **1996**, *6*, 349.
49. Arendt, E.K.; Zannini, E. *10-Teff Cereal Grains for the Food and Beverage Industries*; Woodhead Publishing Series in Food Science, Technology and Nutrition; Woodhead Publishing: Cambridge, UK, 2013; pp. 351–368.
50. Dingeo, C.; Difonzo, G.; Paradiso, V.M.; Rizzello, C.G.; Pontonio, E. Teff type-I sourdough to produce gluten-free muffin. *Microorganisms* **2020**, *8*, 1149. [CrossRef]
51. Bultosa, G.; Hall, A.N.; Taylor, J.R.N. Physico-chemical characterization of grain teff (*Eragrostis tef* (Zucc.) Trotter) starch. *Starch* **2002**, *54*, 461–468. [CrossRef]
52. Campo, E.; del Arco, L.; Urtasun, L.; Oria, R.; Ferrer-Mairal, A. Impact of sourdough on sensory properties and consumers preference of gluten-free breads enriched with teff flour. *J. Cereal Sci.* **2016**, *67*, 75–82. [CrossRef]
53. Wolter, A.; Hager, A.S.; Zannini, E.; Galle, S.; Gänzle, M.G.; Waters, D.M.; Arendt, E.K. Evaluation of exopolysaccharide producing *Weissella cibaria* MG1 strain for the production of sourdough from various flours. *Food Microbiol.* **2014**, *37*, 44–50. [CrossRef]
54. Barretto, R.; Buenavista, R.M.; Rivera, J.L.; Wang, S.; Vara Prasad, P.V.; Siliveru, K. Teff (*Eragrostis tef*) processing, utilization and future opportunities: A review. *Int. J. Food Sci. Technol.* **2021**, *56*, 3125–3137. [CrossRef]
55. Moroni, A.V.; Arendt, E.K.; dal Bello, F. Biodiversity of lactic acid bacteria and yeasts in spontaneously-fermented buckwheat and teff sourdoughs. *Food Microbiol.* **2011**, *28*, 497–502. [CrossRef] [PubMed]
56. Harth, H.; van Kerrebroeck, S.; Vuyst, L. Impact of process conditions on the microbial community dynamics and metabolite production kinetics of teff sourdough fermentations under bakery and laboratory conditions. *Food Sci. Nutr.* **2018**, *6*, 1438–1455. [CrossRef] [PubMed]
57. De Vuyst, L.; van Kerrebroeck, S.; Harth, H.; Huys, G.; Daniel, H.-M.; Weckx, S. Microbial ecology of sourdough fermentations: Diverse or uniform? *Food Microbiol.* **2014**, *37*, 11–29. [CrossRef] [PubMed]
58. Ogunsakin, O.A.; Banwo, K.; Ogunremi, O.R.; Sanni, A.I. Microbiological and physicochemical properties of sourdough bread from sorghum flour. *Int. Food Res. J.* **2015**, *22*, 2610–2618.
59. Awika, J.M.; Rooney, L.W. Sorghum phytochemicals and their potential impact on human health. *Phytochemistry* **2004**, *65*, 1199–1221. [CrossRef]
60. Dykes, L.; Rooney, L.W. Phenolic compounds in cereal grains and their health benefits. *Cereal Foods World* **2007**, *52*, 105–111. [CrossRef]
61. AACC International. *AACC Approved Methods of Analysis*, 11th ed.; AACC International: St. Paul, MN, USA, 2010.
62. Petkova, M.; Stefanova, P.; Gotcheva, V.; Kuzmanova, I.; Angelov, A. Microbiological and physicochemical characterization of traditional Bulgarian sourdoughs and screening of lactic acid bacteria for amylolytic activity. *J. Chem. Technol. Metal.* **2020**, *55*, 921–934.
63. Petkova, M.; Stefanova, P.; Gotcheva, V.; Angelov, A. Isolation and characterisation of lactic acid bacteria and yeasts from typical Bulgarian sourdoughs. *Microorganisms* **2021**, *9*, 1346. [CrossRef]
64. Vasileva, I.; Denkova, R.; Chochkov, R.; Petkova, N.; Teneva, D.; Denkova, Z.; Desev, T.; Denev, P.; Slavov, A. Effect of lavender (*Lavandula angustifolia*) and melissa (*Melissa officinalis*) waste on quality and shelf life of bread. *Food Chem.* **2018**, *253*, 13–21. [CrossRef]
65. ICC. *Standard Methods of the International Association for Cereal Science and Technology*; ICC: Vienna, Austria, 2021.
66. Novotni, D.; Čukelj, N.; Smerdel, B.; Bituh, M.; Dujmić, F.; Ćurić, D. Glycemic index and firming kinetics of partially baked frozen gluten-free bread with sourdough. *J. Cereal Sci.* **2012**, *55*, 120–125. [CrossRef]
67. Kim, W.M.; Gyu-Hee Lee, G.H. Comparison of imported wheat flour bread making properties and korean wheat flour bread making properties made by various bread making methods. *J. Korean Soc. Food Sci. Nutr.* **2015**, *44*, 434–441. [CrossRef]
68. Lönner, C.; Preve-Akesson, K. Effects of lactic acid bacteria on the properties of sour dough bread. *Food Microbiol.* **1989**, *6*, 19–35. [CrossRef]
69. Zlateva, D.; Chochkov, R. Effect of *Spirulina platensis* on the crumb firming of wheat bread during storage. *Ukr. Food J.* **2019**, *8*, 851–860. [CrossRef]

70. *ISO 6658:2017*; Sensory Analysis—Methodology—General Guidance. ISO: Geneva, Switzerland, 2017.
71. Bower, J.A. Statistics for food science V: ANOVA and multiple comparisons. *Nutr. Food Sci.* **1998**, *98*, 41–48. [CrossRef]
72. Puncha-arnon, S.; Uttapap, D. Rice starch vs. rice flour: Differences in their properties when modified by heat–moisture treatment. *Carbohydr. Polym.* **2013**, *91*, 85–91. [CrossRef]
73. Gebru, Y.A.; Sbhatu, D.B.; Kim, K.P. Nutritional composition and health benefits of teff (*Eragrostis tef* (Zucc.) Trotter). *J. Food Qual.* **2020**, *6*, 9595086. [CrossRef]
74. Di Monaco, R.; Torrieri, E.; Pepe, O.; Masi, P.; Cavella, S. Effect of sourdough with exopolysaccharide (EPS)-producing lactic acid bacteria (LAB) on sensory quality of bread during shelf life. *Food Bioproc. Technol.* **2015**, *8*, 691–701. [CrossRef]
75. Antognoni, F.; Mandrioli, R.; Bordoni, A.; di Nunzio, M.; Viadel, B.; Gallego, E.; Laure, D. Integrated evaluation of the potential health benefits of einkorn-based breads. *Nutrients* **2017**, *9*, 1232. [CrossRef]
76. Ua-Arak, T.; Jakob, F.; Vogel, R.F. Influence of levan-producing acetic acid bacteria on buckwheat-sourdough breads. *Food Microbiol.* **2017**, *65*, 95–104. [CrossRef]
77. Hadaegh, H.; Seyyedain Ardabili, S.; Tajabadi Ebrahimi, M.; Chamani, M.; Azizi Nezhad, R. The impact of different lactic acid bacteria sourdoughs on the quality characteristics of toast bread. *J. Food Qual.* **2017**, *11*, 7825203. [CrossRef]
78. Topuzova, Y.A.; Karadzhov, G.I.; Chonova, V.M. Basic raw materials used for production of gluten-free bakery and confectionery products. *Sci. Works UFT* **2012**, *59*, 439–443.
79. Alaunyte, I.; Stojceska, V.; Plunkett, A.; Ainsworth, P.; Derbyshire, E. Improving the quality of nutrient-rich Teff (*Eragrostis tef*) breads by combination of enzymes in straight dough and sourdough breadmaking. *J. Cereal Sci.* **2012**, *55*, 22–30. [CrossRef]
80. Hager, A.S.; Arendt, E.K. Influence of hydroxypropylmethylcellulose (HPMC), xanthan gum and their combination on loaf specific volume, crumb hardness and crumb grain characteristics of gluten-free breads based on rice, maize, teff and buckwheat. *Food Hydrocol.* **2013**, *32*, 195–203. [CrossRef]
81. Neela, S.; Fanta, S.W. Injera (An ethnic, traditional staple food of Ethiopia): A review on traditional practice to scientific developments. *J. Ethn. Food* **2020**, *7*, 32. [CrossRef]
82. Arendt, E.K.; Morrissey, A.; Moore, M.M.; dal Bello, F. Gluten-Free Breads. In *Gluten-Free Cereal Products and Beverages*; Arendt, E.K., dal Bello, F., Eds.; Elsevier: Amsterdam, The Netherlands, 2008; pp. 289–319.
83. Galassi, E.; Taddei, F.; Ciccoritti, R.; Nocente, F.; Gazza, L. Biochemical and technological characterization of two C4 gluten-free cereals: *Sorghum bicolor* and *Eragrostis tef*. *Cereal Chem.* **2020**, *97*, 65–73. [CrossRef]
84. Arendt, E.K.; dal Bello, F. Functional Cereal Products for Those with Gluten Intolerance. In *Technology of Functional Cereal Products*; Hamaker, B.R., Ed.; Woodhead Publishing: Cambridge, UK, 2008; pp. 446–475.
85. Duodu, K.G.; Taylor, J.R.N. The Quality of Breads Made with Non-Wheat Flours. In *Breadmaking*; Cauvain, S.P., Ed.; Woodhead Publishing: Cambridge, UK, 2012; pp. 754–782.
86. Marti, A.; Marengo, M.; Bonomi, F.; Casiraghi, M.C.; Franzetti, L.; Pagani, M.A.; Iametti, S. Molecular features of fermented teff flour relate to its suitability for the production of enriched gluten-free bread. *LWT* **2017**, *78*, 296–302. [CrossRef]
87. Yilmaz, H.O.; Arslan, M. Teff: Nutritional compounds and effects on human health. *Acta Sci. Med. Sci.* **2018**, *2*, 15–18.
88. Olojede, A.O.; Sannic, A.I.; Banwo, K. Rheological, textural and nutritional properties of gluten-free sourdough made with functionally important lactic acid bacteria and yeast from Nigerian sorghum. *LWT Food Sci. Technol.* **2020**, *120*, 108875. [CrossRef]
89. Rico, D.; Ronda, F.; Villanueva, M.; Perez Montero, C.; Martin-Diana, A.B. Development of healthy gluten-free crackers from white and brown teff (*Eragrostis tef Zucc.*) flours. *Heliyon* **2019**, *5*, e02598. [CrossRef]
90. Chanapamokkhot, H.; Thongngam, M. The chemical and physico-chemical properties of sorghum starch and flour. *Kasetsart J.* **2007**, *41*, 343–349.
91. Pandya, T.S.; Srinivasan, R. Effect of hammer mill retention screen size on fiber separation from corn flour using the Elusieve process. *Ind. Crops Prod.* **2012**, *35*, 37–43. [CrossRef]
92. Zannini, E.; Pontonio, E.; Waters, D.M.; Arendt, E.K. Applications of microbial fermentations for production of gluten-free products and perspectives. *Appl. Microbiol. Biotechnol.* **2012**, *93*, 473–485. [CrossRef] [PubMed]
93. Ćuric, D.; Novotni, D.; Tusak, D.; Bauman, I.; Gabric, D. Gluten-free bread production by the corn meal and soybean flour extruded blend usage. *Agric. Consp. Sci.* **2007**, *72*, 227–232.
94. Wolter, A.; Hager, A.S.; Zannini, E.; Czerny, M.; Arendt, E.K. Impact of sourdough fermented with *Lactobacillus plantarum* FST 1.7 on baking and sensory properties of gluten-free breads. *Eur. Food Res. Technol.* **2014**, *239*, 1–12. [CrossRef]
95. Nami, Y.; Gharekhani, M.; Aalami, M.; Hejazi, M.A. Lactobacillus-fermented sourdoughs improve the quality of gluten-free bread made from pearl millet flour. *J. Food Sci. Technol.* **2019**, *56*, 4057–4067. [CrossRef]
96. Edema, M.O.; Emmambux, M.N.; Taylor, J.R.N. Improvement of fonio dough properties through starch modification by sourdough fermentation. *Starch* **2013**, *65*, 730–737. [CrossRef]
97. Schober, T.J.; Bean, S.R.; Boyle, D.L. Gluten-free sorghum bread improved by sourdough fermentation: Biochemical, rheological, and microstructural background. *J. Agric. Food Chem.* **2007**, *55*, 5137–5146. [CrossRef]
98. Sahin, A.W.; Wiertz, J.; Arendt, E.K. Evaluation of a new method to determine the water addition level in gluten-free bread systems. *J. Cereal Sci.* **2020**, *93*, 102971. [CrossRef]
99. Tafti, A.G.; Peighardoust, S.H.; Behnam, F.; Bahrami, A.; Aghagholizadeh, R.; Ghamari, M.; Rafat, S.A. Effects of spray-dried sourdough on flour characteristics and rheological properties of dough. *Czech J. Food Sci.* **2013**, *31*, 361–367. [CrossRef]

100. Moore, M.M.; dal Bello, F.; Arendt, E.K. Sourdough fermented by *Lactobacillusplantarum* FST 1.7 improves the quality and shelf life of gluten-free bread. *Eur. Food Res. Technol.* **2008**, *226*, 1309–1316. [CrossRef]
101. Wolter, A.; Hager, A.S.; Zannini, E.; Czerny, M.; Arendt, E.K. Influence of dextran-producing *Weissella cibaria* on baking properties and sensory profile of gluten-free and wheat breads. *Int. J. Food Microbiol.* **2014**, *172*, 83–91. [CrossRef]
102. Axel, C.; Rocker, B.; Brosnan, B.; Zannini, E.; Furey, A.; Coffey, A.; Arendt, E.K. Application of *Lactobacillus amylovorus* DSM19280 in gluten-free sourdough bread to improve the microbial shelf life. *Food Microbiol.* **2015**, *47*, 36–44. [CrossRef]
103. Di Cagno, R.; Rizzello, G.C.; de Angelis, M.; Cassone, A.; Giuliani, G.; Benedusi, A.; Limitone, A.; Surico, R.F.; Gobbetti, M. Use of selected sourdough strains of *Lactobacillus* for removing gluten and enhancing the nutritional properties of gluten-free bread. *J. Food Prot.* **2008**, *71*, 1491–1495. [CrossRef]
104. Falade, A.T.; Emmambux, M.N.; Buys, E.M.; Taylor, J.R.N. Improvement of maize bread quality through modification of dough, rheological properties by lactic acid bacteria fermentation. *J. Cereal Sci.* **2014**, *60*, 471–476. [CrossRef]
105. Moghaddam, M.F.T.; Jalali, H.; Nafchi, A.M.; Nouri, L. Evaluating the effects of lactic acid bacteria and olive leaf extract on the quality of gluten-free bread. *Gene Reports* **2020**, *21*, 100771. [CrossRef]
106. Corsetti, A.; Settanni, L. Lactobacilli in sourdough fermentation. *Food Res. Int.* **2007**, *40*, 539–558. [CrossRef]
107. Hansen, A.; Schieberle, P. Generation of aroma compounds during sourdough fermentation: Applied and fundamental aspects. *Trends Food Sci. Technol.* **2005**, *16*, 85–94. [CrossRef]
108. Bourne, M.C. Texture profile analysis. *Food Technol.* **1978**, *32*, 62–66.
109. Poutanen, K.; Flander, L.; Katina, K. Sourdough and cereal fermentation in a nutritional perspective. *Food Microbiol.* **2009**, *26*, 693–699. [CrossRef]
110. Tieking, M.; Gänzle, M.G. Exopolysaccharides from cereal-associated lactobacilli. *Trends Food Sci. Technol.* **2005**, *16*, 79–84. [CrossRef]
111. Belz, M.C.E.; Axel, C.; Arendt, E.K.; Lynch, K.M.; Brosnan, B.; Sheehan, E.M.; Coffey, A.; Zannini, E. Improvement of taste and shelf life of yeasted low-salt bread containing functional sourdoughs using *Lactobacillus amylovorus* DSM 19280 and *Weisella cibaria* MG1. *Int. J. Food Microbiol.* **2019**, *302*, 69–79. [CrossRef]

foods

Review

Gluten-Free Bread and Bakery Products Technology

Zuzana Šmídová * and Jana Rysová

Food Research Institute Prague, Radiová 7, 102 00 Prague, Czech Republic; jana.rysova@vupp.cz
* Correspondence: zuzana.smidova@vupp.cz

Abstract: Gluten, a protein fraction from wheat, rye, barley, oats, their hybrids and derivatives, is very important in baking technology. The number of people suffering from gluten intolerance is growing worldwide, and at the same time, the need for foods suitable for a gluten-free diet is increasing. Bread and bakery products are an essential part of the daily diet. Therefore, new naturally gluten-free baking ingredients and new methods of processing traditional ingredients are sought. The study discusses the use of additives to replace gluten and ensure the stability and elasticity of the dough, to improve the nutritional quality and sensory properties of gluten-free bread. The current task is to extend the shelf life of gluten-free bread and bakery products and thus extend the possibility of its distribution in a fresh state. This work is also focused on various technological possibilities of gluten-free bread and the preparation of bakery products.

Keywords: gluten-free products; bread; bakery products; cereals; enzymes; sourdough

Citation: Šmídová, Z.; Rysová, J. Gluten-Free Bread and Bakery Products Technology. *Foods* **2022**, *11*, 480. https://doi.org/10.3390/foods11030480

Academic Editors: Maria Papageorgiou and Theodoros Varzakas

Received: 10 January 2022
Accepted: 2 February 2022
Published: 7 February 2022

1. Introduction

A gluten-free diet is the only treatment for people suffering from gluten intolerance. Gluten consumption leads to a range of gluten-related disorders, such as coeliac disease, dermatitis herpetiformis (cutaneous manifestation of coeliac disease), gluten ataxia and non-coeliac gluten sensitivity [1]. A gluten-free diet requires the use of gluten-free cereals—corn, rice, sorghum, millet, teff—and pseudo-cereals—buckwheat, quinoa, amaranth, canihua—but also other foods that are naturally gluten-free—potatoes, tapioca, nuts, oilseeds, legumes, fruits and vegetables [2]. The main challenges for food technologists are bread, bakery products, pastry and pasta. Because of the absence of gluten, other substances needed to maintain the texture, volume, satisfactory crumb, shelf life and sensory quality must be used. These include the use of hydrocolloids, sourdough or enzyme preparations. The use of them is intended to change the recipe and the production technology.

Gluten-free bread and other gluten-free bakery products are very unusual for a consumer accustomed to classic wheat or wheat–rye bread. Toth et al. (2020) have shown that 70.8% of the asked consumers were dissatisfied with gluten-free breads due their texture and taste [3]. Gluten-free breads usually have a less flexible crumb, which hardens faster, and which is easy to crumble. The taste of these products is also different, depending, of course, on the ingredients used. Gluten-free products are easily accepted by people who have been suffering from gluten intolerance since childhood. The acceptance of a gluten-free diet, and at the same time, the acceptance of gluten-free bread in adults diagnosed with gluten intolerance later in life, is more difficult.

According to Codex Standard 118-1979 [4], gluten represents a protein fraction from wheat, rye, barley, oats or their hybrids and derivatives that some people are intolerant to and that is insoluble in water and 0.5 M sodium chloride solution. Water-insoluble prolamins and glutelins (collectively referred to as gluten) usually make up 70–80% of cereal grain proteins. They are the most important cereal proteins from a technological point of view. In this sense, gluten is a specific structure, a viscoelastic gel that gives wheat dough and bakery products its unique properties. This gel is formed after the addition of water and kneading, when the wheat proteins gliadins and glutenins swell, and with the simultaneous access of oxygen as a complex, a three-dimensional viscoelastic system (gluten in the original technological sense) is created which ensures the required viscoelastic properties of the dough. The result is a three-dimensional sufficiently strong and flexible continuous network capable of maintaining a large volume of gas, and thus ensuring the sufficient volume, shape and texture of the products.

Prolamins and glutelins are represented by a number of related proteins with somewhat different amino acid composition and structure (e.g., gliadin proteins are usually up to several dozen for each wheat variety). In particular, the prolamins of wheat (as well as spelt, Khorasan, einkorn and emmer), rye and barley and their hybrids (Triticale, Tritordeum) cause a disease in predisposed individuals, which is called celiac disease. The relationship between oat prolamins and celiac disease is still the subject of debate. Wheat gliadin fractions are divided into four subfractions: α-, β-, γ- and ω-gliadins (Figures 1 and 2). All subfractions of α-/β-/γ-/ω—gliadins can cause celiac disease in predisposed individuals, and ω-5 gliadins allergic reaction. Celiac disease is caused by two amino acid sequences: ProSer-Gln-Gln (PSQQ) and Gln-Gln-Gln-Pro (QQQP) [5].

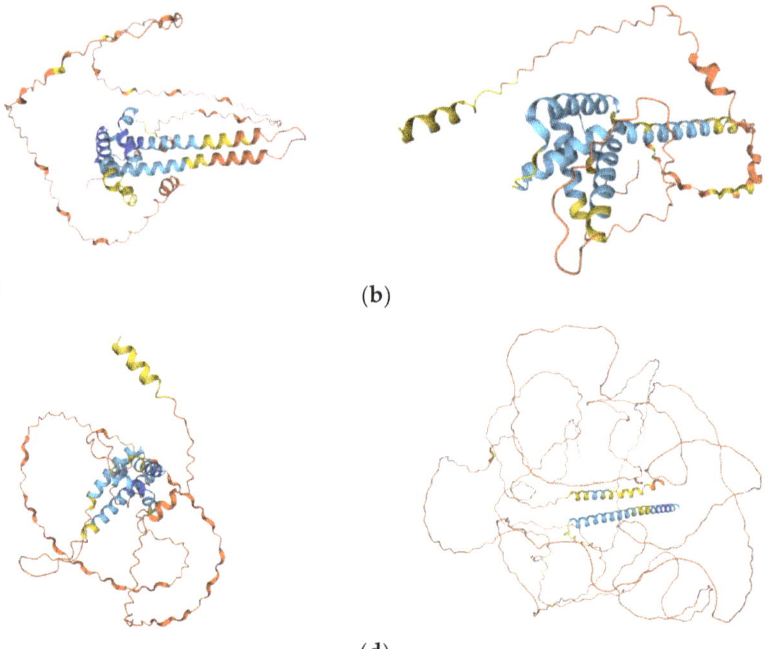

(a) (b) (c) (d)

Figure 1. Gluten structure from the UNIPROT database https://www.uniprot.org/ (accessed date 31 January 2022) [6]. The picture shows (**a**) glutenin LMW subunit; (**b**) α,β-gliadin; (**c**) γ-gliadin; (**d**) glutenin HMW subunit. AlphaFold produces a per-residue confidence score (pLDDT) between 0 and 100. Some regions with low pLDDT may be unstructured in isolation. The colours represent the model confidence: dark blue–Very high (pLDDT > 90); light blue–Confident (90 > pLDDT > 70); yellow –Low (70 > pLDDT > 50); red–Very low (pLDDT < 50).

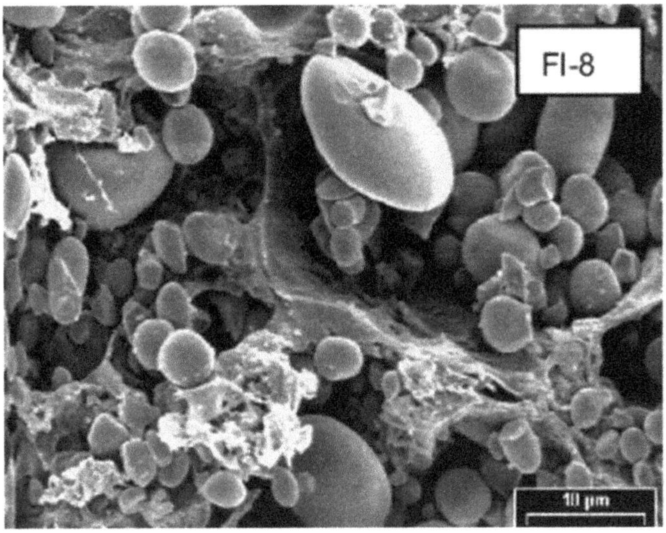

Figure 2. Scanning electron micrograph of the gluten dough prepared in Brabender Farinograph from a wheat flour after 8 min. of mixing (FI-8, magnification: 2600×). Adapted from [7].

2. Raw Materials for Gluten-Free Bread and Bakery Products

The specific technological properties of typical gluten cereals in the production of gluten-free bread and bakery products need to be replaced. Rice, corn, or sorghum and other gluten-free cereals are the basis of the diet in many countries around the world. For the preparation of bread and bakery products from gluten-free raw materials, it is necessary to ensure the volume and cohesion of the dough. Rapid staling of these products is a big problem. The nutritional value of wheat and gluten-free bread can be different. Table 1 shows the comparison of nutritional values between wheat flour and gluten-free bread baking mixtures (or flour, respectively) and Table 2 the composition of fresh gluten-containing and gluten-free buns. The nutritional values strong depend on the raw material composition of these products and are not uniform.

Table 1. Nutritional values of wheat flour and gluten-free bread baking mixtures. The values were obtained from the product packages or available on https://itesco.cz/ (accessed on 24 January 2022).

Flour Nutritional Values	Wheat Flour per 100 g	Gluten-Free 1 per 100 g	Gluten-Free 2 per 100 g	Gluten-Free 3 per 100 g	Gluten-Free 4 per 100 g	Gluten-Free 5 per 100 g
Energy (kJ)	1430	1517	919	1490	1497	1475
Energy (kcal)	337	362	219	356	358	351
Fats (g)	1	1.9	4.4	0.7	5.6	0.9
of which saturates (g)	0.2	0.5	1.9	0.1	0.6	0.2
Carbohydrates (g)	69	81.9	42	84	66	80
of which sugars (g)	2	3.8	<0.5	<0.5	0.8	1.4
Proteins (g)	12	3.2	2.3	2.4	7.2	2.7
Fiber (g)	2	-	1.1	-	6.0	4.4
Salt (g)	<0.005	0.2	1.4	1.5	2.5	0.83

Table 2. Nutritional values of wheat buns and gluten-free buns. These values were obtained from the product packages or available on https://itesco.cz/ (accessed on 24 January 2022).

Fresh Bun Nutritional Values	Conventional per 100 g	Gluten-Free per 100 g
Energy (kJ)	1352	1144
Energy (kcal)	320	272
Fats (g)	5.4	8.9
of which saturates (g)	1.6	1.8
Carbohydrates (g)	55.8	42
of which sugars (g)	1.2	3.9
Proteins (g)	10.0	4.4
Fiber (g)	2.9	3.1
Salt (g)	1.5	1.3

In addition to basic gluten-free ingredients such as gluten-free flours and starches, technologically and nutritionally functional ingredients such as hydrocolloids of cereal and non-cereal origin [8], fruit or vegetable fiber [9,10], flax and chia seeds [11,12], psyllium [13], modified starches (e.g., [14]) and proteins [15] from many sources need to be added to achieve sufficient bread volume, crumb softness and shelf life. The addition of fiber, through its hydration, affects the quality of the bread. Besides the beneficial health effects, fiber improves texture, specific volume, apparent viscosity, consistency, texture, sensory quality and shelf life. This is due to the ability to bind water, form a gel and thicken [8]. The key parameters are fiber length, polymerization degree, soluble/insoluble fiber ratio and fiber interactions with other ingredients [16–19]. Soluble fiber improves dough; coarse fiber reduces gas retention [13,20]. The use of enzyme preparations improves the colour of crumbs, supports the production of flavors, increases specific volume and prevents starch retrogradation [21]. Krishna et al. (2019) [11] evaluated the effect of 1% (total base) addition of flax seed powders from flax (*Linum usitatissimum*) and acacia seed powder on the pasting properties, texture and volume of gluten-free bread. The addition of all seed powders reduced crumb hardness by 30–65% and increased specific loaf volume by 50%. These textural improvements were caused by water absorption capacity and emulsifying ability. A darker bread crumb was observed after flax addition, whereas after acacia addition, dark particles were visible. Scanning electron microscopy of these breads showed the absence of holes in the pore surface and viscoelastic starch–protein network. Steffolani et al. (2004) [12] observed a reduction in specific volume and an increase in bread hardness after the addition of chia seed or flour into rice breads, when the effect was more evident with the flour than with seeds. Chia addition minimized weight loss during baking. Chia flour addition led to a darker crust and crumb. No significant differences between the different breads in acceptability were noted; however, chia seed breads presented better texture than controls. Additionally, Sandri et al. (2017) [22] prepared chia-containing rice breads with acceptable sensory properties when the best formulations were prepared from rice flour blends with 5, 10 and 14% whole chia flour. The overall acceptability scores were 8.7, 8.1 and 7.9 out of the 10-point scale, and were very similar to their white gluten-free bread and wheat bread counterparts. The addition of chia flour was acceptable up to 14%. The use of 5–14% whole chia flour increased the levels of lipids, proteins and dietary fiber compared with the white gluten-free breads.

Moreover, there is a trend to use fermentation processes in bread baking so that the products resemble sensory properties of sourdough bread, e.g., the use of sourdough made from gluten-free flour (e.g., [23]). In addition to organic acids, the lactic bacteria also produce free amino acids as precursors of flavours, contemporary degrade phytates and starch and change the fiber solubility. The resulting organic acids and antimicrobial substances extend the shelf life of the bread. Some lactic acid bacteria produce exopolysaccharides, which can affect the rheological properties of the dough [24].

Another possibility of raw material modification is the use of controlled germination. Germinated seeds are characterized by improved taste and nutritional properties. Germination activates seed enzymes, the partial degradation of storage substances into simpler sugars or peptides occurs and the availability of minerals increases. Similar to fermentation, germination also produces a number of secondary metabolites and leads to the degradation of antinutritive substances [25–27].

An integral part of gluten-free dough is emulsifiers, which enable easier processing of the dough and soften the crumb. Lecithin, mono- and diglycerides of fatty acids and esters of fatty acids with lactic acid are used for this purpose. Milk, egg yolk (except for bread), soy protein, sunflower and lupine flour can also serve as emulsifiers [28,29].

The basic raw materials for gluten-free bread and bakery production are gluten-free flours or native starches to which additional ingredients must be added that "substitute gluten" and ensure optimal dough properties. Flours represent more complex materials in comparison with starches, which also include proteins and a low amount of lipids, as well as some minor components such as fiber, vitamins and minerals. Therefore, they are more convenient. For gluten-free products, there is only scarce information concerning flour requirements. These flours differ in starch characteristic (e.g., amylose and amylopectin ratio), in protein amount, and in particle size and their distribution. As far as gluten-free wheat starch present on the market is concerned, it does not have harmful effects on most celiacs; however, celiac people are still unwilling to consume products with wheat-based ingredients [30].

3. Gluten-Free Dough Specifications

Gluten-free dough is a very complex semi-liquid system that contains polysaccharides and other structure-forming components, viscosity-increasing substances and dough-stabilizing substances. It is characterized by high density and low elasticity. Gluten-free dough contains more water than conventional wheat dough. The amount of water depends on the nature of the basic raw materials, their ability to absorb water and the granulation of the raw materials. Additionally, the kneading, its length and its speed are very important. Prolonged kneading increases the specific volume of bread [31].

When baking, the proteins are denatured with increasing temperature, and the starch gelatinization occurs. A sufficiently strong and flexible spatial structure should be created to maintain the expanding gas bubbles and not collapse during baking or cooling of the product. However, gluten-free flours and starches alone do not create such a structure. Therefore, hydrocolloid addition is necessary because of their swelling and water binding capacity [8,32]. Proper hydration affects the conformation of polymer molecules and the rheological properties of the dough. It also determines the texture and softness of the crumb and the crunchiness of the crust. The less hydrated dough provides a small volume of bread, the more hydrated dough can be processed better and the fermentation takes place better in it. Without the addition of other raw materials, the product is irregular in shape, not very cohesive and the crumb is not sufficiently supple and flexible. If the gluten-free recipe contains less protein, the product has usually light crust. There are not enough amino acids available to enter the Maillard reaction [33,34].

After baking, the crust on the surface is firm and crunchy, whereas the crumb retains moisture. After a certain storage time, the moisture in both parts of the bakery product begins to equalize. The water retained in the crumb diffuses to the surface of the product; the crust softens and may deform. On the contrary, through this redistribution process the crumb loses water, thus reducing the flexibility and suppleness of the bread (Table 3). Starch tends to return from an amorphous state to a crystalline form. Starch is recrystallized (retrograded) and bread is staling. The crumb is now brittle and incoherent or hard [35]. Rapid staling and thus limited shelf life are a big problem with gluten-free products. Short shelf life limits the possibility of the sale of fresh gluten-free bread and is one of the reasons for using dry mixtures for home baking of gluten-free bread and bakery products [33].

Table 3. The table shows the generalized differences between wheat dough and bread and between gluten-free dough and bread. Adapted and modified according to [34,36–40].

	Wheat Flours	**Gluten-Free Flours**
	Raw materials	
swelling	good	better
	Dough	
repeated kneading	yes	no
stickiness	no/small	typically high
	Dynamic oscillation rheometry	
G′storage modul	lower	higher
G′′loss modul	lower	higher
phase angle tg(d)	higher	lower
	Extensograph	
extensibility	high	poor
extensibility resistance	high	mostly lower
R/E ratio	mostly lower	mostly higher
area under the curve (extensibility energy)	high	very low
	Farinograph	
development time	low	different according to the raw material
stability	high	different according to the raw material
degree of softening	not a clear trend	not a clear trend
water binding	mostly lower	mostly higher
	Bread	
volume	high	low
crust color	darker	light
crust	crunchy	more moist, dense
crumb elasticity	good	low
porosity	good	low
pore size	large	small
staling rate	slow	faster
crust moisture	optimal	more moist
crumbliness	low	significantly higher
hardness	soft	higher

The viscoelastic properties of different doughs significantly influence the volume, and the crumb texture of gluten-free baked goods [41]. One of the most important factors affecting the quality of bread and bakery products is the flour particle size. For bread production, bigger particle size is more suitable and particles below 80–100 μm should not be used if gluten-free bread with high volume and soft crumb is desired. Flours with very large particles may result in breads with sandy texture; therefore, 200 μm is the maximum particle size. Flours with larger particles have been proven to reduce the dough gas retention capacity as well as the final bread volume [30].

3.1. Proteins in Gluten-Free Dough and Products

Proteins improve the nutritional value of gluten-free products. The choice of flour and possibly another source of protein affects the rheological properties of the dough and the water binding in the dough. Proteins interact with starch and lipids and together contribute to the stability of the dough and the structure of the product. They also give the impression of full product flavour. Proteins can be of plant origin (legumes, soya, gluten-free cereals, rapeseed, canola, sunflower, potato), animal origin (whey, egg, casein, caseinate) or microorganism-, algae-, seaweed- and insect-based [15].

Conventional proteins represent egg and milk proteins. Eggs are very useful in forming the structural network, but they are not usually used in bread. Milk proteins, including caseinates and whey protein concentrates, are sources of calcium and can bind moisture satisfactorily. They have a positive effect on the colour and volume of bread and bakery products. For example, protein-rich gluten-free bread with the addition of 15% whey protein concentrate and 3% of HPMC were prepared by Rustagi et al. (2018) [42]. However, many celiacs do not tolerate lactose and must omit milk from their diets [43].

The source of proteins are naturally gluten-free cereals—rice, corn, teff, sorghum, Job's tears (Figure 3). These cereals contain prolamins too (an ethanol-soluble protein fraction), but the molecules of these proteins do not contain amino acid sequences that are toxic to people with celiac disease. Rice is often used as the basis for gluten-free bread formulations. Gluten-free oats have also very good properties [44]. The zein protein, a prolamin from corn, behaves similar to gluten when heated to 35–40 °C. Both corn zein and sorghum kafirin increase the plasticity of the dough [31,45].

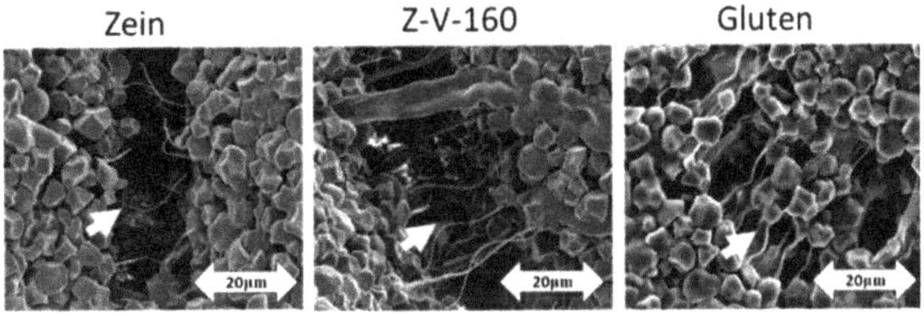

Figure 3. Images of dough samples containing starch with zein, thermally treated-zein (zein heated in vacuum at 160 °C, Z-V-160) and gluten. Adapted from [46]. The white arrows in the image represent the scale-they indicate the distance of 20 μm.

Cereal flours are combined in gluten-free bread and bakery products formulations with flour from other crops and with starches. A combination of cereals with legumes is nutritionally advantageous. Pseudocereals also contain proteins with a preferred amino acid composition [8]. This category includes buckwheat, amaranth and quinoa. For example, Föste et al. (2013) [47] used various buckwheat milling fractions, rice and corn flour and fermented buckwheat brans for gluten-free bread preparation. The specific bread volume, porosity and crumb texture can be improved by using buckwheat flour. Soya is a traditional gluten-free ingredient. Soy protein has a very advantageous amino acid composition, participates in interactions with other substances, binds water well and slows down the staling of bread. Legume flours from pea, chickpea, lupine, lentil and bean are rich in proteins with a high lysine content. They significantly affect the dough quality. Lupine and soya flour show emulsifying properties because of the lecithin content. Legumes reduce the glycemic index of food products. The disadvantage of legumes is their typical taste [35,48–50]. Recently, other flours, such as from nuts and seeds, have appeared in the range of bakery products. Typically, walnut flour and peanut flour are relatively expensive, but very suitable for some formulations. Fat in walnut flour or other nuts flours (except for coconut) contains polyunsaturated fatty acids, and nut proteins are also of very high quality due to their composition. Coconut flour binds water very well, too [51,52].

An alternative solution is the use of insect proteins, e.g., cricket flour which improves the texture of gluten-free bread [31].

The use of proteins in the form of protein concentrates or isolates from different sources in gluten-free baked goods leads to the quality and nutritional profile improvement [15]. Through the comparison of plant- and animal-based proteins, Gorissen et al. [53] reported lower content of essential amino acids in plant-based protein isolates than in animal-based proteins. Differences in the composition of the amino acid spectrum of gluten-free raw materials are known, especially in the composition of essential amino acids. Cereal proteins are deficient in lysine, and some cereals also in threonine and tryptophan. Therefore, from this point of view, it is recommended to combine different types of plant proteins and thus optimize the ratio of amino acids. The combination of different ingredients makes easier to ensure the presence of other nutrients, such as vitamins or minerals.

3.2. *Starch in Gluten-Free Dough and Products*

Starch, together with flours from gluten-free crops, is one of basic ingredients in gluten-free bread and bakery products. It is involved in the formation of the crumb structure, responsible for the volume and colour of the product. It is also used as a thickening, gelling, stabilizing, moisture retention and anti-staling agent [54]. According to Abdel-Aal (2009) [55], starch influences gluten-free products in three ways: it enhances crumb softness, ensures dough consistency and affects starch gelatinization. Starch is stored in starch grains of various sizes and shapes according to its plant source. Individual starches differ in their composition, size and shape depending on the plant species and the interactions between genes and environment [54]. The amylose starch fraction forms single chains, whereas amylopectin is branched with a significantly larger molecule. When heated in suspension/dough, the starch grains swell, are partially solubilized [56], and gradually lose their cohesiveness. Starch gelatinization occurs at a temperature of 50–70 ° C, when their chains are released, and a viscous solution is formed from the suspension. Upon cooling, the viscosity increases, new bonds are formed between the molecules and a gel is formed. During storage, the gel further changes, loses water and eventually retrogrades. Amylose retrogradation proceeds faster than the same process for amylopectin. It follows that by choosing the type of starch, it is possible to partially influence the staling of the bread [54]. Starch behavior may be affected by bound lipids.

Native starches are the most commonly used in gluten-free products, e.g., potato, corn, rice and tapioca, and pea starch has also appeared. Specially prepared gluten-free wheat starch is also used for its properties ensuring an optimal bread texture [15,57]. Modified starches are produced for food purposes have a wide range of physical properties according to the purpose they are used. Starches can be modified by heating of the starch solution or by heating in the dry state; the heating can be performed by drying or extrusion. Chemical modification of starch is also possible [58]. For gluten-free products, starches with good water absorption and slow retrogradation are selected. Specially modified starches are suitable for frozen products. A new modification is the so-called superheated starch, prepared by heating the starch suspension to high temperatures until dissolved and then cooling to form a spreadable gel with a creamy consistency [59,60]. Additionally, various types of banana flour can be applied as well as the direct use of bananas in the dough [61].

There are significant differences in granular structure among various types of starches, which affect their ability to produce high quality gluten-free baked goods. When the baked goods are based on starch, they show higher volume, lower hardness and a lighter crust since Maillard reactions are reduced. The starch addition results in softer, and resilient crumbs. The type of starch also influences the quality of the baked goods. For the specific gluten-free formulations different mixtures of flours and starches must be optimized [30,41].

Preventing the retrogradation of starch and thus prolonging the shelf life of gluten-free bread and bakery products can be achieved in several ways:

(a) Using enzyme preparations.
(b) Application of hydrocolloids.
(c) Using sourdough fermentation.
(d) Suitable packaging method.

3.2.1. Use of Enzyme Preparations

The most common enzyme preparations use amylases, which improve the colour of crumbs and support the production of flavors. Amylases partially degrade amylopectin and thus modify the starch recrystallization process [62–64]. Transglutaminase improves the dough viscoelasticity and decrease crumb hardness, and cyclodextrinase also enhances dough viscoelasticity, leading to improvement in shape index and crumb firmness [65].

Cyclodextrin glycosyltransferases form cyclic structures from starch with different affinities for water outside and inside the ring. Lactase and tyrosinase create crosslinks of non-starch polysaccharides with proteins with the use of phenolic substances [25]. Oxidases such as lipoxygenase, sulfhydryl oxidase, glucose oxidase and peroxidase stabilize the dough; for example, glucose oxidase added to rice bread improved volume and reduced stiffness. Proteases and peptidases improve the interaction between protein molecules and starch and reduce the viscosity of the dough [31,66]. Microbial transglutaminase, which forms covalent bonds between the free epsilon-amino group of lysine and the amide group of glutamine, is used to promote the formation of a spatial network of gas bubble trapping molecules. Transglutaminase supports the rheological and viscoelastic properties of the dough [67–71]. Silva et al. (2020) [72] tested gluten-free bread from red rice flour and cassava flour, with the addition of transglutaminase and chitosan at concentrations of 0%, 1% and 2%. Bread with chitosan and transglutaminase showed lighter brown coloration because of incomplete Maillard reaction and low specific volumes, probably related to chitosan interference with yeast fermentation. With the use of chitosan, viscosity increased. Bread containing chitosan had a lower rate of staling due to water retention.

The use of enzymes influences the quality of the gluten-free baked goods, and the effect depends on the type of the flour used. Some enzymes have positive effects on product volume and delay staling [30].

3.2.2. Use of Hydrocolloids

The application of hydrocolloids is crucial for the quality of gluten-free bread. Hydrocolloids swell and form a gel. This heated gel thickens the mass of dough forming the walls of gas bubbles, preventing the loss of gas released during whipping, leavening or from raising agents. After baking, hydrocolloids stabilize the crumb structure, bind water, and prevent rapid starch retrogradation. They stabilize the product during freezing. Due to the higher water binding, the recipes with the addition of hydrocolloids contain higher doses of water [32,73].

Vegetable gums (guar gum, locust bean gum, arabic gum, tara gum, carob, konjac gum), beta-glucans, pentosans and arabinoxylans, cellulose derivatives (methyl cellulose MC, carboxymethyl cellulose CMC, hydroxypropylmethyl cellulose HPMC), microbial exopolysaccharides (xanthan, dextrins) and seaweed polysaccharides (agar, carrageenans) are used to produce gluten-free bread and bakery products [42,74–76]. Flax flour or ground chia seeds are sometimes used to increase viscosity.

Cellulose derivatives, especially CMC and HPMC, are among the most used hydrocolloids in gluten-free dough. They can interact with other raw materials in the matrix. They are most often combined with other types of thickeners, proteins, and emulsifiers [66]. For example, Liu et al. (2018) [77] compared the effect of HPMC, CMC, xanthan gum and pectin on the behavior of steamed potato dough. The addition of 2% HPMC increased mostly the specific volume of bread and porosity and reduced the stiffness of the crumb by almost 29%. The addition of hydrocolloids significantly reduced the content of both readily available and slowly available starch and, conversely, increased the content of resistant starch. Model experiments with HPMC and rice flour on Mixolab examined the effect of HPMC dose (1–3%) and water dose (90–110%) on the rheological properties of gluten-free dough and crumb quality. The optimal dose is 2.2% HPMC and 110% water) [78]. Lazaridou et al. (2007) [79] showed that 1% CMC and 2% pectin led to breads with improved breads volume, porosity and crumb elasticity. With the use of HPMC, Hager et al. (2013) [80] observed an increased volume in corn and teff breads, a decreased size of rice breads and a positive effect on the crumb hardness of each bread.

Salehi (2019) [41] dealt with the application of HPMC, CMC and other hydrocolloids in rice flour dough. HPMC in combination with carrageenan forms a softer crumb. To slow down the aging of bread, the addition of CMC or HPMC 0.1–0.5% is recommended.

Kaur and Chopra (2018) [61] tested 74% corn starch bread with tapioca, rice flour and 2.2% HPMC. Belorio and Gómez (2020) [81] tested the use of different types of hydrocolloids (HPMC, xanthan and psyllium) in rice and corn bread and the effect of water levels. The water dose has been optimized to form a thermoreversible gel that provides a sufficient volume of bread. For corn bread with added HPMC, the optimal hydration was 80%; for rice bread with HPMC, the hydration was higher—100%.

Fruit and vegetable pomace containing fiber and antioxidants can also be used [9,41,82]. Djeghim et al. (2021) [9] observed the addition of various by-products with gluten-free bread formulations based on corn and chickpea flours (2/1 *w*/*w*)—orange and apple pomace, tomato peel, pepper peel, prickly pear peel and prickly pear seed peel on the dough rheology and properties of gluten-free breads. They found out that the addition of the above-mentioned by-products significantly improved the specific volume of gluten-free bread, with values increasing from 1.48 to 2.50 cm^3/g, and increased the maximum dough height, the total CO_2 production and CO_2 retention coefficient.

The effect of apple, orange and carrot pomace powders, on dough rheology and quality characteristics of the rice sweet bakery were studied by Kirbas et al. (2019) [10]. With an increase in the content of pomace powders, the dough elasticity, specific gravity and apparent viscosity increased. The addition of pomace powder increased crumb hardness and decreased the specific volume of the rice-based sweet bakery products. The addition of 5% of orange pomace powder had the highest acceptance scores with respect to the colour, flavour, texture, appearance and acceptability of the products.

The comparison of breads and bakery products prepared from gluten-free flour alone and with various hydrocolloids (gums) showed that the incorporation of hydrocolloids led to a significant improvement in the texture, volume, color, appearance, flavor and overall acceptability. Different hydrocolloids have slightly different effects on rheology, texture and other properties, thus affecting the resulting quality of various types of gluten-free breads and bakery products. For example, xanthan gum is able to maintain unchanged texture parameters during storage; the addition of xanthan, carrageenan and guar gums decrease dough extensibility, whereas arabic gum and HPMC lead to increased extensibility [41]. HPMC is preferred to other hydrocolloids since it provides gluten-free products with appropriate physical characteristics, higher specific volumes of products and better sensory properties [30]. Moreover, the use of hydrocolloids is the easiest way to raise the content of dietary fiber in gluten-free baked goods [8].

3.2.3. Microbial Fermentation in Gluten-Free Bread Production

The use of sourdough is a traditional procedure in conventional baking technology. In the preparation of gluten-free bread, starter cultures began to be applied later, because gluten-free raw materials have a specific composition different from rye flour; therefore, the classical culture of rye sourdough bacteria and yeasts may not grow sufficiently in gluten-free substrates. Although it is possible to gradually "dilute" rye flour with gluten-free raw material during repeated fermentation so that the proportion of rye is reduced to a minimum value, such a process would take a long time, and there would still be the danger of the presence of gluten traces [83]. Therefore, suitable strains of microorganisms capable of fermenting rice, buckwheat, sorghum or corn flour are sought. The choice of a suitable starting culture will significantly affect the resulting properties of the dough and the product. Additionally, the oilseed, chia and flaxseed sourdoughs can be used [84]. During fermentation, the dough is acidified. At the same time, the enzymes naturally contained in the flour are also activated and break down high molecular weight substances, and thus make them more accessible. The activity of the cereal grain's own enzymes is combined with the action of microbial enzymes. Substances affecting the taste and smell of the products are formed. Fermentation increases the swelling of carbohydrates and improves the viscoelastic properties of the dough. The fermentation products include organic acids with a predominance of lactic and acetic acids, but some strains also produce propionic acid [85–89]. These acids significantly increase the shelf life of the bakery products. For example, Kaur and Chopra (2018) [61] deal with the use of teff flour and rice sourdough as a possible combination. Bacterial strains of sourdough microflora, for example *Lactobacillus reuterii* or *Weisella cibaria*, are able to produce the exopolysaccharides fructan, levan, dextran or reuteran. These polysaccharides naturally increase the viscosity of the dough and thus contribute to the formation of the product texture. The presence of these polysaccharides reduces the hardness of the crumb, improves its porosity, and slows down the staling of the bread [25,31]. Additions of dried sourdough are also being applied. The advantage of dried sourdough is its standardized quality, the disadvantage is the possible inactivation of living microbial strains during the drying process [90]. A non-traditional sourdough using *Lactobacillus sanfranciscensis* for fermentation of chia, quinoa and hemp flour to produce gluten-free corn/rice bread was tested by Jagelaviciute and Cizeikiene (2021) [87]. This sourdough showed a decreased pH, specific volume and rate of bread staling and, on the other hand, increased bread porosity compared with bread made only with chia, quinoa or hemp seed flour. The use of non-fermented chia and hemp flour increased the firmness and the rate of bread staling, whereas use of non-traditional hemp and quinoa sourdough reduced the rate of bread staling.

The use of sourdough in gluten-free baked goods leads to products with improved technological and nutritional properties [23], which are softer, tend to stale more slowly and have a delayed mould spoilage rate and thus a prolonged shelf life. Sourdough also brings nutritional benefits because it makes minerals more available and its presence leads to the production of exopolysaccharides, which function as hydrocolloids [30].

3.3. Gluten-Free Bread and Bakery Products Spoilage

Because gluten-free bread contains more water, it has a higher water activity. It is not usually baked using sourdough; therefore, the possibility of infestation by mould and other microorganisms is a significant problem. Mould species involved in bread and bakery products spoilage have been identified; they are represented by fungi of the genus *Penicillium, Cladosporiun, Neurospora or Rhizopus, Aspergillus, Fusarium, Mucor, Endomyces, Chrysonilia*, etc. [91,92]. Mould contamination leads to off-flavour generation and mycotoxins production, endangers human health and causes economic losses and consumer dissatisfaction [93]. Baked goods can also be attacked by yeasts, e.g., of the genera *Pichia, Candida* or *Torulaspora* and bacteria of the genus *Bacillus* (e.g., *B. subtilis, B. amyloliquefaciens, B. licheniformis, B. cereus*), with *B. amyloliquefaciens* as the main species causing rope spoilage [91,94]. The same microorganisms cause spoilage of gluten-free products [91].

To ensure the shelf life of bread and bakery products, various physical methods can be applied. Ultraviolet light, infrared treatment, microwave heating and ultra-high-pressure treatments can be used for bakery products preservation [95]. The disadvantages include low penetration ability of ultraviolet light, higher cost of infrared treatment or condensation problems asssociated with microwave heating [96]. The application of modified atmosphere or gamma irradiation alone does not give advantageous outputs, in practice, more types of protective factors will have to be combined. The use of antimicrobial compounds extracted from plants—biopreservation—provides very promising results and is considerably efficient in slowing down the growth of fungi [93].

For the distribution of finished bread, buns and other bakery products, it is necessary to choose packaging material with good barrier properties. It is possible to directly use packaging with antimicrobial and antioxidant properties, ethanol emitting or carbon dioxide emitting packaging, moisture absorbing packaging or packaging ensuring chemical preservation of the product, for example by potassium acetate, calcium propionate or potassium sorbate (Figure 4) [96–98]. Barrier packaging using oxygen absorbers and the use of modified active packaging (MAP) is a way to extend the shelf life of products. Modified atmosphere packaging [99] is now commonly used for food packaging. The use of natural essential oils is also tested for wheat bread and other bakery products' shelf life extension; the use of this treatment can be proposed in gluten-free bakery products, too [91,100–102].

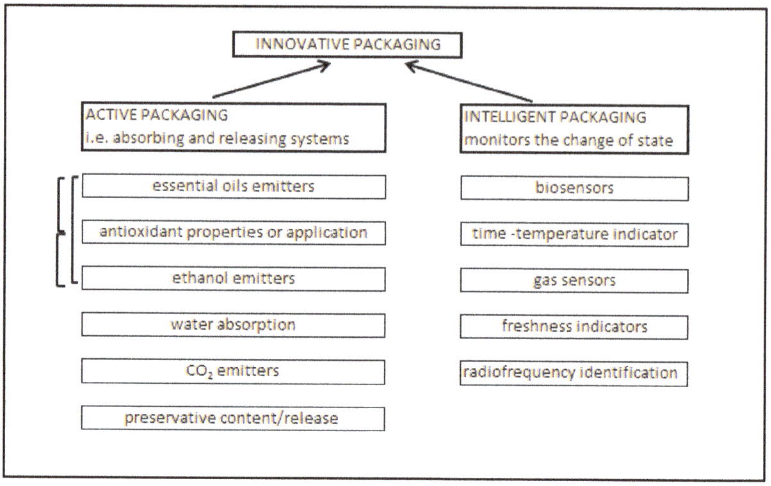

Figure 4. Innovative food packaging systems adapted and modified from [96,101]. Synergies between particular packaging types are marked with brackets.

The use of gluten-free sourdough improves the microbial stability of bread and bakery products as well as their taste [103,104] and represents another tool of biopreservation. In addition to lactic acid, sourdough bacteria also produce acetic and propionic acids with antimicrobial properties. Axel et al. (2016) [103] found out that the addition of *Lactobacillus reuteri* R29 containing sourdough extended the shelf life by 2 days for rice and quinoa bread compared with controls. Similar results were achieved with *Lactobacillus amylovorus* DSM19280-inoculated quinoa sourdough bread [104]. Some lactic acid bacteria in sourdough also form bacteriocins directed against competing microorganisms, and thus improve the shelf life of bakery products [105–107].

Many possibilities exist as to how to avoid spoilage in gluten-free baked goods. To prevent spoilage of these products and prolong their shelf life, it is necessary to combine several types of protective measures and thus use their synergistic effect.

3.4. *New Technologies in Gluten-Free Dough and Bread Preparation*

Recently, several technological processes have been tested to influence the properties of gluten-free dough and improve baking. Treatment of the dough with a high pressure (pascalization) of 100–1000 MPa reduces the temperature of starch gelatinization and change the properties of proteins, including crosslinking. Starch swells and gelatinizes without granules degradation; the extent of swelling depends on the intensity and length of pascalization. This changes the viscoelastic properties of the dough, increases its flexibility, but sometimes also its viscosity [25,31]. The experiments were also performed using ultrasound and micromilling to reduce flour particles. However, no positive effects on bread volume and porosity have been found [31,45].

The properties of gluten-free dough can also be influenced by heating of dry ingredients before dough preparation. Protein denaturation and partial gelatinization of the starch occurs, which increases the flexibility of the dough and the ability to retain gas. The dough viscosity, resistance and stiffness increase, as well as the dough volume [31,108]. Microwave heating was used to heat the rice flour with a moisture content of 20–30%. Proteins denatured after opening their three-dimensional structure. The specific volume and elasticity of bread has significantly improved, and the staling of bread has slowed down [109]. The "Instant controlled pressure drop" technology based on the heating of gluten-free flour for a short time under reduced pressure was tested on a mixture of rice and bean flour. The temperature was in the range of 100–165 °C, and the pressure of

5 kPa and the heating time of 20–60 s were used. The appropriate heating conditions of gluten-free raw materials were determined so that the bread baked from the heat-treated mixture resembled the control wheat bread [110].

Microwave and infrared technologies are also tested in the gluten-free bread baking process. Microwave heating would be cost effective and fast, but the resulting product had low volume, a solid crumb and was rapidly subject to the staling process. The poorer energy penetration into the bread mass is the disadvantage of infrared heating technology, but the resulting product was better sensory evaluated. The so-called "jet-impingement" using hot air convection heating on the surface of the bread was also tested. Homogeneous heat transfer occurred, but the process was energy consuming. The disadvantage was the formation of a solid crumb and dense texture, loss of water and aroma. The starch gelatinization was not complete and starch digestibility was thus decreased. The combination of both methods was recommended for this reason [111,112].

Ohmic heating was the other tested technology. The food material was heated up by its resistance during the electric current passage. The advantage of this process would be the homogeneity of the heating [66]. Another possibility of heating is the partial baking under the reduced pressure. In the first stage, bread was baked at a normal pressure at 180 °C, and after the formation of a solid crust preventing the bread collapsing, it was baked at a pressure reduced to 60 kPa. No changes in bread volume or stiffness were observed, but product moisture was lost and the crust colour was affected. After vacuum baking, other types of starch crystals were formed in the bread and the bread tended to grow stale more slowly [113].

The new technologies improving the gluten-free dough and products quality and shelf life have been still evolving. Some of them have proven to be appropriate. It has been shown that some promising methods do not give completely satisfactory results. Other technologies need to be combined. Research in this area will certainly continue. In practice, however, the economic side of the process would be crucial. Gluten-free bakery products should be not only of high quality, but also be affordable for customers.

4. Clean Label vs. Gluten-Free Products

Green label means that the products do not contain additives. New consumer demands are moving in this direction [114]. However, in gluten-free products, this is very difficult because of the poor baking properties of gluten-free raw materials, missing texturing properties of wheat gluten, etc. Only some of the used additives are appropriate for clean label products. A solution can be achieved using reformulations [115]. To mimic the gluten function hydrocolloids are used which are all classified as food additives and have their E numbers in accordance with EU regulation No. 1333/2008 [116]. Psyllium and beta-D-glucan can be used as gluten replacements in clean label formulations [117]. As far as enzymes are concerned, there are some exceptions they do not have E numbers.

Chia seeds, buckwheat flour and flax flour absorb water very well; lupine flour or soy flour can serve as natural emulsifiers, thus they do not have E numbers. An appropriately chosen starting culture for the preparation of sourdough improves the shelf life and taste of the bakery products, but can also modify the rheology properties of the dough due to the production of exopolysaccharides [118]. The sourdough can also be used in clean label formulations. Flours modified by heating or extrusion are tested instead of E numbers labeled thickeners [119].

5. Conclusions

Gluten intolerance is becoming more common in the population, and patients with this intolerance must follow a gluten-free diet. Bread and other bakery products are staple foods and pose a problem in a gluten-free diet due to their short shelf life and the need to replace gluten. Naturally gluten-free cereals and pseudocereals, but also milled legumes, seeds and nuts, are being increasingly used for the preparation of gluten-free baked goods. The additions of hydrocolloids are traditionally used in gluten-free product formulations.

Recently, amylase, transglutaminase and other enzymatic preparations have been applied to bakery gluten-free mixtures. The use of fermentation for native sourdough preparation or the addition of dried sourdough to dry baking mixes improves the taste and shelf life of gluten-free bread. The quality of gluten-free dough and bread is improved by the addition of modified starches and protein isolates or concentrates. Rheological properties of gluten-free dough, the texture and sensory quality of gluten-free bread also improve the utilization of sourdough with specific microbial strains selected for the gluten-free raw materials. Newly tested baking technologies could improve the texture and slow down the staling of these products. New packaging materials and packaging methods can affect the shelf life of gluten-free bread and pastries. In the future, it is possible to anticipate the use of other non-traditionally processed gluten-free raw materials as well as new technologies for the sourdough preparation and bread baking.

Author Contributions: Both authors contributed to the article equally, for 50% and 50% in data searching, writing—original draft preparation, writing—review and editing, and funding acquisition. All authors have read and agreed to the published version of the manuscript.

Funding: This research was funded by the Ministry of Agriculture of the Czech Republic, institutional support RO0318, METROFOOD-CZ project MEYS, grant number LM2018100 and METROFOOD-PP (EU Horizon 2020 Grant agreement No. 871083).

Data Availability Statement: Not applicable.

Conflicts of Interest: The authors declare no conflict of interest. The funders had no role in the design of the study; in the collection, analyses, or interpretation of data; in the writing of the manuscript, or in the decision to publish the results.

References

1. Al-Toma, A.; Volta, U.; Auricchio, R.; Castillejo, G.; Sanders, D.S.; Cellier, C.; Mulder, C.J.; Lundin, K. European Society for the Study of Coeliac Disease (ESsCD) guideline for coeliac disease and other gluten-related disorders. *United Eur. Gastroenterol. J.* **2019**, *7*, 583–613. [CrossRef]
2. Gabrovská, D.; Hálová, I.; Chrpová, D.; Ouhrabková, J.; Sluková, M.; Vavreinová, S.; Faměra, O.; Kohout, P.; Pánek, J.; Skřivan, P. *Cereals in Human Nutrition (Obiloviny v Lidské Výživě)*, 1st ed.; Federation of the Food and Drink Industries of the Czech Republic: Prague, Czech Republic, 2015; pp. 44–49.
3. Toth, M.; Vatai, G.; Koris, A. Consumers' Acceptance, Satisfaction in Consuming Gluten-free Bread: A Market Survey Approach. *Int. J. Celiac Dis.* **2020**, *8*, 44–49.
4. Codex Standard 118-1979. Available online: http://www.fao.org/fao-who-codexalimentarius/sh-proxy/en/?lnk=1&url=https%253A%252F%252Fworkspace.fao.org%252Fsites%252Fcodex%252FStandards%252FCXS%2B118-1979%252FCXS_118e_2015.pdf (accessed on 9 January 2022).
5. Wieser, H. Chemistry of gluten proteins. *Food Microbiol.* **2007**, *24*, 115–119. [CrossRef]
6. Gluten Structure from the Database UNIPROT. Available online: https://www.uniprot.org/ (accessed on 22 January 2022).
7. Gomez, A.; Ferrero, C.; Calvelo, A.; Añón, M.; Puppo, M. Effect of Mixing Time on Structural and Rheological Properties of Wheat Flour Dough for Breadmaking. *Int. J. Food Prop.* **2011**, *14*, 583–598. [CrossRef]
8. Culetu, A.; Susman, I.E.; Duta, D.E.; Belc, N. Nutritional and Functional Properties of Gluten-Free Flours. *Appl. Sci.* **2021**, *11*, 6283. [CrossRef]
9. Djeghim, F.; Bourekoua, H.; Różyło, R.; Bieńczak, A.; Tanaś, W.; Zidoune, M.N. Effect of By-Products from Selected Fruits and Vegetables on Gluten-Free Dough Rheology and Bread Properties. *Appl. Sci.* **2021**, *11*, 4605. [CrossRef]
10. Kirbas, Z.; Kumcuoglu, S.; Tavman, S. Effects of apple, orange and carrot pomace powders on gluten-free batter rheology and cake properties. *J. Food Sci. Technol.* **2019**, *56*, 914–926. [CrossRef]
11. Krishna, K.R.; Bejkar, M.; Du, S.; Serventi, L. Flax and wattle seed powders enhance volume and softness of gluten-free bread. *Food Sci. Technol. Int.* **2019**, *25*, 66–75. [CrossRef] [PubMed]
12. Steffolani, E.; de la Hera, E.; Pérez, G.; Gómez, M. Effect of Chia on Gluten-Free Bread Quality. *J. Food Qual.* **2014**, *37*, 309–317. [CrossRef]
13. Fratelli, C.; Santos, F.G.; Muniz, D.G.; Habu, S.; Braga, A.R.C.; Capriles, V.D. Psyllium improves the quality and shelf life of gluten-free bread. *Foods* **2021**, *10*, 954. [CrossRef] [PubMed]
14. Genevois, C.E.; de Escalada Pla, M.F. Soybean by-products and modified cassava starch for improving alveolar structure and quality characteristics of gluten-free bread. *Eur. Food Res. Technol.* **2021**, *247*, 1477–1488. [CrossRef]
15. Skendi, A.; Papageorgiou, M.; Varzakas, T. High Protein Substitutes for Gluten in Gluten-Free Bread. *Foods* **2021**, *10*, 1997. [CrossRef]

16. Korus, J.; Achremowicz, B. Fiber preparations of different origin used as additives in baking gluten-free breads. *Food Sci. Technol. Qual.* **2004**, *1*, 65–73.
17. Arslan, M.; Rakha, A.; Xiaobo, Z.; Mahmood, M.A. Complimenting gluten free bakery products with dietary fiber: Opportunities and constraints. *Trends Food Sci. Technol.* **2019**, *83*, 194–202. [CrossRef]
18. Morreale, F.; Benavent-Gila, Y.; Rosell, C.M. Inulin enrichment of gluten free breads: Interaction between inulin and yeast. *Food Chem.* **2019**, *278*, 545–551. [CrossRef] [PubMed]
19. Drabinska, N.; Zielinski, H.; Krupa-Kozak, U. Technological benefits of inulin-type fructans application in gluten-free products—A review. *Trends Food Sci. Technol.* **2016**, *56*, 149–157. [CrossRef]
20. Tsatsaragkou, K.; Protonotariu, S.; Mandala, I. Structural role of fibre addition to increase knowledge of non-gluten bread. *J. Cereal Sci.* **2016**, *67*, 58–67. [CrossRef]
21. Azizi, S.; Azizi, M.H.; Moogouei, R.; Rajaei, P. The effect of Quinoa flour and enzymes on the quality of gluten-free bread. *Food Sci. Nutr.* **2020**, *8*, 2373–2382. [CrossRef] [PubMed]
22. Sandri, L.T.B.; Santos, F.G.; Fratelli, C.; Capriles, V.D. Development of gluten-free bread formulations containing whole chia flour with acceptable sensory properties. *Food Sci Nutr.* **2017**, *5*, 1021–1028. [CrossRef]
23. Ramos, L.; Alonso-Hernando, A.; Martínez-Castro, M.; Morán-Pérez, J.A.; Cabrero-Lobato, P.; Pascual-Maté, A.; Téllez-Jiménez, E.; Mujico, J.R. Sourdough Biotechnology Applied to Gluten-Free Baked Goods: Rescuing the Tradition. *Foods* **2021**, *10*, 1498. [CrossRef]
24. Ketabi, A.; Soleimanian-Zad, S.; Kadivar, M.; Sheikh-zeinoddin, M. Production of microbial exopolysaccharides in the sourdough and its effects on the rheological properties of dough. *Food Res. Int.* **2008**, *41*, 948–951. [CrossRef]
25. Deora, N.V.; Deswal, A.H.N.; Mishra, H.N. Alternative approaches towards gluten-free dough development: Recent trends. *Food Eng. Rev.* **2014**, *6*, 89–104. [CrossRef]
26. Horstman, S.W.; Atzler, J.J.; Heitmann, M.; Zannini, E.; Lynch, K.M.; Arendt, E.K. A comparative study of gluten-free sprouts in the gluten-free bread-making process. *Eur. Food Res. Technol.* **2019**, *245*, 617–629. [CrossRef]
27. Ouazib, M.; Garzon, R.; Zaidi, F.; Rosell, C.M. Germinated, toasted and cooked chickpea as ingredients for breadmaking. *J. Food Sci. Technol.* **2016**, *53*, 2664–2672. [CrossRef] [PubMed]
28. Nunes, M.H.B.; Moore, M.M.; Ryan, L.A.M.; Arendt, E.K. Impact of emulsifiers on the quality and rheological properties of gluten-free breads and batters. *Eur. Food Res. Technol.* **2009**, *228*, 633–642. [CrossRef]
29. Medvid, I.; Shydlovska, O.; Ishchenko, T. Influence of the combination of emulsifiers on the properties of rice gluten-free dough and the quality of bread. *Food and Environ. Saf.* **2021**, *20*, 172–181.
30. Gómez, M.; Sciarini, L.S. Gluten-Free Bakery Products and Pasta. In *Advances in the Understanding of Gluten Related Pathology and the Evolution of Gluten-Free Foods*; Arranz, E., Fernández-Bañares, F., Rosell, C.M., Rodrigo, L., Peña, A.S., Eds.; OmniaScience: Barcelona, Spain, 2015; pp. 565–604.
31. Capelli, A.; Oliva, N.; Cini, E. A systematic review of gluten-free dough and bread: Dough rheology, bread characteristics, and improvement strategies. *Appl. Sci.* **2020**, *10*, 6559. [CrossRef]
32. Anton, A.A.; Artfield, S.D. Hydrocolloids in gluten-free breads: A review. *Int. J. Food Sci. Nutr.* **2008**, *59*, 11–23. [CrossRef]
33. McCarthy, D.F.; Gallagher, E.; Gormley, T.R.; Schober, T.J.; Arendt, E.K. Application of response surface methodology in the development of gluten-free bread. *Cereal. Chem.* **2005**, *82*, 609–615. [CrossRef]
34. Demirkesen, I.; Mert, B.; Sumnu, G.; Sahin, S. Rheological properties of gluten-free bread formulations. *J. Food Engineer.* **2010**, *96*, 295–303. [CrossRef]
35. Conte, P.; Fadda, C.; Drabinska, N.; Krupa-Kozak, U. Technological and nutritional challenges, and novelty in gluten-free breadmaking: A review. *Pol. J. Food Nutr. Sci.* **2019**, *69*, 5–21. [CrossRef]
36. Burešová, I.; Bureš, D.; Čurečkova, K. Comparison of Gluten-Free Dough Ability to Produce Leavening Gas During Baking and its Impact on Crumb Characteristics. *Kvasny Prum.* **2017**, *63*, 8–10. [CrossRef]
37. Arendt, E.K.; Da Bello, F. *Gluten-Free Cereal products and Beverages*; Department of Food and Nutritional Sciences University College Cork Ireland: Cork, Ireland; Elsevier: Amsterdam, The Netherlands, 2008.
38. Burešová, I. Evaluation of Rheological Characteristics of Gluten-Free Doughs (Metody Hodnocení Reologických Vlastností Bezlepkového Těsta). Ph.D. Thesis, Faculty of Science, Palacký University Olomouc, Olomouc, Czech Republic, 2015. Available online: https://theses.cz/id/z89z21/buresova-iva-2015.pdf (accessed on 28 January 2022).
39. Arendt, E.K. *Advances in Gluten Free Cereal Research*; School of Food and Nutritional Sciences, University College Cork: Cork, Ireland, 2016; Available online: https://www.ksla.se/wp-content/uploads/2016/02/Elke-Arendt.pdf (accessed on 29 January 2022).
40. Burešová, I.; Buňka, F.; Kráčmar, S. Rheological characteristics of gluten-free dough. *J. Microbiol. Biotech. Food Sci.* **2014**, *3*, 195–198.
41. Salehi, F. Improvement of gluten-free bread and cake properties using natural hydrocolloids: A review. *Food Sci. Nutr.* **2019**, *7*, 3391–3402. [CrossRef]
42. Rustagi, S.; Khan, S.; Choudhary, S.; Pandey, A.; Khan, M.K.; Kumari, A.; Singh, A. Hydroxypropyl methylcellulose and whey protein concentrate as technological improver in formulation of gluten-free protein rich bread. *Curr. Res. Nutr Food Sci Jour.* **2018**, *6*, 211–221.
43. Lerner, A.; Wusterhausen, P.; Ramesh, A.; Torsten, M. Celiac Disease and Lactose Intolerance. *Int. J. Celiac Dis.* **2018**, *6*, 68–70.

44. Huettner, E.; Arendt, E. Recent advances in gluten-free baking and the current status of oats. *Trends Food Sci. Technol.* **2010**, *21*, 303–331. [CrossRef]
45. Espinoza-Herrera, J.; Martínez, L.M.; Serna-Saldívar, S.O.; Chuck-Hernández, C. Methods for the modification and evaluation of cereal proteins for the substitution of wheat gluten in dough systems. *Foods* **2021**, *10*, 118. [CrossRef]
46. Federici, E. Physical and chemical treatments of zein to improve gluten-free bread quality. A Dissertation Submitted to the Faculty of Purdue University In Partial Fulfillment of the Requirements for the degree of Doctor of Philosophy. Department of Food Science, West Lafayette, Indiana May 2021. Available online: https://www.google.com/url?sa=t&rct=j&q=&esrc=s&source=web&cd=&ved=2ahUKEwir0rf3yuT1AhV-7rsIHUp8D4EQFnoECBgQAQ&url=https%3A%2F%2Fhammer.purdue.edu%2Farticles%2Fthesis%2FPHYSICAL_AND_CHEMICAL_TREATEMENTS_OF_ZEIN_TO_IMPROVE_GLUTEN-FREE_BREAD_QUALITY%2F14079554%2F1%2Ffiles%2F26640344.pdf&usg=AOvVaw3O_ie9adAd3_SLq1-sknJF (accessed on 13 January 2022).
47. Föste, M.; Elgeti, D.; Jekle, M.; Becker, T. Manufacture of gluten-free breads—A question of the substrate? *Bak. + Biscuit* **2013**, *6*, 46–49.
48. Sciarini, L.S.; Ribotta, P.D.; León, A.E.; Pérez, G.T. Influence of gluten-free flours and their mixtures on batter properties and bread quality. *Food Bioprocess. Technol.* **2010**, *3*, 577–585. [CrossRef]
49. Foschia, M.; Horstmann, S.W.; Arendt, E.K.; Zannini, E. Legumes as Functional Ingredients in Gluten-Free Bakery and Pasta Products. *Annu. Rev. Food Sci. Technol.* **2017**, *8*, 75–96. [CrossRef] [PubMed]
50. Melini, F.; Melini, V.; Luziatelli, F.; Ruzzi, M. Current and forward-looking approaches to technological and nutritional improvements of gluten-free bread with legume flours: A critical review. *Compr. Rev. Food Sci. Food Saf.* **2017**, *16*, 1101–1122. [CrossRef] [PubMed]
51. Aditiya, U.J.; Changqi, L.; Shridhar, K.S. Functional properties of select seed flours. *LWT—Food Sci. Technol.* **2015**, *60*, 325–331.
52. Burbano, J.J.; Cabezas, D.M.; Correa, M.J. Effect of walnut flour addition on rheological, thermal and microstructural properties of a gluten free-batter. *LWT—Food Sci. Technol.* **2022**, *154*, 112819. [CrossRef]
53. Gorissen, S.H.M.; Crombag, J.J.R.; Senden, J.M.G.; Waterval, W.A.H.; Bierau, J.; Verdijk, L.B.; van Loon, L.J.C. Protein content and amino acid composition of commercially available plant-based protein isolates. *Amino Acids* **2018**, *50*, 1685–1695. [CrossRef]
54. Horstmann, S.; Lynch, K.M.; Arendt, E.K. Starch characteristics linked to gluten-free products. *Foods* **2017**, *6*, 29. [CrossRef]
55. Abdel-Aal, E.-S.M. 11 Functionality of Starches and Hydrocolloids in Gluten-Free Foods. In *Gluten-Free Food Science and Technology*; Gallagher, E., Ed.; Wiley-Blackwell: Oxford, UK, 2009; p. 200.
56. Hug-Iten, S.; Escher, F.; Conde-Petit, B. Structural Properties of Starch in Bread and Bread Model Systems: Influence of an Antistaling α-Amylase. *Cereal Chem.* **2001**, *78*, 421–428. [CrossRef]
57. Mancebo, C.M.; Merino, C.; Martínez, M.M.; Gómez, M. Mixture design of rice flour, maize starch and wheat starch for optimization of gluten free bread quality. *J. Food Sci. Technol.* **2015**, *52*, 6323–6333. [CrossRef]
58. Ziobro, R.; Korus, J.; Witczak, M.; Juszczak, L. Influence of modified starches on properties of gluten-free dough and bread. Part II: Quality and staling of gluten-free bread. *Food Hydrocoll.* **2012**, *29*, 68–74. [CrossRef]
59. Hu, X.; Guo, B.; Liu, C.; Yan, X.; Chen, J.; Luo, S.; Liu, Y.; Wang, H.; Yang, R.; Zhong, Y.; et al. Modification of potato starch by using superheated steam. *Carbohydr. Polym.* **2018**, *198*, 375–384. [CrossRef]
60. Witczak, M.; Ziobro, R.; Juszczak, L.; Korus, J. Starch and starch derivatives in gluten-free systems—A review. *J. Cereal Sci.* **2016**, *67*, 46–57. [CrossRef]
61. Kaur, R.S.; Chopra, C.S. Gluten-free products for celiac susceptible people. *Front. Nutr.* **2018**, *5*, 116.
62. Palabiyik, I.; Yildiz, O.; Toker, O.S.; Cavus, M.; Ceylan, M.M.; Yurt, B. Investigating the addition of enzymes in gluten-free flours—The effect on pasting and textural properties. *LWT—Food Sci. Technol.*, **2016**, *69*, 633–641. [CrossRef]
63. Motahar, S.F.S.; Ariaeenejad, S.; Salami, M.; Emam-Djomeh, Z.; Mamaghani, A.S.A. Improving the quality of gluten-free bread by a novel acidic thermostable α-amylase from metagenomics data. *Food Chem.* **2021**, *352*, 129307. [CrossRef] [PubMed]
64. Haghighat-Kharazi, S.; Kasaai, M.R.; Milani, J.M.; Khajeh, K. Microencapsulation of α-amylase in beeswax and its application in gluten-free bread as an anti-staling agent. *Food Sci. Nutr.* **2020**, *8*, 5888–5897. [CrossRef] [PubMed]
65. Ngemakwe, P.N.; Le Roes-Hill, M.; Jideani, V. Advances in gluten-free bread technology. *Food Sci. Technol. Int.* **2014**, *21*, 256–276. [CrossRef]
66. Bender, D.; Schoenlechner, R. Innovative approaches towards improved gluten-free bread properties. *J.Cereal Sci.* **2020**, *91*, 102904. [CrossRef]
67. Huang, W.; Li, L.; Wang, F.; Wan, J.; Tilley, M.; Ren, C.; Wu, S. Effects of transglutaminase on the rheologicaland Mixolab thermomechanical characteristics of oat dough. *Food Chem.* **2010**, *121*, 934–939. [CrossRef]
68. Ogilvie, O.; Roberts, S.; Sutton, K.; Larsen, N.; Gerrard, J.; Domigan, L. The use of microbial transglutaminasein a bread system: A study of gluten protein structure, deamidation state and protein digestion. *Food Chem.* **2021**, *340*, 127903. [CrossRef]
69. Dłużewska, E.; Marciniak-Lukasiak, K.; Kurek, N. Effect of transglutaminase additive on the quality of gluten-free bread. *CyTA–J. Food* **2015**, *13*, 80–86. [CrossRef]
70. Tomić, J.; Torbica, A.; Belović, M. Effect of non-gluten proteins and transglutaminase on dough rheological properties and quality of bread based on millet (*Panicum miliaceum*) flour. *LWT* **2020**, *118*, 108852. [CrossRef]
71. Diowksz, A.; Sadowska, A. Impact of sourdough and transglutaminase on gluten-free buckwheat bread quality. *Food Biosci.* **2021**, *43*, 101309. [CrossRef]

72. Silva, H.A.; Paiva, E.G.; Lisboa, H.M.; Duarte, E.; Cavalcanti-Mata, M.; Gusmão, T.; de Gusmão, R. Role of chitosan and transglutaminase on the elaboration of gluten-free bread. *J. Food Sci. Technol.* **2020**, *57*, 1877–1886. [CrossRef] [PubMed]
73. Zhao, F.; Li, Y.; Li, C.; Ban, X.; Cheng, L.; Hong, Y.; Gu, Z. Co-supported hydrocolloids improve the structure and texture quality of gluten-free bread. *LWT* **2021**, *152*, 112248. [CrossRef]
74. Azghar, A.; Zia, M. Effects of xanthan gum and guar gum on the quality and storage stability of gluten free frozen dough bread. *Amer. J. Food Nutr.* **2016**, *6*, 107–112.
75. Manik, L.C.M.; Nur, M. The recent development of gluten-free bread quality using hydrocolloids. In *IOP Conference Series: Earth and Environmental Science*; IOP Publishing: Bristol, UK, 2021; Volume 733, p. 012101.
76. Ren, Y.; Linter, B.R.; Linforth, R.; Foster, T.J. A comprehensive investigation of gluten free bread dough rheology, proving and baking performance and bread qualities by response surface desigh and principal component analysis. *Food Funct.* **2020**, *11*, 5333–5345. [CrossRef]
77. Liu, X.; Mu, T.; Sun, H.; Zhang, M.; Chen, J.; Fauconnier, M.L. Influence of different hydrocolloids on dough thermo-mechanical properties and in vitro starch digestibility of gluten-free steamed bread based on potato flour. *Food Chem.* **2018**, *239*, 1064–1074. [CrossRef]
78. Morreale, F.; Garzón, R.; Rosell, C.M. Understanding the role of hydrocolloids viscosity and hydration in developing gluten-free bread. A study with hydroxypropylmethylcellulose. *Food Hydrocoll.* **2018**, *77*, 629–635. [CrossRef]
79. Lazaridou, A.; Duta, D.; Papageorgiou, M.; Belc, N.; Biliaderis, C.G. Effects of hydrocolloids on doughrheology and bread quality parameters in gluten-free formulations. *J. Food Eng.* **2007**, *79*, 10331047. [CrossRef]
80. Hager, A.S.; Arendt, E.K. Influence of hydroxypropylmethylcellulose (HPMC), xanthan gum and theircombination on loaf specific volume, crumb hardness and crumb grain characteristics of gluten-free breadsbased on rice, maize, teff and buckwheat. *Food Hydrocoll.* **2013**, *32*, 195–203. [CrossRef]
81. Belorio, M.; Gómez, M. Effect of hydration on gluten-free breads made with hydroxypropyl methylcellulose in comparison with psyllium and xanthan gum. *Foods* **2020**, *9*, 1548. [CrossRef]
82. Zoghi, A.; Mirmahdi, R.S.; Mohammadi, M. The role of hydrocolloids in the development of gluten-free cereal-based products for coeliac patients: A review. *Int. J. Food Sci. Technol.* **2021**, *56*, 3138–3147. [CrossRef]
83. Marti, A.; Bottega, G.; Franzetti, L.; Morandin, F.; Quaglia, L.; Pagani, M.A. From wheat sourdough to gluten-free sourdough: A non-conventional process for producing gluten-free bread. *Int. J. Food Sci. Technol.* **2015**, *50*, 1268–1274. [CrossRef]
84. Maidana, S.D.; Finch, S.; Garro, M.; Savoy, G.; Gänzle, M.; Vignolo, G. Development of gluten-free breads started with chia and flaxseed sourdoughs fermented by selected lactic acid bacteria. *LWT* **2020**, *125*, 109189. [CrossRef]
85. Moroni, A.V.; Dal Bello, F.; Arendt, E.K. Sourdough in gluten-free bread-making: An ancient technology to solve a novel issue? *Food Microbiol.* **2009**, *26*, 676–684. [CrossRef] [PubMed]
86. Rinaldi, M.; Paciulli, M.; Caligiani, A.; Scazzina, F.; Chiavaro, E. Sourdough fermentation and chestnut flour in gluten-free bread: A shelf-life evaluation. *Food Chem.* **2017**, *224*, 144–152. [CrossRef]
87. Jagelaviciute, J.; Cizeikiene, D. The influence of non-traditional sourdough made with quinoa, hemp and chia flour on the characteristics of gluten-free maize/rice bread. *LWT* **2021**, *137*, 110457. [CrossRef]
88. Olojede, A.O.; Sanni, A.I.; Banwo, K.; Adesulu-Dahunsi, A.T. Sensory and antioxidant properties and in-vitro digestibility of gluten-free sourdough made with selected starter cultures. *LWT* **2020**, *129*, 109576. [CrossRef]
89. Puerta, P.; Garzón, R.; Rosell, C.M.; Fiszman, S.; Laguna, L.; Tárrega, A. Modifying gluten-free bread's structure using different baking conditions: Impact on oral processing and texture perception. *LWT* **2021**, *140*, 110718. [CrossRef]
90. Różyło, R.; Rudy, S.; Krzykowski, A.; Dziki, D.; Gawlik-Dziki, U.; Różyło, K.; Skonecki, S. Effect of adding fresh and freeze-dried buckwheat sourdough on gluten-free bread quality. *Int. J. Food Sci. Technol.* **2014**, *50*, 313–322. [CrossRef]
91. Melini, V.; Melini, F. Strategies to extend bread and GF bread shelf-life: From sourdough to antimicrobial active packaging and nanotechnology. *Fermentation* **2018**, *4*, 9. [CrossRef]
92. Pitt, J.I.; Hocking, A.D. *Fungi and Food Spoilage, 3rd ed*; Springer: New York, NY, USA, 2009; pp. 401–404.
93. Garcia, M.V.; Copetti, M.V. Alternative methods for mould spoilage control in bread and bakery products. *Int. Food Res. J.* **2019**, *26*, 737–749.
94. Valerio, F.; De Bellis, P.; Di Biase, M.; Lonigro, S.L.; Giussani, B.; Visconti, A.; Lavermicocca, P.; Sisto, A. Diversity of spore-forming bacteria and identification of *Bacillus amyloliquefaciens* as a species frequently associated with the ropy spoilage of bread. *Int. J. Food Microbiol.* **2012**, *156*, 278–285. [CrossRef] [PubMed]
95. Axel, C.; Zannini, E.; Arendt, E. Mold spoilage of bread and its biopreservation: A review of current strategies for bread shelf life extension. *Crit. Rev. Food Sci. Nutr.* **2016**, *57*, 3528–3542. [CrossRef]
96. Qian, M.; Liu, D.; Zhang, X.; Yin, Z.; Ismail, B.B.; Ye, X.; Guo, M. A review of active packaging in bakery products: Applications and future trends. *Trends Food Sci. Technol.* **2021**, *114*, 459–471. [CrossRef]
97. Romão, B.; Botelho, R.B.A.; Alencar, E.R.; Nunes da Silva, V.S.; Bertoldo Pacheco, M.T.; Puppin Zandonadi, R. Chemical composition and glycemic index of gluten-free bread commercialized in Brazil. *Nutrients* **2020**, *12*, 2234. [CrossRef]
98. Roman, L.; Belorio, M.; Gomez, M. Gluten-free breads: The gap between research and commercial reality. *Compr. Rev. Food Sci. Food Saf.* **2019**, *18*, 690–702. [CrossRef]
99. Kurek, M.A.; Wyrwisz, J.; Karp, S. Effect of modified atmosphere packaging on the quality of wheat bread fortified with soy flour and oat fibre. *Food Meas.* **2019**, *13*, 1864–1872. [CrossRef]

100. Pasqualone, A. Bread Packaging: Features and Functions. In *Flour and Breads and their Fortification in Health and Disease Prevention*; Preedy, V.R., Watson, R.R., Eds.; Academic Press: London, UK, 2019; pp. 211–222.
101. Valková, V.; Ďúranová, H.; Galovičová, L.; Vukovic, N.L.; Vukic, M.; Kačániová, M. In Vitro antimicrobial activity of lavender, mint, and rosemary essential oils and the effect of their vapours on growth of *Penicillium* spp. in a bread model system. *Molecules* **2021**, *26*, 3859. [CrossRef]
102. Galovičová, L.; Borotová, P.; Valková, V.; Vukovic, N.L.; Vukic, M.; Štefániková, J.; Ďúranová, H.; Kowalczewski, P.Ł.; Čmiková, N.; Kačániová, M. *Thymus vulgaris* essential oil and its biological activity. *Plants* **2021**, *10*, 1959. [CrossRef] [PubMed]
103. Axel, C.; Brosnan, B.; Zannini, E.; Furey, A.; Coffey, A.; Arendt, E.K. Antifungal sourdough lactic acid bacteria as biopreservation tool in quinoa and rice bread. *Int. J. Food Microbiol.* **2016**, *239*, 86–94. [CrossRef] [PubMed]
104. Axel, C.; Röcker, B.; Brosnan, B.; Zannini, E.; Furey, A.; Coffey, A.; Arendt, E.K. Application of *Lactobacillus amylovorus* DSM19280 in gluten-free sourdough bread to improve the microbial shelf life. *Food Microbiol.* **2015**, *47*, 36–44. [CrossRef]
105. Bartkiene, E.; Lele, V.; Ruzauskas, M.; Domig, K.J.; Starkute, V.; Zavistanaviciute, P.; Bartkevics, V.; Pugajeva, I.; Klupsaite, D.; Juodeikiene, G.; et al. Lactic acid bacteria isolation from spontaneous sourdough and their characterization including antimicrobial and antifungal properties evaluation. *Microorganisms* **2020**, *8*, 64. [CrossRef] [PubMed]
106. Zangeneh, M.; Khorami, S.; Khalegh, M. Bacteriostatic activity and partial characterization of the bacteriocin produced by L. plantarum sp. isolated from traditional sourdough. *Food Sci. Nutr.* **2020**, *11*, 6023–6030. [CrossRef]
107. Quattrini, M.; Liang, N.; Fortina, M.G.; Xiang, S.; Curtis, J.; Gänzle, M. Exploiting synergies of sourdough and antifungal organic acids to delay fungal spoilage of bread. *Int. J. Food Microbiol.* **2019**, *302*, 8–14. [CrossRef]
108. Jeong, D.; Hong, J.S.; Liu, Q.; Choi, H.; Chung, H. The effects of different levels of heat-treated legume flour on nutritional, physical, textural, and sensory properties of gluten-free muffins. *Cereal Chem.* **2020**, *98*, 392–404. [CrossRef]
109. Villanueva, M.; Harasym, J.; Muñoz, J.M.; Ronda, F. Rice flour physically modified by microwave radiation improves viscoelastic behavior of doughs and its bread-making performance. *Food Hydrocoll.* **2019**, *90*, 472–481. [CrossRef]
110. Boulemkahel, S.; Benatallah, L.; Besombes, C.; Allaf, K.; Zidoune, M.N. Impact of instant controlled pressure drop (DIC) treatment on the technological quality of gluten-free bread based on rice-field bean formula using design of experiments. *Afr. J. Food Sci.* **2021**, *15*, 121–130.
111. Chhanwal, N.; Bhushette, P.R.; Anandharamakrishnan, C. Current perspectives on non-conventional heating ovens for baking process—A review. *Food Bioprocess. Technol.* **2019**, *12*, 1–15. [CrossRef]
112. Rosell, C.M.; Aalami, M.; Mahdavi, S.A. Innovative Gluten-Free Products. In *Innovative Processing Technologies for Healthy Grains*; Pojic, M., Tiwari, U., Eds.; Wiley-Blackwell: Hoboken, NJ, USA, 2020; pp. 177–198. [CrossRef]
113. Simsek, S.T. Evaluation of partial-vacuum baking for gluten-free bread: Effects on quality attributes and storage properties. *J. Cereal Sci.* **2020**, *91*, 102891. [CrossRef]
114. Do Nascimento, K.d.O.; do Nascimento Dias Paes, S.; Ivanilda, M.A. A Review 'Clean Labeling': Applications of Natural Ingredients in Bakery Products. *J. Food Nutr. Res.* **2018**, *6*, 285–294. [CrossRef]
115. Kajzer, M.; Diowksz, A. The clean label concept: Novel approaches in gluten-free breadmaking. *Appl. Sci.* **2021**, *11*, 6129. [CrossRef]
116. Regulation (EC) No 1333/2008 of the European Parliament and of the Council of 16 December 2008 on Food Additives. Available online: https://eur-lex.europa.eu/legal-content/PL/TXT/?uri=CELEX:32008R1333 (accessed on 15 December 2021).
117. Karp, S.; Wyrwisz, J.; Kurek, M.A.; Wierzbicka, A. The use of high-in-β-glucan oat fibre powder as a structuring agent in gluten-free yeast-leavened cake. *Food Sci. Technol. Int.* **2019**, *25*, 618–629. [CrossRef] [PubMed]
118. Montemurro, M.; Pontonio, E.; Rizzello, C.G. Design of a "Clean-Label" gluten-free bread to meet consumers demand. *Foods* **2021**, *10*, 462. [CrossRef] [PubMed]
119. Carcelli, A.; Masuelli, E.; Diantom, A.; Vittadini, E.; Carini, E. Probing the Functionality of Physically Modified Corn Flour as Clean Label Thickening Agent with a Multiscale Characterization. *Foods* **2020**, *9*, 1105. [CrossRef] [PubMed]

foods MDPI

Review

The Role of Hydrocolloids in Gluten-Free Bread and Pasta; Rheology, Characteristics, Staling and Glycemic Index

Alina Culetu [1], Denisa Eglantina Duta [1], Maria Papageorgiou [2] and Theodoros Varzakas [3,*]

[1] National Institute of Research & Development for Food Bioresources, IBA Bucharest, 6 Dinu Vintila Street, 021102 Bucharest, Romania; alinaculetu@gmail.com (A.C.); denisa.duta@bioresurse.ro (D.E.D.)
[2] Department of Food Science and Technology, International Hellenic University, P.O. Box 141, 57400 Thessaloniki, Greece; mariapapage@ihu.gr
[3] Department of Food Science and Technology, University of the Peloponnese, 24100 Kalamata, Greece
* Correspondence: t.varzakas@uop.gr; Tel.: +30-2721045279

Abstract: Hydrocolloids are important ingredients controlling the quality characteristics of the final bakery products. Hydrocolloids are frequently used in gluten-free (GF) recipes, mimicking some rheological properties of gluten, improving dough properties, delaying starch retrogradation and improving bread texture, appearance and stability. Hydrocolloids addition increases viscosity and incorporation of air into the GF dough/batter. Besides their advantages for the technological properties of the GF bread, hydrocolloids addition may impact the glycemic index (GI) of the final product, thus answering the demand of people requiring products with low GI. This review deals with the application of hydrocolloids in GF bread and pasta with a focus on their effect on dough rheology, bread hardness, specific volume, staling and GI.

Keywords: gluten-free; hydrocolloids; dough rheological properties; texture; volume; sensory; glycemic index; staling; bread; pasta

Citation: Culetu, A.; Duta, D.E.; Papageorgiou, M.; Varzakas, T. The Role of Hydrocolloids in Gluten-Free Bread and Pasta; Rheology, Characteristics, Staling and Glycemic Index. *Foods* **2021**, *10*, 3121. https://doi.org/10.3390/foods10123121

Academic Editor: Moshe Rosenberg

Received: 6 November 2021
Accepted: 13 December 2021
Published: 16 December 2021

1. Introduction

Hydrocolloids are a group of water-soluble polysaccharides with different chemical structures, high molecular weight and hydrophilic long-chain molecules. Hydrocolloids' addition has a positive impact on gluten-free (GF) cereal-based products because they improve the structure, volume, texture, taste and overall quality of the final products as well as a shelf-life extension [1–3].

The use of hydrocolloids in GF applications depends on their colloidal properties, the ability to increase the water-binding capacity, viscosity, hydration rate and the effect of temperature on hydration because, for most hydrocolloids, the viscosity decreases with increasing temperature [1]. Hydrocolloids also improve the development and retention of gases during fermentation.

Hydrocolloids are classified according to their origin, as shown in Figure 1. Different types of hydrocolloids were used in GF products, including hydroxypropyl methylcellulose (HPMC), xanthan gum (XG), guar gum (GG), locust bean gum, psyllium, carrageenan, pectin, carboxymethyl cellulose (CMC), konjac gum, gelatine, agarose, agar, β-glucan, gum arabic (GA) and alginate [4–6].

Furthermore, hydrocolloids addition represents the easiest way to increase the dietary fiber content of GF bakery products. In general, GF products are characterized by a much lower nutritional value due to the fact that they lack important nutrients, such as vitamins, proteins, minerals and dietary fiber. One of these ingredients used in the food industry, classified as dietary fiber, is β-glucan, a non-starch polysaccharide that is located in the walls of endosperm cells of oats and barley. Moreover, psyllium, a natural bioactive soluble fiber that can be used as hydrocolloid replacer due to its water-holding, gel-forming and structure building properties, received attention in GF preparations in the last years.

Psyllium is able to control crumb texture, as it is interchangeable with other commonly used hydrocolloids (XG, GG, HPMC) [7].

Figure 1. Classification of the main hydrocolloids according to their origin.

In the present manuscript, the impact of hydrocolloids addition into the formulation of GF bread and pasta products, with a focus on the dough rheology, hardness, specific volume, staling, glycemic index and sensory characteristics, are reviewed.

A comparison of articles from the Web of Science database by using the terms "gluten-free bread/pasta/noodles/cake/cookie/muffin/biscuit" in the article title AND "hydrocolloid" as well as the exact name of each of the following hydrocolloids in the abstract: XG, HPMC, GG, psyllium, pectin, CMC, locust bean gum, β-glucan, carrageenan, alginate, GA (document type: articles and review articles; language: English; no other exclusion criteria), showed a significantly higher number of papers published for bread as compared with the other GF products, followed by those addressing pasta products (Figure 2). This is explained by the fact that the gluten absence is critical in GF breads in regard to the bread structure, which makes it more challenging to find new approaches to improve the bread properties. Figure 3a shows the number of publications for GF bread according to the name of the hydrocolloids, while the papers' distribution over time is shown in Figure 3b.

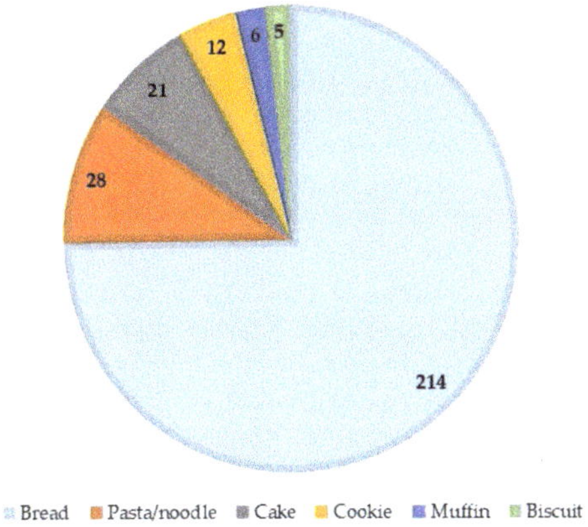

Figure 2. Number of publications dealing with hydrocolloid applications in different GF products. Results were obtained on 19 October 2021 on the Web of Science database.

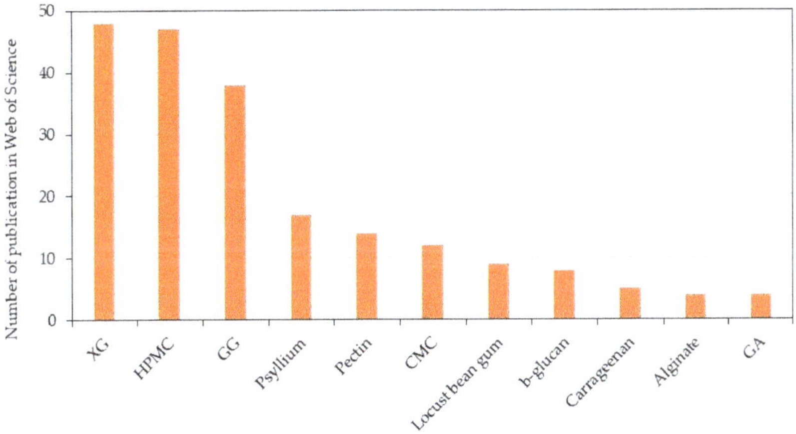

(**a**)

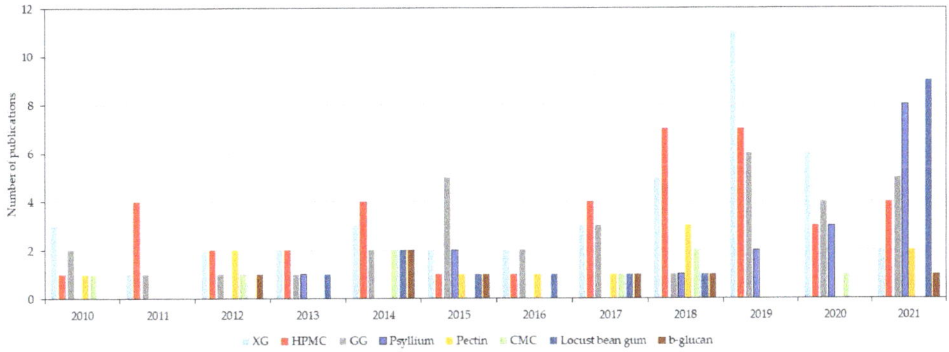

(**b**)

Figure 3. (**a**) Number of publications by name of the hydrocolloid in GF bread applications. (**b**) Number of publications by name of the hydrocolloid in GF bread applications over time (in the last 11 years). Results were obtained on 19 October 2021 on the Web of Science database.

XG and HPMC are the most frequently employed hydrocolloids for GF breads, mainly for their impact to increase the volume and porosity as well to produce softer products, followed by GG and psyllium. A higher frequency was noted in the last 3 years for psyllium application in GF bread. Psyllium is a promising addition to improve GF bread, enhancing the volume, structure, texture, appearance and acceptability of GFB, in addition to increasing the dietary fiber content and decreasing the glycemic response of GF bread [7].

2. Hydrocolloids in GF Bread

In GF doughs, hydrocolloids are used to create a viscoelastic network in order to balance the lack of gluten. Comprehensive reviews about the impact of the hydrocolloids on dough handling, technological and nutritional properties of GF breads underlined their function as structuring agents, mimicking the gluten network because of the ability to bind water [2,4,5,8]. In addition, hydrocolloids bring positive effects on the viscoelastic properties of the GF dough and bread texture [8].

A recent review stated that HPMC is the most favorable hydrocolloid in GF bread manufacturing [9]. HPMC forms a gel network on heating and shows lower variability than other hydrocolloids [10]. The presence of HPMC in the GF system makes the starch granules adhere to one another, and there is more space to entrap water in the system [4]. HPMC, together with the components from the rice flour, form hydrophilic bonds that are beneficial to the water absorption and contribute to the stability and homogeneity of the GF dough [11]. Factors that are related to HPMC functionality were related to the type of flours used, the presence of other ingredients and the percent of methoxyl groups contained in the HPMC molecule [12]. Besides the HPMC addition and hydration levels, Morreale et al. [11] pointed out the importance of HPMC viscosity to obtain GF rice breads with optimal quality.

The charge and the molecular weight of the hydrocolloids are amongst the main factors that influence bread quality [4,13]. The polar charge has an effect on the water affinity. Negatively charged hydrocolloids are more prone to build intermolecular hydrogen bonds with water, while uncharged hydrocolloids have intramolecular hydrogen bonds that reduce the interactions with water [4]. In a GF bread formulation based on potato starch, Horstmann et al. [13] suggested that negatively charged hydrocolloids such as sodium alginate and pectin create repulsive forces with negatively charged phosphate groups of the potato starch, delaying the pasting and gelatinization of starch granules, leading to lower viscosity and therefore to higher bread volume due to the high gas cell expansion. On the other hand, hydrocolloids with a neutral charge and higher molecular weight, such as GG and locust bean gum, create hydrogen bonds with leached amylose that leads to higher viscosity, thus lowering the elasticity and decreasing bread volume due to limiting gas expansion. Moreover, the molecular weight affects the water holding capacity of hydrocolloids [4,14]. Funami et al. [14] correlated higher water holding capacity for hydrocolloids with a higher molecular weight. Because of the higher molecular weight of certain hydrocolloids (XG, CMC, agarose and β-glucan) and due to increasing concentration, Lazaridou et al. [15] attributed the reduced loaf volume in GF bread formulation based on rice flour, corn starch and sodium caseinate.

Besides the factors mentioned above, the impact of hydrocolloids on the bread quality also depends on the level of the hydrocolloid used, the type of flour and other ingredients, as well as on the interaction with other components in the GF system [2]. Regarding the presence of other ingredients, it was shown that protein addition at certain levels of addition causes antagonistic interaction with the hydrocolloids. For example, in a formulation with rice flour-cassava starch and 5% HPMC, the addition of soy protein isolate (1%, 2%, 3%) and egg white solids (5% and 10%) reduced dough stability by lowering the hydrocolloid functionality, modifying the available water within the dough, weakening the interactions between hydrocolloid and starch and, consequently, reducing the foam stability [16]. Besides HPMC, other hydrocolloids such as XG and methylcellulose were reported to be used together with rich protein sources in GF formulations [17].

Dough hydration in GF bread is an important feature of final product quality. The correct volume of water is significant for strengthening the three-dimensional dough structure [11]. It is generally known that the greater the hydration, the higher the increment of the bread volume; there is a maximum hydration level after which the dough collapses during the baking process [18]. Recently, Sahin et al. [19] proved that Farinograph was a better tool in establishing the optimal amount of water in GF rice breads with different hydrocolloids as compared to the common method that uses the calculation based on the water hydration capacity of the individual ingredients: flour, starch and hydrocolloids. The authors stated that the advantage of the Farinograph method is that it takes into account the temperature changes during mixing and its effect on hydration, simulating the real process. Moreover, the Farinograph method provides data for dough stability and development time.

The following sub-sections deal with the effect of hydrocolloids on dough rheology, bread crumb hardness, bread specific volume, bread staling and glycemic index.

2.1. *Effect of Hydrocolloids on Dough Rheology*

The rheological behavior of dough is an important topic that has drawn significant attention in the research community, as rheology is linked to baking properties and bread quality. For example, it a correlation was found between the rheological properties of dough samples, and the firmness of GF bread as higher viscoelastic values of dough resulted in bread with lower hardness [20].

Hydrocolloids improve dough development and gas retention by an increase in viscosity, which will permit the production of improved GF breads [21].

Rheological investigation of the hydrocolloids effect on GF dough is achieved not only by empirical methodologies such as farinograph, alveograph, extensograph and Mixolab determinations but also with typical rheometers through creep-recovery and oscillation tests, which include strain and frequency sweeps that allow evaluating the viscoelastic dough properties [15,22,23]. The rheometer measures the deformation energy stored in the sample during a shear process, which represents the elastic component (G′—storage modulus), while the deformation energy used up and lost during shearing represents the viscous component (G″—loss modulus) of the dough. In GF bread, an equilibrium between elastic and viscous properties is needed [15]. Atypical viscoelastic behavior is achieved when G′ values are higher than G″ values, which enables gas cell expansion.

Mancebo et al. [24] stated that the creep-recovery test might estimate the bread quality characteristics better than the oscillatory test because the low deformations used in the latter do not correspond to the real processing and baking conditions.

Table 1 presents some results published in the literature with the effect of hydrocolloids addition on the rheological dough properties and the type of rheological test used.

The correct selection of the hydrocolloid and the amount of water in the recipe can lead to dough properties such as the wheat-containing one. In order to obtain high-quality GF bread, a high water content of up to 150% is needed [20]. Investigating different types of hydrocolloids, Sabanis and Tzia [10] found that XG required 10% more water than HPMC, GG and carrageenan in formulations based on corn starch and rice flour due to its higher water-binding capacity. Moreover, when increasing HPMC, GG and carrageenan addition levels from 1% to 2%, the water increased from 75% to 85%. In rice flour and cornstarch-based doughs prepared with different water amounts (130–150%), Lazaridou et al. [15] reported a decrease in elastic modulus as the water amount increased.

Many research on GF dough formulations underlined that dough samples present viscoelastic properties up to 0.1% strain level and the decrease in linearity was very significant beyond 1% strain level, which indicates the breakdown of the GF dough structure [15,25]. Similarly, with GF, wheat doughs showed linear viscoelasticity at strain levels lower than 0.1–0.25% [26,27], while other systems have different viscoelastic regions; for example, zein suspensions had a linear viscoelastic region below 0.003% strain level [28].

The addition of hydrocolloids to GF dough formulations showed increased elastic and viscous moduli. The elastic and viscous moduli of GF cornbread dough are increased with hydrocolloids addition, denoting a stronger dough structure formed by entrapping gas and retaining water, thus leading to higher viscosity [25]. The authors found a higher increase for HPMC than guar-based doughs. The higher increase in moduli values produced by HPMC addition compared to other hydrocolloids was explained by its capacity to form a foam that enables it to entrap gas inside the dough structure [4].

The oscillatory and creep tests showed that the elasticity and resistance to deformation of GF dough formulations supplemented with hydrocolloids followed the order: XG > CMC > pectin > agarose > β-glucan [15]. The higher elasticity shown by XG was attributed to its property to form a weak gel at low shear rates.

Sciarini et al. [29] used rheology at large deformation (resistance to penetration) and small deformation (frequency sweep) to study the hydrocolloids effect on GF dough prepared with rice flour, cassava starch and soy. The first method gives information about dough resistance, and XG showed the highest resistance, followed by CMC, alginate and carrageenan. The higher resistance given by XG was explained by its capacity to embrace a helix conformation in aqueous media, which changes the molecule into a rigid form. Regarding the frequency sweep tests, carrageenan was the only hydrocolloid, which showed a significant increase in both elastic and viscous dynamic moduli compared with a control dough; XG, alginate and CMC were similar to control.

Peressini et al. [30] found that XG and propylene glycol alginate (PGA) enhanced the storage modulus of a rice–buckwheat dough, with greater effect for PGA. The rheological properties and crumb quality of dough were improved through the use of PGA, which is modified alginate characterized as amphiphilic with special surface activity and emulsifying capacity [30,31]. A mixture of hydrocolloids improves both the structure and texture of the GF bread than the use of a single hydrocolloid. Zhao et al. [31] stated that co-supported hydrocolloids (HPMC–PGA) improve the overall quality of GF bread; namely, HPMC acted as a skeleton, and PGA served as a supporting matrix. The dough structure was enhanced by the rearrangement of polysaccharide polymers.

In a formulation made with a mixture of rice and buckwheat flour, HPMC or CMC showed a reducing strength and extension of the 3D network in the dough rheological behavior. HPMC addition also showed a modification of the dough thermal behavior [23].

It is known that hydrocolloids and starches that come from various botanical sources differ in functionality and properties related to granule size, composition or morphology that influence gelatinization, respectively. Thus, in GF sorghum bread formulations, the interaction between hydrocolloids (XG, HPMC and locust bean gum) and starches (potato, tapioca and rice) revealed that the best combinations in terms of bread quality were between potato starch (xanthan, tapioca starch) HPMC and rice starch (xanthan). Doughs with lower viscosities produced loaves with better crumb grain characteristics [32].

Studying the interaction between different hydrocolloids, Mancebo et al. [24] found no synergic effects between HPMC and psyllium in GF rice bread. Both hydrocolloids increased viscoelastic moduli, but only psyllium reduced the pasting temperature and compliance values, indicating higher dough strength [24]. Psyllium has very similar rheological characteristics with XG, both being responsible for weak gelling properties. Psyllium shows important hydration capacity and gel-forming properties, able to entrap CO_2 [18].

By adding 5.5% psyllium to a formulation based on chickpea flour, an increase in consistency was shown during the initial stages of mixing at the beginning of heating related to protein network weakening as measured by the Mixolab technique [33]. A favorable dough consistency explained the increased cohesiveness and springiness of the crumb, which are desirable outcomes in the GF bread-making process.

Table 1. Effect of hydrocolloids on dough rheology.

Type of Hydrocolloid	Level Used *	Other Ingredients	Type of the Rheological Test	Effect	References
GG	1%	chestnut flour with 4% chia flour	Creep-recovery (rheometer)	Improved the dough elasticity by 65.9%	[34]
HPMC	2%			Improved the dough elasticity by 64.8%	
Tragacanth gum	1%			Improved dough elasticity by 45.8%	
XG–GG (mix)	0.5%	100% rice flour, 8% sugar, 8% shortening, 2% salt, 1% instant yeast, 150% water	Frequency sweep	Increased elastic and viscous moduli	[20]
CMC	1%	70% rice flour, 30% buckwheat flour, 85% water	Frequency sweep	Increased complex modulus, improved the internal structure, increased the crumb porosity, similar to the standard wheat bread	[23]
HPMC	1%	70% rice flour, 30% buckwheat flour, 85% water			
HPMC	1%	70% rice flour, 30% buckwheat flour, 100% water			
HPMC	1–1.5%	75% corn starch, 25% rice flour, 2% yeast, 4% sunflower oil, 4% sucrose, 2% salt, 75–85% water	Shear properties, Power law	Improved viscosity	[10]
GG					
Carrageenan					
XG					
HPMC	5.5%	22.2% corn meal, 77.8% corn starch, 5.5% sugar, 2.2% salt, 1.1% yeast, 83.3% water	Strain and frequency sweep measurements	Increased elastic and viscous moduli	[25]
XG	4%	90% sorghum flour, 10% potato starch, 100% water, 6% sugar, 3% baking powder, 1.5% salt	RVA	Lowered viscosity 2.8 vs. 3.4 cP (control)	[32]
HPMC	3%	90% sorghum flour, 10% tapioca starch, 100% water, 6% sugar, 3% baking powder, 1.5% salt		3.3 vs. 3.4 cP (control)	
XG	3%	90% sorghum flour, 10% rice starch, 100% water, 6% sugar, 3% baking powder, 1.5% salt		3.0 vs. 3.4 cP (control)	
Psyllium and HPMC	0–4% and 2–4%	100% rice flour, 3% yeast, 1.8% salt, 10% oil, 5% sugar, 90–110% water	Dynamic oscillatory and creep-recovery test	Psyllium incorporation reduced the pasting temperature and compliance values and increased elastic and viscous moduli	[24]

Table 1. *Cont.*

Type of Hydrocolloid	Level Used *	Other Ingredients	Type of the Rheological Test	Effect	References
XG	0.5–1.5%	60% rice flour, 40% buckwheat flour, 1.5% salt, 4.4% oil, 5.3% yeast, 80–90% water	Frequency sweep test	Elastic modulus from 4 to 22 times higher than control	[30]
PGA	0.5–1.5%			Elastic modulus from 1.5 to 3 times higher than control	
XG	0.5%	45% rice flour, 45% cassava starch, 10% soy flour, 2% salt, 2% shortening, 3%yeast, 75% water	Large deformation and frequency sweep	Resistance: 35.6 vs. 46.3 g (control)	[29]
Carrageenan	0.5%			Increased moduli Elastic: 60.8 vs. 29.7 kPa (control) Viscous: 12.9 vs. 6.8 kPa (control)	
XG, CMC	1% and 2%	rice flour, corn starch, sodium caseinate, fresh yeast, sunflower oil, salt, sugar, 140–150% water	Oscillation measurements	Increased elasticity	[15]

* based on flour weight basis.

2.2. Effect of Hydrocolloids on Bread Hardness

Bread crumb hardness is an important textural attribute as it is associated with the perception of consumers for freshness as well as for its relation with product shelf life. Bread crumb texture is influenced by the ingredients and recipe used. Usually, hydrocolloid addition tends to decrease bread hardness. The type of hydrocolloid, concentration and interaction are the factors that contribute to the hardness of the bread crumb [13]. As shown in Table 2, different hydrocolloids decreased the hardness of GF bread.

Rice bread prepared with different types of hydrocolloids showed a softer crumb than control samples without addition, and the hardness increases with the following order: mix XG–GG < HPMC < guar < XG ≈ mix locust bean gum-XG < pectin < locust bean gum. The combination of hydrocolloids with an emulsifier such as DATEM further lowered the hardness values and improved bread quality regarding the specific volume and sensory properties [20].

However, Calle et al. [35] showed the highest value for hardness in the case of breads prepared with HPMC, XG and GG, but they attributed this increase to the type of flour used, a rhizome flour from *Colocasia* spp. On the same level of hydrocolloids addition (2.5% reported to the amount of millet flour and tapioca starch), Chakraborty et al. [36] showed that XG decreased the bread hardness as compared to other hydrocolloids, varying as follows: GG > GA > tragacanth > XG [36]. On one side, XG was shown to have a softening effect over crumb hardness [36,37], while other studies found an increase in crumb hardness [10,15]. In line with the results of Lazaridou et al. [15] for rice-based GF bread, Peressini et al. found elevation with XG level in the crumb firmness of rice–buckwheat bread [30].

Differences may appear from the bread manufacturing process and especially from the amount of water used. Encina-Zelada et al. [38] also showed that higher levels of XG (3.5%) at a constant water level (90%) led to an increased crumb hardness of bread formulated with 50% rice, 30% maize and 20% quinoa flours. By increasing the water content (to 110%), the hardness and consistency were decreased, producing bread with higher specific volume and softer crumbs; however, the high amount of water yielded stickier and less viscous doughs.

Table 2. Effect of hydrocolloids on bread hardness compared to control.

Type of Hydrocolloid	Level Used *	Other Ingredients *	Hardness, g or N **	References
Carrageenan	0.5%	40% rice flour, 40% corn flour, 20% soy flour, 2% salt, 2% shortening, 3% compressed yeast, 158% water (flour basis).	818 vs. 720 g	[40]
Alginate	0.5%		723 vs. 720 g	
XG	0.5%		402 vs. 720 g	
CMC	0.5%		639 vs. 720 g	
Gelatine	0.5%		730 vs. 720 g	
HPMC	2%	100% potato flour, 70% water, 1% yeast	28.9 vs. 58.3 N	[41]
CMC	1%		32.7 vs. 58.3 N	
XG	2%		24.1 vs. 58.3 N	
Apple pectin	1%		33.6 vs. 58.3 N	
HPMC	2%	100% rhizome flour, 227% water, 1.5% salt, 3% yeast, 2% sugar, 2% oil	316 vs. 263 g	[35]
HPMC, XG, GG	0.29%, 0.21%, 0.50%		323 vs. 263 g	
XG	1.5%	58.3% corn starch, 25% rice flour, 16.7% soy flour, 3.3% pre-gelatinized corn starch, 3.3% vegetable oil, 1.7% egg white, 1.6% salt, 1.6% sugar, 1.3% yeast, 0.42% sodium stearoyl lactylate	5.1 vs. 26.2%	[37]
XG, CMC	1%, 1%		5.7 vs. 26.2%	
HPMC	1.5%	75% corn starch, 25% rice flour, 2% yeast, 4% sunflower oil, 4% sucrose, 2% salt, 80% water	2.96 vs. 4.9%	[10]
GG	1.5%		3.46 vs. 4.9%	
Carrageenan	1.5%		3.94 vs. 4.9%	
GG	5%	100% fresh cheese, 50% tapioca starch, 20% pre-cooked corn flour, 10% margarine, 6% sugar, 97% milk	16.5 vs. 20.0%	[39]
XG	0.5%	45% rice flour, 45% cassava starch, 10% soy flour, 2% salt, 2% shortening, 3% yeast, 75% water	162 vs. 249 g	[29]
CMC	0.5%		113 vs. 249 g	
Carrageenan	0.5%		132 vs. 249 g	
Alginate	0.5%		141 vs. 249 g	
GG	1.9%	50% rice flour, 15% corn flour, 30.6% cornstarch, 4.4% potato starch, 1.6% salt, 5.1% yeast, 5.9% oil, 83.6% g water	2.91 vs. 6 N	[42]
HPMC	2.3%		1.86 vs. 6 N	

* based on flour weight basis. ** vs. control: no hydrocolloid addition.

The capacity of the hydrocolloids to bind water helps to avoid water loss during bread storage. Sabanis and Tzia [10] found that the crumb hardness increases in the following order: HPMC < GG < carrageenan.

At a higher concentration of GG, the hardness of GF cheese bread decreased. A mixture of GG and HPMC led to an increase in bread hardness, which was explained by the water competition among the hydrocolloids and between the hydrocolloids and tapioca starch, the main GF ingredient [39].

In rice–buckwheat GF bread, the addition of XG or PGA improved crumb hardness by increasing the amount of water in the dough and, accordingly, the moisture content of the crumb because water has a plasticizing effect on the texture properties of the crumb cell walls [30]. Propylene glycol alginate breads showed greater improvement in terms of increased specific volume, decreased crumb firmness and crumb structure than XG breads. The positive effects of PGA were explained by a combined effect of low dough viscosity and elasticity produced by the polymer and the capacity to form elastic films at the gas and liquid interface, thus protecting the gas cells from instability [30].

By investigating the interactions between HPMC, psyllium and water in rice bread, no significant changes were recorded for specific bread volume when HPMC addition increased from 2% to 4% at different hydration levels between 90 and 110%. An opposite effect was observed in the case of increasing psyllium addition level from 0 to 4% when bread volume decreased and hardness increased. This outcome was diminished at higher water addition levels [24].

2.3. Effect of Hydrocolloids on Bread Specific Volume

Depending on the type and level of hydrocolloid addition used and the type of formulation, the effect of hydrocolloids over the specific volume of GF breads is different. There is no general correlation between the hydrocolloid concentration and the bread volume. For example, GF formulations based on potato starch containing pectin, HPMC and XG, did not show any significant effect over the specific volume when higher levels of hydrocolloid were used; while, in formulations with locust bean gum, GG and sodium alginate, the volume was dependent on the hydrocolloid level employed [13]. Thus, bread with the highest volume was obtained using 1% XG [40], while an opposite effect was reported by Lazaridou et al. [15] when using 1% and 2% XG (Table 3). The negative effect of XG on bread volume was explained by the hydrogen bonds that are formed between the negatively charged carboxyl groups present in the XG forms and water and starch and at higher levels of gum addition, leading to a rigid gel formation [36]. XG at high levels of addition produces doughs with too high resistance and consistency, which cause limited gas cell expansion during proofing [15,30]. The swelling of the starch granules is different in the presence of XG, and the granules are covered by a gum layer that limits the swelling at high temperatures [30]. Mezaize et al. [42] also reported that the incorporation of 0.6% XG into GF bread based on rice and cornflour and potato starch did not change the volume as compared to control, as XG addition makes the dough system too rigid to incorporate gases. On the other hand, the addition of 1.9% GG and 2.3% HPMC, respectively, increased the specific volume as compared to 0.6% XG.

Table 3. Effect of hydrocolloids on the bread specific volume as compared to control.

Type of Hydrocolloid	Level Used	Other Ingredients *	Specific Volume, cm^3/g **	References
Carrageenan Alginate XG CMC Gelatine	0.5% 0.5% 0.5% 0.5% 0.5%	40% rice flour, 40% corn flour, 20% soy flour, 2% salt, 2% shortening, 3% compressed yeast, 158% water	2.6 vs. 2.4 2.5 vs. 2.4 2.9 vs. 2.4 2.6 vs. 2.4 2.5 vs. 2.4	[40]
HPMC CMC XG Apple pectin	2% 1% 2% 1%	100% potato flour, 70% water, 1% yeast	2 vs. 1.25 1.75 vs. 1.25 1.85 vs. 1.25 1.6 vs. 1.25	[41]
HPMC HPMC GG	1.5% 2% 1.5%	75% corn starch, 25% rice flour, 2% yeast, 4% sunflower oil, 4% sucrose, 2% salt, 80% water for 1.5% hydrocolloid/85% water for 2% hydrocolloid	2.9 vs. 2.68 2.85 vs. 2.68 2.85 vs. 2.68	[10]
GG	2.5%	100% fresh cheese, 50% tapioca starch, 20% pre-cooked corn flour, 10% margarine, 6% sugar, 68% milk	2.4 vs. 2.1	[39]
XG CMC Carrageenan Alginate	0.5% 0.5% 0.5% 0.5%	45% rice flour, 45% cassava starch, 10% soy flour, 2% salt, 2% shortening, 3%yeast, 75% water	1.86 vs. 1.98 2.14 vs. 1.98 2.38 vs. 1.98 1.99 vs. 1.98	[29]
GG HPMC	1.9% 2.3%	50% rice flour, 15% corn flour, 30.6% cornstarch, 4.4% potato starch, 1.6% salt, 5.1% yeast, 5.9% oil, 83.6% water	2.82 vs. 2.47 3.33 vs. 2.47	[42]
CMC Agarose β-glucan	1% 1% 1%	rice flour, corn starch, sodium caseinate, fresh yeast, sunflower oil, salt, sugar, 140% water	2.67 vs. 2.19 2.62 vs. 2.19 2.68 vs. 2.19	[15]
Pectin	2%	rice flour, corn starch, sodium caseinate, fresh yeast, sunflower oil, salt, sugar, 150% water	2.52 vs. 2.21	[15]

* based on flour weight basis. ** vs. control: no hydrocolloid addition.

Another example was in the case of rice–buckwheat bread, where a level of addition of 0.5% XG gave the maximum bread volume, and a further increase in the gum concentration led to lower volume [30]. There should be a balance between the water level and the hydrocolloid concentration. Thus, to obtain higher bread volume, Peressini et al. [30] increased water level and decreased XG level. In GF formulations based on maize starch, 2% XG and 2% psyllium produced breads with similar specific volume but higher when compared to breads with 2% HPMC [18].

Sciarini et al. [40] stated that in formulations with high water content, batter consistency is strongly associated with bread volume. In their study, Lazaridou et al. [15] also reported that 1% addition of CMC, agarose and β-glucan in GF formulation significantly increased the loaf volume.

In GF cheese breads based on tapioca starch and pre-cooked corn flour, GG increased the specific loaf volume, while the mixture of GG and HPMC did not produce higher loaf volume [39].

Another study showed that HPMC was much more effective than GG in a corn-based GF bread formulation [25]. Mainly, the volume of HPMC breads was almost 1.2–1.6 times bigger than that of the control, and the increment is higher than that obtained for GG. Moreover, the addition of HPMC improved the quality of breads, which were characterized by a crumb structure more aerated, elastic and fine [25]. Breads with higher specific volume were found using HPMC and maize starch than other formulations with rice flour, which was explained by the presence of proteins that leads to a higher consistency than in the case of rice flours [18,24]. The specific volume of bread prepared with rice and corn flours and potato starch increased at 2.3% HPMC and 1.9% GG addition, respectively [42].

With the aim to investigate the most commonly used GF flours in bread manufacturing, Hager and Arendt [12] found that the volume of teff and maize breads was positively influenced by HPMC addition, the volume of rice bread decreased, and for the buckwheat bread, no effect was recorded. XG decreased the bread volume for all types of flour used. On the other hand, HPMC reduced the hardness of all the breads, while XG had a diverse role: decreasing for maize bread, increasing for teff and buckwheat breads and no effect for rice bread.

2.4. *Staling of GF Bread in the Presence of Hydrocolloids*

The fast-staling process in GF bread is an important issue. Crumb textural parameters—hardness/firmness and resilience—are used to measure crumb staling. To predict the bread shelf-life, kinetic models (i.e., Avrami model) that describe the crumb hardness are employed [7].

One of the aims of the hydrocolloids addition to bakery products is to improve their shelf life by retaining the moisture content and retarding the process of staling [40]. Bread staling rate is evidence of the product's shelf life and plays a significant role in the consumers' acceptability. Hydrocolloids influence the starch retrogradation in bread by diminishing the loss and diffusion of water from the crumb. Starch retrogradation and bread hardness are delayed as a consequence of higher moisture content in the bread [37].

Staling rate was calculated, reporting the difference between crumb hardness at 24 h and at 2 h after baking [19]. The staling rate of rice bread prepared with different hydrocolloids decreased in the following order: GG > locust bean gum ≈ sodium alginate > XG [19].

Increasing the level of XG from 5 to 15 g/kg flour in a GF formulation made from corn starch, rice flour, soy flour and pre-gelatinized corn starch decreased staling during storage, while CMC-containing formulae showed no significant difference after 3 days of storage at 17–20 °C [37]. Another study confirmed that the staling rate was slower in the presence of 1% XG or 1% CMC in a formulation with rice, corn and soy flours after bread storage at room temperature [40]. Formulations with the highest water content and lower moisture loss had the minimum staling. The hydrogen bonding between hydrocolloids and starch retards starch retrogradation [10].

Guar gum may also delay bread staling as it was observed in the GF cheese bread during storage for 6 days at room temperature due to its hydrophilic character that prevents water release and polymer aggregation. The mechanism proposed was based on a possible inhibition of amylopectin retrogradation as GG preferentially binds to starch [39].

Sciarini et al. [29] observed the following trend for the staling rate (related to the crumb-hardening) of bread based on rice flour, cassava starch, full-fat active soy and hydrocolloids: control > XG > carrageenan > alginate > CMC.

Moreover, the bread staling was faster with GG than sodium caseinate at a 1.5% level of addition in GF potato flour-based bread formulations because of its excessive moisture accumulation, but both hydrocolloids were effective in reducing the rate of staling when compared to the control bread. Besides the positive effect of the hydrocolloids on bread staling, benefits over the bread staling can be brought by the use of potato flour in the bread formulation due to its higher starch content and longer amylopectin side-chains, which contribute to the retaining of moisture in the bread during storage when compared to other cereals [43].

Psyllium is an effective anti-staling agent that significantly delays bread staling due to its higher capacity to bind water, limiting the water mobility, which decreases starch hydration, gelatinization and retrogradation thus, influencing the crumb hardening kinetics [7,44]. A reduction in bread staling was reported with a 17.14% psyllium addition and 117.86% water to a formulation consisting of 75% rice flour, 25% cassava starch, 25% whole egg, 10.5% whole milk powder, 6% white cane sugar, 6% soy oil, 2% salt, 0.8% dry yeast, and 0.1% calcium propionate. The authors found 75% softer crumbs in the psyllium-enriched GF bread [44]. In wholegrain buckwheat/carob-based GF bread (90.7%/7.3%), 2% psyllium addition delayed crumb hardening during 10 days of storage [7]. The staling effect was also attributed to the types of flour used (i.e., buckwheat and carob). Other studies found that chickpea flour in combination with psyllium reduced and delayed GF bread staling after 7 days of storage [33,45]. The higher fiber content from psyllium addition contributed to a greater crumb springiness and cohesiveness that inhibited the bread from crumbling during storage [33].

2.5. Estimated Glycemic Index of GF Bread in the Presence of Hydrocolloids

Celiac disease is associated with a high incidence of type I diabetes, and patients must maintain a constant glycemic control while adhering to a strict GF diet [46]. The glycemic index is influenced by several factors such as starch granule, bread structure and viscoelasticity. It was previously reported that the glycemic index of GF bread is much higher compared to the traditional bread, exerting an influence over chronic diseases [47–49]. The strategies to reduce the glycemic response of starchy gluten-free products refers to the replacement of flours and starches with alternative raw materials (characterized by an increased content of dietary fiber, protein and resistant starch), the addition of viscous dietary fibers and application of different processing conditions such as grain germination, sourdough fermentation or hydration level [50–52].

The use of high amounts of pure starches and rice flour in GF products determines higher glycemic index values (i.e., above 80) [53]. In GF rice bread, de la Hera et al. [54] underlined that the more compact the structure of the bread, the lower the glycemic response. In breads with higher amounts of water (90–110%), the estimated glycemic index was higher. Other alternative GF raw materials, such as *Colocasia esculenta* (a rhizome) flour, either thermally treated or in mixtures with hydrocolloids, contribute to the reduction in the glycemic index (i.e., below 30) [35].

There are few papers investigating the effect of hydrocolloids addition on the glycemic index of GF breads (Table 4). Liu et al. [41] showed that hydrocolloids addition (HPMC, CMC, XG and apple pectin) significantly reduced the rapidly digestible starch and the estimated glycemic index of the gluten-free bread based on potato flour compared to control bread. The hydrocolloid forms a layer around the starch granules, retarding the enzymatic hydrolysis and thus acting as a barrier to the enzyme attack or to the release of the products of hydrolysis [41,55]. Hydrocolloids addition modifies the starch gelatinization properties, influencing the starch digestibility. Higher percentages of hydrocolloid addition contribute to viscosity changes that cover the starch surface, preventing the α-amylase access [41]. The authors explained these phenomena for HPMC, CMC or apple pectin additions, while XG showed an opposite effect attributed to its higher molecular weight. Higher molecular weight was reported to enhance the viscosity of the liquid in the upper digestive tract, reducing the in vitro starch digestion and the glycemic response [56]. It was also reported that the addition of certain hydrocolloids (sodium carboxymethyl cellulose and XG) decreased the glycemic index of wheat-based bread [57].

Under simulated gastric and intestinal conditions, it was shown that the addition of guar gum in waxy maize starch reduced the glycemic response parameters, namely, by almost 25% in the starch hydrolysis and by 15% at the end of in vitro intestinal digestion [58]. The decreasing effect of gums over the post-prandial glycemia after ingestion of starchy foods was attributed to the gum's capacity to induce high viscosity in the gut lumen [59]. The authors found that the consumption by a non-diabetic group of subjects of wholemeal bread made with 15% guar addition produced a significantly lower blood glucose level at 30 min compared to control bread. In addition, the plasma insulin responses at 30 and 60 min were lower in the case of 10 and 15% guar additions compared to the control.

Recently, Montemurro et al. [60] formulated a "clean-label" gluten-free bread using natural hydrocolloids (a mixture of psyllium, flaxseed and chia flours as structuring agents), rice and maize flour fortified with quinoa flour and chestnut dough containing exopolysaccharides, which showed similar in vitro glycemic index (a value of 85 calculated with wheat bread as reference) as compared to other commercial GF breads. A lower estimated glycemic index (55.2) was obtained for a GF potato steam bread, containing 4.84% pregelatinized potato flour, 1.68% HPMC, 5.87% egg white protein, and 69.69% water based on potato flour [61]. The value was much lower compared to the value of 73.6 for the wheat steamed bread [61].

Table 4. Glycemic index for GF bread containing hydrocolloids.

Type of Hydrocolloid	Level Used	Other Ingredients	GI Value	Method *	References
None	-	100% potato flour, 70% water, 1% yeast	73.3	in vitro starch digestibility glucose	[41]
Apple pectin	0.5%		65.1		
	1%		**64.8**		
	2%		65.1		
HPMC	0.5%		65.0		
	1%		60.5		
	2%		**58.9**		
CMC	**0.5%**		**66.2**		
	1%		68.4		
	2%		66.6		
XG	**0.5%**		**62.7**		
	1%		**62.7**		
	2%		63.3		
None	-	100% flour (50% *Colocasia* flour blended with 50% pre-treated *Colocasia* flour), 227% water, 1.5% salt, 3% compressed yeast, 2% sugar, 2% oil	23.9	in vitro starch digestibility white bread	[35]
HPMC	2%		**23.1**		
HPMC + XG + GG	0.29 + 0.21 + 0.50%		26.2		
HPMC	1.68%	100% potato flour, 4.84% pregelatinized potato flour, 5.87% egg white protein, 69.69% water	55.2	in vitro starch digestibility	[61]
None	-	75% rice flour, 25% cassava starch, 25% whole egg, 10.5% whole milk powder, 6% white cane sugar, 6% soy oil, 2% salt, 0.8% dry yeast, 117.86% water	66.5	in vivo white wheat bread	[62]
Psyllium	17.14%		**50**		
XG + CMC	0.3%, 0.3%	75% chickpea flour, 25% cassava starch, 6% white cane sugar, 2% salt, 0.8% dry yeast, 0.1% calcium propionate, 25% whole eggs, 6% soybean oil, 125% water	79.2	in vivo rice bread	[63]
Psyllium	5.5%		**74.6**		

* Refers to the method used to determine the glycemic index and the type of the standard food used for comparison. Bold represents the lowest GI in the corresponding study.

Because different compounds (among them, fat, protein, dietary fiber, hydrocolloids, starch type) may interfere in the glycemic analysis, it is relatively difficult to compare the glycemic values between breads. Moreover, the method used plays an important role in the calculation of the glycemic index.

Some researches focused on evaluating the influence of psyllium on the post-prandial glycemic response of GF bread [62,63]. The addition of 17.14% psyllium to a GF bread formulation based on rice flour and cassava starch exhibited a decrease in the glycemic index by 25% compared to a control bread without psyllium addition [62]. Similarly, the combination of chickpea and 5.5% psyllium in gluten-free bread-making reduced the glycemic index by 25% [63].

Besides the reduction in the glycemic response, psyllium addition enhanced the bread volume, appearance and sensory acceptability score, yielding softer crumbs as well as higher dietary fiber content.

3. Hydrocolloids in GF Pasta/Noodles

Pasta/noodles represent one of the most consumed GF products due to their versatility to be produced in different shapes, from various ingredients: legumes, pseudocereals, etc. Hydrocolloids play a crucial role in obtaining fresh and cooked pasta. The dough rheology during mixing, heating and cooling is influenced by the hydration during pasta preparation. The addition of hydrocolloids may affect pasta color, hardness and firmness. Table 5 presents the effect of hydrocolloid addition on some GF pasta.

Table 5. Effect of hydrocolloids in GF pasta.

Type of Hydrocolloid/ Obtained Product	Level Used	Other Ingredients	Type of the Rheological Test	Effect	References
XG, GG, CMC/noodles	0.5%	Tiger nut flour	Mixolab rheological behavior: mixing, heating and cooling consistency, extrusion force	Improved dough extensibility; XGgavehigher firmness, reduced adhesiveness, increased chewiness and resilience	[64]
Gellan Gum, CMC, Pectin PEC, Agar, Tapioca starch, Guar seed flour and Chitosan/spaghetti	2.0%	Maize flour and naked oat	Elongation and shear viscosity (capillary rheometer)	Improved cooking quality and texture properties (adhesiveness, cooking loss, hardness). Chitosan: reduced glycemic index. CMC and agar: reducing the blood cholesterol.	[65]
XG/noodle	5%	Rice flour, glutinous rice flour	Pasting properties (RVA); Frequency sweep test (controlled-stress rheometer); Dough development characteristics: water absorption, development time, stability, softening (DoughLab equipment)	Enhanced tensile strength, peak viscosity, gel strength. Increased chewiness and hardness.	[66]

Table 5. *Cont.*

Type of Hydrocolloid/ Obtained Product	Level Used	Other Ingredients	Type of the Rheological Test	Effect	References
GG, gum acacia and gum tragacanth/pasta	0.5–1%	Amaranth flour	Pasting properties (RVA)	GG and gum tragacanth: increased peak, trough, breakdown and final viscosities. Gum acacia: reverse trend.	[67]
GG, XG, sodium alginate	1% and 2%	Proso millet flour	Frequency sweep tests (controlled stress rheometer)	Improved dough rheology (increased viscosity and elasticity at 2% addition) Improved pasta network strength by GG and XG addition	[68]

Sensory attributes of GF pasta are influenced by the nature of the raw ingredients used and the addition of hydrocolloids. GF tiger nut noodles made with XG and an adapted amount of water showed the best quality, considering the lowest cooking losses obtained and higher firmness values. Colour was differently affected by hydrocolloids addition, observing a decrease in luminosity, although significant only when hydration was adapted in the presence of XG, GG or CMC [64]. The authors stated that GG, XG and CMC, increased the noodles diameter while the level of hydration influenced the rheological behavior due to the high ability to retain water.

Padalino et al. [65] evaluated the following sensory attributes of GF spaghetti: color, homogeneity, odor, overall quality for noncooked spaghetti and elasticity, firmness, bulkiness, adhesiveness, color, homogeneity, odor, taste and overall quality for cooked spaghetti. The best overall quality was obtained by the addition of 2% CMC or chitosan. Moreover, pasta based on maize and oat flours with added chitosan as hydrocolloid showed an increased content of water-insoluble fibers, which is beneficial for reducing the glycemic index; spaghetti with CMC and agar, on the other hand, returned an increased water-soluble fiber content, which makes them recommended for reducing the blood cholesterol level.

Pasta prepared with 1.0% GG and amaranth flour showed higher sensory scores for firmness, texture, taste and overall quality of pasta [67].

In GF pasta with cassava starch and cornflour, XG improved dough handling and a level of addition of 0.6% had the highest potential to improve the pasta capacity to prevent structure disintegration, showing the lowest cooking loss and the lowest values for firmness, cohesiveness, chewiness, springiness and cutting force as well as a non-adhesive mouthfeel [69].

De Arcangelis et al. [70] prepared innovative GF pasta with the highest cooking quality and texture using a combination of 0.1% PGA, 0.5% monoglycerides of fatty acids and the gelatinization of a mixture of flours (buckwheat, maize and rice).

4. Conclusions

Hydrocolloids are widely used in GF systems to increase: dough handling properties, viscosity and incorporation of air into the GF dough/batter, overall quality and to extend the shelf-life of final products as a result of their structure-building and water-binding properties. Most of the hydrocolloids benefits are explained by their property to increase the water-holding ability of the dough system due to high molecular weight that helps to create a more stable structure.

In GF bread, hydrocolloids are used as gluten replacements and stabilizing agents. Furthermore, hydrocolloids can delay the release of digested carbohydrates and, thus, decrease the glycemic bread index. Among the hydrocolloids that reduced in vitro starch digestibility and estimated glycemic index are: HPMC, CMC, XG, apple pectin or psyllium, depending on the addition level in the GF formulations.

The positive effect that the hydrocolloids addition brings to the GF dough matrix depends not only on the type and concentration used but also on the interactions with the flour and other ingredients as well as on the process parameters (temperature, pH). XG and HPMC are the most employed hydrocolloids for GF breads. In GF pasta, hydrocolloid addition is used to improve dough handling, cooking quality and texture, as well as to obtain higher sensory scores. There is a lower number of publications that study the impact of hydrocolloids on the batter rheology of GF sweet products as compared to GF bread.

Author Contributions: Conceptualization, A.C., D.E.D., M.P. and T.V.; writing—original draft preparation, A.C. and D.E.D.; writing—review and editing, A.C., D.E.D., M.P. and T.V.; supervision, M.P. and T.V.; project administration, A.C. and D.E.D.; funding acquisition, A.C. and D.E.D. All authors have read and agreed to the published version of the manuscript.

Funding: This work was achieved through the Core programme (PN 19 02), supported by the Ministry of Research, Innovation and Digitalization, Romania, project number 19 02 02 02.

Data Availability Statement: Data are contained within the article.

Conflicts of Interest: The authors declare no conflict of interest.

Abbreviations

CMC	carboxymethyl cellulose
GA	gum arabic
GF	gluten-free
GG	guar gum
GI	Glycemic Index
G′	elastic (storage) modulus
G″	viscous (loss) modulus
HPMC	hydroxypropyl methylcellulose
PGA	propylene glycol alginate
RVA	Rapid Visco Analyzer
XG	xanthan gum

References

1. Padalino, L.; Conte, A.; Del Nobile, M.A. Overview on the general approaches to improve gluten-free pasta and bread. *Foods* **2016**, *5*, 87. [CrossRef]
2. Zoghi, A.; Mirmahdi, R.S.; Mohammadi, M. The role of hydrocolloids in the development of gluten-free cereal-based products for coeliac patients: A review. *Int. J. Food Sci. Technol.* **2021**, *56*, 3138–3147. [CrossRef]
3. Gao, Y.P.; Janes, M.E.; Chaiya, B.; Prinyawiwatkul, W.; Brennan, M.A.; Brennan, C.S.; Prinyawiwatkul, W. Gluten-free bakery and pasta products: Prevalence and quality improvement. *Int. J. Food Sci. Technol.* **2018**, *53*, 19–32. [CrossRef]
4. Anton, A.A.; Artfield, S.D. Hydrocolloids in gluten-free breads: A review. *Int. J. Food Sci. Nutr.* **2008**, *59*, 11–23. [CrossRef]
5. Salehi, F. Improvement of gluten-free bread and cake properties using natural hydrocolloids: A review. *Food Sci. Nutr.* **2019**, *7*, 3391–3402. [CrossRef] [PubMed]
6. Xu, J.; Zhang, Y.; Wang, W.; Li, Y. Advanced properties of gluten-free cookies, cakes, and crackers: A review. *Trends Food Sci. Technol.* **2020**, *103*, 200–213. [CrossRef]
7. Filipčev, B.; Pojić, M.; Šimurina, O.; Mišan, A.; Mandić, A. Psyllium as an improver in gluten-free breads: Effect on volume, crumb texture, moisture binding and staling kinetics. *LWT* **2021**, *151*, 112156. [CrossRef]
8. Mir, S.A.; Shah, M.A.; Naik, H.R.; Zargar, I.A. Influence of hydrocolloids on dough handling and technological properties of gluten-free breads. *Trends Food Sci. Technol.* **2016**, *51*, 49–57. [CrossRef]
9. Cappelli, A.; Oliva, N.; Cini, E. A systematic review of gluten-free dough and bread: Dough rheology, bread characteristics, and improvement strategies. *Appl. Sci.* **2020**, *10*, 6559. [CrossRef]
10. Sabanis, D.; Tzia, C. Effect of hydrocolloids on selected properties of gluten-free dough and bread. *Food Sci. Technol. Int.* **2010**, *17*, 279–291. [CrossRef] [PubMed]
11. Morreale, F.; Garzón, R.; Rosell, C.M. Understanding the role of hydrocolloids viscosity and hydration in developing gluten-free bread. A study with hydroxypropylmethylcellulose. *Food Hydrocoll.* **2018**, *77*, 629–635. [CrossRef]
12. Hager, A.-S.; Arendt, E.K. Influence of hydroxypropylmethylcellulose (HPMC), xanthan gum and their combination on loaf specific volume, crumb hardness and crumb grain characteristics of gluten-free breads based on rice, maize, teff and buckwheat. *Food Hydrocoll.* **2013**, *32*, 195–203. [CrossRef]

13. Horstmann, S.W.; Axel, C.; Arendt, E.K. Water absorption as a prediction tool for the application of hydrocolloids in potato starch-based bread. *Food Hydrocoll.* **2018**, *81*, 129–138. [CrossRef]
14. Funami, T.; Kataoka, Y.; Omoto, T.; Goto, Y.; Asai, I.; Nishinari, K. Food hydrocolloids control the gelatinization and retrogradation behavior of starch. 2a. Functions of guar gums with different molecular weights on the gelatinization behavior of corn starch. *Food Hydrocoll.* **2005**, *19*, 15–24. [CrossRef]
15. Lazaridou, A.; Duta, D.; Papageorgiou, M.; Belc, N.; Biliaderis, C.G. Effects of hydrocolloids on dough rheology and bread quality parameters in gluten-free formulations. *J. Food Eng.* **2007**, *79*, 1033–1047. [CrossRef]
16. Crockett, R.; Ie, P.; Vodovotz, Y. Effects of soy protein isolate and egg white solids on the physicochemical properties of gluten-free bread. *Food Chem.* **2011**, *129*, 84–91. [CrossRef]
17. Skendi, A.; Papageorgiou, M.; Varzakas, T. High Protein Substitutes for Gluten in Gluten-Free Bread. *Foods* **2021**, *10*, 1997. [CrossRef]
18. Belorio, M.; Gómez, M. Effect of hydration on gluten-free breads made with hydroxypropyl methylcellulose in comparison with psyllium and xanthan gum. *Foods* **2020**, *9*, 1548. [CrossRef] [PubMed]
19. Sahin, A.S.; Wiertz, J.; Arendt, E.K. Evaluation of a new method to determine the water addition level in gluten-free bread systems. *J. Cereal Sci.* **2020**, *93*, 102971. [CrossRef]
20. Demirkesen, L.; Mert, B.; Sumnu, G.; Sahin, S. Rheological properties of gluten-free bread formulations. *J. Food Eng.* **2010**, *96*, 295–303. [CrossRef]
21. Capriles, V.D.; Arêas, J.A.G. Novel approaches in gluten-free breadmaking: Interface between food science, nutrition, and health. *Compr. Rev. Food Sci. Food Saf.* **2014**, *13*, 871–890. [CrossRef]
22. Torbica, A.; Hadnađev, M.; Dapčević, T. Rheological, textural and sensory properties of gluten-free bread formulations based on rice and buckwheat flour. *Food Hydrocoll.* **2010**, *24*, 626–632. [CrossRef]
23. Baldino, N.; Laitano, F.; Lupi, F.R.; Curcio, S.; Gabriele, D. Effect of HPMC and CMC on rheological behavior at different temperatures of gluten-free bread formulations based on rice and buckwheat flours. *Eur. Food Res. Technol.* **2018**, *244*, 1829–1842. [CrossRef]
24. Mancebo, C.M.; San Miguel, M.Á.; Martínez, M.M.; Gómez, M. Optimisation of rheological properties of gluten-free doughs with HPMC, psyllium and different levels of water. *J. Cereal Sci.* **2015**, *61*, 8–15. [CrossRef]
25. Ozturk, O.K.; Mert, B. The effects of microfluidization on rheological and textural properties of gluten-free corn breads. *Food Res. Int.* **2018**, *105*, 782–792. [CrossRef]
26. Phan-Thien, N.; Safari-Ardi, M. Linear viscoelastic properties of flour-water doughs at different water concentrations. *J. Non Newton. Fluid Mech.* **1998**, *74*, 137–150. [CrossRef]
27. Fanari, F.; Desogus, F.; Scano, E.A.; Carboni, G.; Grosso, M. The effect of the relative amount of ingredients on the rheological properties of semolina doughs. *Sustainability* **2020**, *12*, 2705. [CrossRef]
28. Zhong, Q.; Ikeda, S. Viscoelastic properties of concentrated aqueous ethanol suspensions of alpha-zein. *Food Hydrocoll.* **2012**, *28*, 46–52. [CrossRef]
29. Sciarini, L.S.; Ribotta, P.D.; León, A.E.; Pérez, G.T. Incorporation of several additives into gluten free breads: Effect on dough properties and bread quality. *J. Food Eng.* **2012**, *111*, 590–597. [CrossRef]
30. Peressini, D.; Pin, M.; Sensidoni, A. Rheology and breadmaking performance of rice-buckwheat batters supplemented with hydrocolloids. *Food Hydrocoll.* **2011**, *25*, 340–349. [CrossRef]
31. Zhao, F.; Li, Y.; Li, C.; Ban, X.; Cheng, L.; Hong, Y.; Gu, Z.; Li, Z. Co-supported hydrocolloids improve the structure and texture quality of gluten-free bread. *LWT* **2021**, *152*, 112248. [CrossRef]
32. Ari Akin, P.; Miller, R.A. Starch–hydrocolloid interaction in chemically leavened gluten-free sorghum bread. *Cereal Chem.* **2017**, *94*, 897–902. [CrossRef]
33. Santos, F.G.; Capriles, V.D. Relationships between dough thermomechanical parameters and physical and sensory properties of gluten-free bread texture during storage. *LWT* **2021**, *139*, 110577. [CrossRef]
34. Moreira, R.; Chenlo, F.; Torres, M.D. Effect of chia (*Sativa hispanica* L.) and hydrocolloids on the rheology of gluten-free doughs based on chestnut flour. *LWT* **2013**, *50*, 160–166. [CrossRef]
35. Calle, J.; Benavent-Gil, Y.; Rosell, C.M. Development of gluten free breads from Colocasia esculenta flour blended with hydrocolloids and enzymes. *Food Hydrocoll.* **2020**, *98*, 105243. [CrossRef]
36. Chakraborty, S.K.; Kotwaliwale, N.; Navale, S.A. Selection and incorporation of hydrocolloid for gluten-free leavened millet breads and optimization of the baking process thereof. *LWT* **2020**, *119*, 108878. [CrossRef]
37. Mohammadi, M.; Sadeghnia, N.; Azizi, M.-H.; Neyestani, T.-R.; Mortazavian, A.M. Development of gluten-free flat bread using hydrocolloids: Xanthan and CMC. *J. Ind. Eng. Chem.* **2014**, *20*, 1812–1818. [CrossRef]
38. Encina-Zelada, C.R.; Cadavez, V.; Monteiro, F.; Teixeira, J.A.; Gonzales-Barron, U. Combined effect of xanthan gum and water content on physicochemical and textural properties of gluten-free batter and bread. *Food Res. Int.* **2018**, *111*, 544–555. [CrossRef]
39. Rodriguez-Sandoval, E.; Cortes-Rodriguez, M.; Manjarres-Pinzon, K. Effect of hydrocolloids on the pasting profiles of tapioca starch mixtures and the baking properties of gluten-free cheese bread. *J. Food Process. Preserv.* **2015**, *39*, 1672–1681. [CrossRef]
40. Sciarini, L.S.; Ribotta, P.D.; León, A.E.; Pérez, G.T. Effect of hydrocolloids on gluten-free batter properties and bread quality. *Int. J. Food Sci. Technol.* **2010**, *45*, 2306–2312. [CrossRef]

41. Liu, X.; Mu, T.; Sun, H.; Zhang, M.; Chen, J.; Fauconnier, M.L. Influence of different hydrocolloids on dough thermo-mechanical properties and in vitro starch digestibility of gluten-free steamed bread based on potato flour. *Food Chem.* **2018**, *239*, 1064–1074. [CrossRef] [PubMed]
42. Mezaize, S.; Chevallier, S.; Le Bail, A.; De Lamballerie, M. Optimization of gluten-free formulations for french-style breads. *J. Food Sci.* **2009**, *74*, E140–E146. [CrossRef]
43. Moradi, M.; Bolandi, M.; Arabameri, M.; Karimi, M.; Baghaei, H.; Nahidi, F.; Eslami Kanafi, M. Semi-volume gluten-free bread: Effect of guar gum, sodium caseinate and transglutaminase enzyme on the quality parameters. *J. Food Meas. Charact.* **2021**, *15*, 2344–2351. [CrossRef]
44. Fratelli, C.; Santos, F.G.; Muniz, D.G.; Habu, S.; Braga, A.R.C.; Capriles, V.D. Psyllium improves the quality and shelf life of gluten-free bread. *Foods* **2021**, *10*, 954. [CrossRef]
45. Santos, F.G.; Aguiar, E.V.; Braga, A.R.C.; Alencar, N.M.M.; Rosell, C.M.; Capriles, V.D. An integrated instrumental and sensory approach to describe the effects of chickpea flour, psyllium, and their combination at reducing gluten-free bread staling. *Food Packag. Shelf Life* **2021**, *28*, 100659. [CrossRef]
46. Cohn, A.; Sofia, A.M.; Kupfer, S.S. Type 1 diabetes and celiac disease: Clinical overlap and new insights into disease pathogenesis. *Curr. Diab. Rep.* **2014**, *14*, 517. [CrossRef]
47. Segura, M.E.M.; Rosell, C.M. Chemical composition and starch digestibility of different gluten-free breads. *Plant. Foods Hum. Nutr.* **2011**, *66*, 224–230. [CrossRef] [PubMed]
48. Romão, B.; Falcomer, A.L.; Palos, G.; Cavalcante, S.; Botelho, R.B.A.; Nakano, E.Y.; Raposo, A.; Shakeel, F.; Alshehri, S.; Mahdi, W.A.; et al. Glycemic index of gluten-free bread and their main ingredients: A systematic review and meta-analysis. *Foods* **2021**, *10*, 506. [CrossRef]
49. Capriles, V.D.; Arêas, J.A.G. Effects of prebiotic inulin-type fructans on structure, quality, sensory acceptance and glycemic response of gluten-free breads. *Food Funct.* **2013**, *4*, 104–110. [CrossRef]
50. Capriles, V.D.; Arêas, J.A.G. Approaches to reduce the glycemic response of gluten-free products: In vivo and in vitro studies. *Food Funct.* **2016**, *7*, 1266–1272. [CrossRef] [PubMed]
51. Bender, D.; Schönlechner, R. Innovative approaches towards improved gluten-free bread properties. *J. Cereal Sci.* **2020**, *91*, 102904. [CrossRef]
52. Ramos, L.; Alonso-Hernando, A.; Martínez-Castro, M.; Morán-Pérez, J.A.; Cabrero-Lobato, P.; Pascual-Maté, A.; Téllez-Jiménez, E.; Mujico, J.R. Sourdough biotechnology applied to gluten-free baked goods: Rescuing the tradition. *Foods* **2021**, *10*, 1498. [CrossRef] [PubMed]
53. Aguiar, E.V.; Santos, F.G.; Krupa-Kozak, U.; Capriles, V.D. Nutritional facts regarding commercially available gluten-free bread worldwide: Recent advances and future challenges. *Crit. Rev. Food Sci. Nutr.* **2021**, 1–13. [CrossRef]
54. de la Hera, E.; Rosell, C.M.; Gomez, M. Effect of water content and flour particle size on gluten-free bread quality and digestibility. *Food Chem.* **2014**, *151*, 526–531. [CrossRef] [PubMed]
55. Chung, H.J.M.; Liu, Q.; Lim, S.T. Texture and in vitro digestibility of white rice cooked with hydrocolloids. *Cereal Chem.* **2007**, *84*, 246–249. [CrossRef]
56. Thondre, P.S.; Monro, J.A.; Mishra, S.; Henry, C.J.K. High molecular weight barley β-glucan decreases particle breakdown in chapattis (Indian flat breads) during in vitro digestion. *Food Res. Int.* **2010**, *43*, 1476–1481. [CrossRef]
57. Ho, L.H.; Tan, T.C.; Aziz, N.A.A.; Bhat, R. In vitro starch digestibility of bread with banana (*Musa acuminata* X *balbisiana* ABB cv. Awak) pseudo-stem flour and hydrocolloids. *Food Biosci.* **2015**, *12*, 10–17. [CrossRef]
58. Dartois, A.; Singh, J.; Kaur, L.; Singh, H. Influence of guar gum on the in vitro starch digestibility—Rheological and microstructural characteristics. *Food Biophys.* **2010**, *5*, 149–160. [CrossRef]
59. Ellis, P.R.; Burley, V.J.; Leeds, A.R.; Peterson, D.B. A guar enriched wholemeal bread reduces postprandial glucose and insulin responses. *J. Hum. Nutr. Diet.* **1988**, *1*, 77–84. [CrossRef]
60. Montemurro, M.; Pontonio, E.; Rizzello, C.G. Design of a "clean-label" gluten-free bread to meet consumers demand. *Foods* **2021**, *10*, 462. [CrossRef]
61. Liu, X.; Mu, T.; Sun, H.; Zhang, M.; Chen, J.; Fauconnier, M.L. Effect of ingredients on the quality of gluten-free steamed bread based on potato flour. *J. Food Sci. Technol.* **2019**, *56*, 2863–2873. [CrossRef]
62. Fratelli, C.; Muniz, D.G.; Santos, F.G.; Capriles, V.D. Modelling the effects of psyllium and water in gluten-free bread: An approach to improve the bread quality and glycemic response. *J. Funct. Foods* **2018**, *42*, 339–345. [CrossRef]
63. Santos, F.G.; Aguiar, E.V.; Rosell, C.M.; Capriles, V.D. Potential of chickpea and psyllium in gluten-free breadmaking: Assessing bread's quality, sensory acceptability, and glycemic and satiety indexes. *Food Hydrocoll.* **2021**, *113*, 106487. [CrossRef]
64. Gasparre, N.; Rosell, C.M. Role of hydrocolloids in gluten free noodles made with tiger nut flour as non-conventional powder. *Food Hydrocoll.* **2019**, *97*, 105194. [CrossRef]
65. Padalino, L.; Mastromatteo, M.; De Vita, P.; Ficco, D.B.M.; Del Nobile, M.A. Effects of hydrocolloids on chemical properties and cooking quality of gluten-free spaghetti. *Int. J. Food Sci. Technol.* **2013**, *48*, 972–983. [CrossRef]
66. Cai, J.; Chiang, J.H.; Tan, M.Y.P.; Saw, L.K.; Xu, Y.; Ngan-Loong, M.N. Physicochemical properties of hydrothermally treated glutinous rice flour and xanthan gum mixture and its application in gluten-free noodles. *J. Food Eng.* **2016**, *186*, 1–9. [CrossRef]
67. Chauhan, A.; Saxena, D.C.; Singh, S. Effect of hydrocolloids on microstructure, texture and quality characteristics of gluten-free pasta. *Food Meas.* **2017**, *11*, 1188–1195. [CrossRef]

68. Romero, H.M.; Santra, D.; Rose, D.; Zhang, Y. Dough rheological properties and texture of gluten-free pasta based on proso millet flour. *J. Cereal Sci.* **2017**, *74*, 238–243. [CrossRef]
69. Milde, L.B.; Chigal, P.S.; Olivera, J.E.; González, K.G. Incorporation of xanthan gum to gluten-free pasta with cassava starch. Physical, textural and sensory attributes. *LWT* **2020**, *131*, 109674. [CrossRef]
70. De Arcangelis, E.; Cuomo, F.; Trivisonno, M.C.; Marconi, E.; Messia, M.C. Gelatinization and pasta making conditions for buckwheat gluten-free pasta. *J. Cereal Sci.* **2020**, *95*, 103073. [CrossRef]

foods

MDPI

Article

Impact of Gluten-Free Sorghum Bread Genotypes on Glycemic and Antioxidant Responses in Healthy Adults

Lorenza Rodrigues dos Reis Gallo [1,*], Caio Eduardo Gonçalves Reis [1], Márcio Antônio Mendonça [2], Vera Sônia Nunes da Silva [3], Maria Teresa Bertoldo Pacheco [3] and Raquel Braz Assunção Botelho [1,*]

[1] Department of Nutrition, College of Health Sciences, University of Brasilia, Brasilia 70910-900, Brazil; caioedureis@gmail.com
[2] College of Agronomy and Veterinary Medicine, University of Brasilia, Brasilia 70910-900, Brazil; marcioamen@gmail.com
[3] Institute of Food Technology, Secretariat of Agriculture and Supply of Sao Paulo, Sao Paulo 13070-178, Brazil; vera.silva@ital.sp.gov.br (V.S.N.d.S.); mtb@ital.sp.gov.br (M.T.B.P.)
* Correspondence: lorenzagallo23@gmail.com (L.R.d.R.G.); raquelbotelho@unb.br (R.B.A.B.); Tel.: +55-61-999864501 (L.R.d.R.G.); +55-61-981378620 (R.B.A.B.)

Abstract: Sorghum is used to provide good quality gluten-free products due to phytochemicals and low glycemic index (GI). This study aimed to determine the chemical composition, the antioxidant activity and capacity, and the glycemic and insulinemic responses of gluten-free (GF) sorghum bread. GF bread samples were produced with three different sorghum genotypes. The samples were analyzed for chemical composition, resistant starch and dietary fiber content; antioxidant activity by ORAC; antioxidant capacity by FRAP; GI; and insulinemic responses. This double-blind, crossover, randomized clinical trial was conducted with 10 healthy men aged 28.0 ± 4.9 years (77.6 ± 11.7 kg and 24.2 ± 2.3 kg/m^{2}). All sorghum bread showed significantly more fiber than rice bread (control). Brown sorghum bread was classified as low GI, bronze and white as medium GI, and control as high GI. Brown sorghum bread presented a low carbohydrate content, a significant amount of fiber, and a significantly lower 3 h AUC glucose response than those of the control, aside from the highest antioxidant activity value ($p \leq 0.001$). Therefore, brown sorghum was superior to other genotypes analyzed in this study, and its production should be encouraged to provide gluten-free products with a better nutritional profile. More research is required to explore the effects of different sorghum genotypes in food products on human health.

Keywords: gluten-free; sorghum; bread; antioxidant activity; resistant starch; dietary fibers

Citation: dos Reis Gallo, L.R.; Reis, C.E.G.; Mendonça, M.A.; da Silva, V.S.N.; Pacheco, M.T.B.; Botelho, R.B.A. Impact of Gluten-Free Sorghum Bread Genotypes on Glycemic and Antioxidant Responses in Healthy Adults. *Foods* **2021**, *10*, 2256. https://doi.org/10.3390/foods10102256

Academic Editors: Maria Papageorgiou and Theodoros Varzakas

Received: 24 August 2021
Accepted: 15 September 2021
Published: 23 September 2021

Publisher's Note: MDPI stays neutral with regard to jurisdictional claims in published maps and institutional affiliations.

1. Introduction

There is a growing demand for gluten-free products (GFP) due to adverse symptoms of gluten such as celiac disease, gluten or wheat allergy, and gluten sensitivity [1,2]. Many wheat flour substitutes are applied to produce GFP, including sorghum (*Sorghum bicolor L.*) [3], a gluten-free (GF) grain with a simple cultivation process, high nutritional profile, and health benefits [4–7]. Sorghum can be incorporated into gluten-free bread formulations, as well as other baked products, such as cakes and cookies, flakes, and pasta [8–12].

Several sorghum genotypes are presented by the genotypes white, cream, yellow, orange, bronze, red, brown, black, and various combinations of these colors [13]. Each genotype presents different characteristics. For example, a white-colored genotype with no pigmented testa presents no significant levels of tannins; a red-colored genotype with pigmented testa presents a moderate tannins content; and a brown-colored genotype with pigmented testa presents a high tannins content [14].

Phenolic compounds of sorghum are usually concentrated in the grain's pericarp [15], and sorghum genotypes containing tannins have a higher antioxidant capacity than sorghum genotypes that do not contain tannins [8]. Moreover, in comparison with white sorghum genotypes, colored sorghum genotypes have a higher phenolic compounds concentration [16]. Therefore, brown-, bronze- and red-pigmented sorghums are rich in phenolic compounds (e.g., flavonoids). They are also rich sources of several phytochemicals, including tannins, phenolic acids, anthocyanins, phytosterols, and policosanols, providing significant antioxidant properties [8,17]. According to Awika and Rooney [8], sorghum genotypes can contain from 0.5 to 68 mg/g of tannins and several different phenolic acids with 3 to 43 mg/100 g of total phenolic content [14].

These sorghum polyphenols are known to function as potent antioxidants, at least in vitro [18]. Despite phenolic compounds' bioavailability after dietary intake being a research topic in recent years, clinical studies are scarce and controversial [7,19]. According to Prior and Wu [19], for some phenolic compounds, there are differences among their primary forms circulating in blood or tissues after oral ingestion and the original forms in the diet. Moreover, some anthocyanins are absorbed intact, and absorption can be saturated. Additionally, the quantities excreted in the urine are less than 0.1% of the total intake, although 60 to 90% of anthocyanin may disappear from the gastrointestinal tract within 4 h after a meal.

Another relevant characteristic of sorghum is the high level of resistant starch (RS), ranging from 2.2 to 6.5 g/100 g [6]. Moreover, sorghum is fiber-rich, containing over 95% of the non-starch polysaccharides [20]. These characteristics provide a slow starch absorption similar to a low-glycemic index food, positively impacting glucose metabolism [12]. Among several studies regarding sorghum products [8–12], only Wolter et al. [11] have calculated the glycemic index (GI) of one genotype of sorghum in bread using in vitro starch digestibility. This resulted in a medium GI (GI = 69). Therefore, there is a lack of information about sorghum products' composition and their impact on glucose metabolism. More studies are needed to assess the effects of sorghum bread consumption on human health [11].

Thus, this study aimed to analyze the chemical composition, antioxidant properties, and the effects on the glucose metabolism of GF sorghum bread made with three different genotypes in healthy adults. Additionally, the study focused on evaluating different sorghum genotypes to produce more options for gluten-free products with better nutritional qualities.

2. Materials and Methods

2.1. Sorghum and Bread Production

According to our group's previous study, three genotypes of sorghum provided by Embrapa (Brazilian Agriculture Research Enterprise) were chosen by the different pericarp colors (white—BRS 501, brown—BR 305, and bronze—BRS 332) and the highest acceptability [21]. One batch of each sorghum genotype was transformed into flour using a Thermomix processor (speed 10 for 3 minutes) (Vorwek TM6, Wuppertal, Germany).

Four bread samples were prepared according to the recipes used by Andrade de Aguiar et al. [21], one as control with commercial white rice flour (produced with *Oryza sativa*) and three with the different types of sorghum flours (made with white, brown, and bronze genotypes). All formulations had the same proportion of ingredients (22.36% of flour, 10.06% of potato starch, 6.04% of whole egg, 5.36% of soy oil, 4.14% of cassava flour, 3.80% of brown sugar, 3.69% of egg whites, 1.12% of dry yest, 0.56% of salt, 0.39% of xanthan gum, and 42.48% of water), only changing the primary flour (rice or sorghum types) for each type of bread. All ingredients used were gluten-free.

Ingredients were weighed to prepare the four different samples of bread. Dry yeast was mixed with 26% of the total water (T—35 °C) and brown sugar and fermented for 10 min. All the other dry ingredients were mixed in a food mixer for 1 min. Then, egg, egg whites, and soy oil were added and mixed with the dry ingredients for one more minute. Hydrated yeast was added to the previous mixture. Each dough was placed in a rectangular cake tin that measured 8.66 (width), ×3.94 (height), 2.56 (depth) inches for a fermentation process of 25 min then was baked in a gas oven (Brastemp, São Paulo, Brazil). Rice bread dough was baked in a pre-heated oven at 180 °C for 45 min and sorghum bread dough for 50 min. After baking, samples were removed from the tin, and each bread slice was stored in freezer bags. For bread chemical composition, samples were analyzed the day after bread production, and for antioxidant analysis, bread slices were frozen at −80 °C until analysis, approximately for two months. For resistant starch analysis, samples were stored in the same freezer bags at room temperature for 24 h before analysis.

2.2. Bread Chemical Composition

For each type of bread, three separate recipes were baked. Then, analyses were conducted in triplicate for moisture (AOAC Official Method 925.09), ash (AACC Official Method 08-03.01), proteins (Kjeldahl, AACC 46-13), and lipids (by extraction with petroleum ether by dragging under pressure with Extractor Ankom Model XT10). Total dietary fiber analysis (AOAC Official Method 985.29) was conducted in dry base samples. Carbohydrates were calculated by difference: 100—(weight in grams (protein + fat + ash + fiber + Resistant Starch (RS)) in 100 g of food). RS was analyzed on wet basis samples by the official analysis methods [22–25], in which 0.1 g of each sample was incubated in a shaking water bath (Yatherm Scientific, Gautama Buddha Nagar, Uttar Pradesh, India) with 4 mL of pancreatic α-amylase and amyloglucosidase (AMG) for 16 h at 37 °C and 200 strokes/min. The reaction ended with the addition of 4 mL of ethanol (99% *v/v*). The samples were centrifuged (Centrifuge Eppendorf 5702 R) at 3000 rpm for 10 min at 4 °C. The centrifugation pellet was re-suspended in 2 mL of ethanol (50% *v/v*), and then, in 6 mL of ethanol, followed by centrifugation under the same conditions twice. RS in the pellet was dissolved in 2 mL of 2 M KOH by vigorously stirring in an ice-water bath over a magnetic stirrer (Warmnest HJ-3) for 20 min. This solution was neutralized with 8 mL of 1.2 M sodium acetate buffer (pH 3.8), and the starch was quantitatively hydrolyzed to glucose by adding 0.1 mL of AMG. This solution was mixed and incubated for 30 min in a water bath at 50 °C with intermittent mixing. Then, samples were centrifuged at 3000 rpm for 10 min, 0.1 mL aliquots (in duplicate) were transferred into glass tubes, 3.0 mL of glucose oxidase/peroxidase reagent (GOPOD) was added and incubated at 50 °C for 20 min. The last step was to measure the absorbance of each solution in a spectrophotometer (Biochrom, Cambridge, United Kingdom) at 510 nm against the reagent blank (0.1 mL of sodium acetate buffer with pH 4.5 + 3 mL of GOPOD) [22–25].

2.3. Antioxidant Activity

The bread antioxidant activity was determined using an oxygen radical absorbance capacity (ORAC) assay quantified by fluorescence, a standardized method for determining antioxidant capacity in foods [26]. The standard curve was generated using the area under the curve (AUC) for different standard concentrations of Trolox (6-hydroxy-2,5,7,8-tetramethylchroman-2-carboxylic acid). The reagents were phosphate buffer pH 7.4, fluorescein $16{,}371.10^{-8}$ mol/L, APPH 178 mmol/L, and Trolox. All extract samples and Trolox standard solutions were pipetted with nine replicates into a black microplate and incubated at 37 °C for 10 min. Then, the peroxyl radical generator 2,2′-azobis (2-amidinopropane) dihydrochloride (AAPH) and the fluorescence were measured at 37 °C every 60 s using a spectrophotometer from 485 and 520 nm. The ORAC values were calculated from the Trolox standard curve with $R^2 = 0.9912$ and were expressed as milligrams of Trolox Equivalent (TE) per 1 g of extract (dry weight basis) [27].

2.4. Individuals

The sample size calculation was performed using the G*Power software (version 3.1.9.2; Dusseldorf University, Dusseldorf, Germany) [28], assuming glucose levels as the primary variable of the study and based on the result of Poquette et al. [29], with a statistical power of 90% and an alpha error of 5% (two-tails). This resulted in a sample size of 9 participants (crossover design).

Participants were recruited through social media and public advertisements. Eligibility criteria included the following: age 18–50 years; no medications that affect glucose metabolism; slight bodyweight fluctuation (≤5 kg in the past three months); willingness to eat all test foods; no self-reported allergy to the foods provided in the study; no self-reported sleep disorders; no cardiovascular, metabolic, and gastrointestinal diseases; no reported family history of type 2 diabetes mellitus in first-degree relatives; and fasting capillary blood glucose 70–100 mg/dL. Participants with metabolic disorders, who were taking any kind of medicine or allergic to the food provided in the study, or did not conclude all experimental sessions, were excluded.

Ethical approval for the study protocol was obtained from the Ethics Committee of Health Sciences School of the University of Brasilia (CAAE 58257416.1.0000.0030). All participants provided written informed consent before participation.

2.5. Clinical Trial Design

This study was a double-blind, crossover, randomized clinical trial in which 13 male individuals were eligible to participate in five experimental sessions with a 3- to 15-day washout period. For screening, participants answered questionnaires based on the recruitment criteria.

All experimental sessions were initiated between 7 and 8 a.m. with participants in 12 h fasting. Participants were instructed not to consume alcohol or perform any non-habitual physical activity 24 h before the sessions. They were also advised to maintain regular dietary intake and physical activity during the study protocol.

At each session, capillary glucose level was assessed by finger stick blood using a glucometer (Accu-check Performa; Roche Diagnostics, Basileia, Suiça) to ensure the food-deprived state (glucose level < 100 mg/dL). Then, an indwelling catheter was placed in the participant's forearm, and blood samples were drawn at baseline (−10) and 15, 30, 45, 60, 90, 120, and at 180 min after consuming glucose or the test bread. Participants were not allowed to eat or drink anything (except water) besides the test meals provided during the study sessions. They were allowed to read, listen to music, watch TV, use the computer, and use the toilet inside the laboratory.

2.6. Anthropometric and Body Composition Measurements

On the first day of the experiment, weight, height, and body fat were measured with 12 h fasting state. Bodyweight was assessed using a weight scale with 150 kg capacity and 50 g precision (OMRON model HN-289) and a stadiometer for height measuring to the nearest 0.1 cm (Wiso model Series 12). Body mass index (BMI) was computed and classified according to WHO [30]. Body fat percentage was measured by tetrapolar bioelectrical impedance (Body Compositon Analyzer—Quantum II, RJL Systems) according to the protocol described by Lukaski et al. [31].

2.7. Test Meals

Five experiment sessions were performed with anhydrous glucose or four bread samples made with rice flour (control) or white, brown, and bronze sorghum, all containing 50 g of available carbohydrate. Test meals were chosen to compare three different genotypes of sorghum and control bread. In each session, the participant consumed the test meal within 10 min in a randomized order. The simple randomization process was performed using a random number table [32]. The researchers were blinded to all experimental treatments, and the participants were blind to the type of bread consumed.

2.8. Antioxidant Capacity

To determine the blood antioxidant capacity, 3 mL of blood was collected in EDTA tubes with GSH 0.65 mmol/L and centrifuged at 4000 rpm for 15 min at 4 °C. The plasma was frozen in liquid nitrogen and stored at −80 °C until analyses. Antioxidant capacity analyses were conducted by FRAP (Ferric Reducing Ability of Plasma) method [33].

2.9. Biochemical Measurements

A volume of 4 mL of blood was collected in a red top vacutainer at each time point. After clotting and centrifugation at 4000 rpm for 15 min at 4 °C (Centrifuge Eppendorf 5702 R), glucose and insulin concentrations were measured by glucose oxidase (ADVIA, model 2400, Siemens Healthcare Diagnostics S.A., São Paulo, Brazil) and electrochemiluminescence methods (ADVIA, model Centaur, Siemens Healthcare Diagnostics S.A., São Paulo, Brazil), respectively. Sensitivity of glucose oxidase was 0.12 mmol/l (within-run CV of 0.41%) and insulin immunoassay was 1.39 pmol/l (within-run CV of 1.9%). The incremental AUC for glucose and insulin (3 h) was calculated, excluding the values below the baseline values, based on the trapezoidal method [31] using Microsoft Excel software, version 2010 (Microsoft Corporation, Washington, USA). The glycemic (GI) and insulinemic index (II) of the sorghum bread samples were determined based on the 2 h AUC response compared with glucose response as standard value (100) [34].

2.10. Statistical Analysis

Levene's and the Shapiro–Wilk tests were performed to determine data homogeneity of variance and normality, respectively. One-way analyses of variance (ANOVA) with Tukey post hoc test was applied to assess food composition, antioxidant status (ORAC and FRAP), glucose, and insulin AUC between glucose and the test meals (control and sorghum bread—white, brown, and bronze). Two-way repeated-measures ANOVA was used to examine the effects of test meals on postprandial glycemic and insulinemic responses with Bonferroni adjustments as post hoc comparisons when significant meal versus time interactions were found. The effect size of glucose and insulin responses and glucose and insulin AUC were calculated using ^{2} Eta squared. Statistical analyses were conducted using Statistical Package for the Social Sciences, version 21 (SAS Institute, Inc., Cary, North Carolina, USA). Differences were considered significant at $p < 0.05$ (two-tailed). The results are presented as mean values and standard deviations.

3. Results

3.1. Individuals Characteristics

Of 13 individuals initially recruited, two did not conclude all the phases, and one was excluded due to a medical condition. The excluded participant had a basal insulin and insulin response higher than the healthy participants. Therefore, data of ten healthy non-celiac males were analyzed. The ten participants that concluded the experiment presented 28.0 ± 4.9 years, 77.6 ± 11.7 kg, 1.78 ± 0.07 m, 24.2 ± 2.3 kg/m^2, 21.36% of body fat, and capillary fasting blood glucose of 89 ± 4.3 mg/dL. All participants were non-smokers.

3.2. Bread Chemical Composition

As shown in Table 1, the control bread presented a statistical difference in moisture compared to sorghum bread samples ($p < 0.05$), and white sorghum bread showed significantly less ashes compared to other samples ($p < 0.05$). For carbohydrates, all samples of sorghum bread presented less carbohydrate content compared to control (rice bread) ($p < 0.05$). Additionally, brown bread presented significantly more carbohydrates compared to bronze ($p = 0.001$). For RS, control and bronze bread samples presented higher values compared to the other sorghum genotypes ($p < 0.05$) with no statistical difference between bronze and control ($p = 0.16$). For fiber, all sorghum bread presented significantly higher values compared with control ($p < 0.05$), with brown presenting the highest content, followed by white and bronze bread (all $p < 0.05$). Bronze bread presented a higher value of protein and lipids than the other bread samples, with a significant difference among all samples for lipids content ($p < 0.05$).

Table 1. Chemical composition (g/100 g) and antioxidant activity of control and sorghum bread.

	Control	Brown	Bronze	White
Moisture	40.97 ± 0.73 [a]	47.42 ± 0.91 [b]	46.94 ± 0.74 [b]	48.80 ± 0.46 [b]
Ashes	1.51 ± 0.00 [b]	1.53 ± 0.00 [b]	1.52 ± 0.02 [b]	1.35 ± 0.00 [a]
Carbohydrate	37.51 ± 0.85 [c]	31.68 ± 0.54 [b]	29.29 ± 0.56 [a]	30.60 ± 0.39 [ab]
Resistant starch	3.05 ± 0.05 [b]	1.77 ± 0.12 [a]	2.75 ± 0.19 [b]	1.55 ± 0.06 [a]
Fiber	3.96 ± 0.03 [a]	5.79 ± 0.03 [d]	4.71 ± 0.13 [b]	5.48 ± 0.03 [c]
Protein	5.36 ± 0.52 [a]	5.42 ± 0.24 [a]	6.13 ± 0.15 [b]	5.36 ± 0.18 [a]
Lipids	7.58 ± 0.00 [a]	6.41 ± 0.08 [b]	8.68 ± 0.00 [d]	6.87 ± 0.00 [c]
ORAC (μmol TE/g)	25.60 ± 2.77 [a]	45.49 ± 2.07 [b]	30.84 ± 0.28 [a]	22.41 ± 3.04 [a]

Values in the same line marked with different letters show statistical significance ($p < 0.05$).

Brown sorghum bread presented significantly higher ORAC antioxidant activity compared to the others ($p \leq 0.001$). Moreover, white and bronze sorghum showed no significant difference from the control bread ($p = 0.38$ and $p = 0.09$, respectively) (Table 1).

3.3. Postprandial Glucose and Insulin Responses

As presented in Table 2, regarding 3 h AUC glucose response, brown bread presented a lower value than the control bread and glucose drink ($p < 0.05$; $^2 = 0.132$). Additionally, there were no significant differences among sorghum bread samples. For 3 h insulin AUC, there were no statistical differences among all bread samples. The glucose drink AUC response was significantly higher than other bread samples ($p < 0.05$).

Table 2. Glycemic and insulinemic index (%) of control and sorghum bread.

	Glucose 3 h AUC	Glycemic Index	Insulin 3 h AUC	Insulin Index
Glucose	2619.75 ± 2094.94 [a]	100	4797.29 ± 3009.89 [a]	100
Control	2098.50 ± 1352.53 [a]	80	3372.05 ± 3255.73 [b]	70
Brown	1144.50 ± 590.67 [b]	44	2379.59 ± 3083.12 [b]	50
Bronze	1571.25 ± 908.22 [ab]	60	2697.02 ± 2890.74 [b]	56
White	1662.75 ± 1362.39 [ab]	63	2094.09 ± 1212.01 [b]	44

Values in the same column marked with different letters show statistical significance ($p < 0.05$).

According to the GI classification (high GI ≥ 70; intermediate GI 56–69; low GI ≤ 55)[30], brown sorghum bread presented a low GI, bronze, and white sorghum an intermediate GI, and a high GI was found only for the control bread. For the insulin index, regarding the bread samples, the control showed a higher score (70) compared to sorghum bread (56, 50 and 44 for bronze, brown and white genotypes, respectively) (Table 2).

There were no significant time versus meal interaction effects at any time point on glycemic and insulinemic responses between tested meals (control and sorghum bread) ($p \geq 0.64$ and $p \geq 0.48$, respectively) (Figure 1).

3.4. Antioxidant Capacity

There was no significant time versus meal effect on blood antioxidant capacity (FRAP) between tested meals (control and sorghum bread samples) with $p \geq 0.504$.

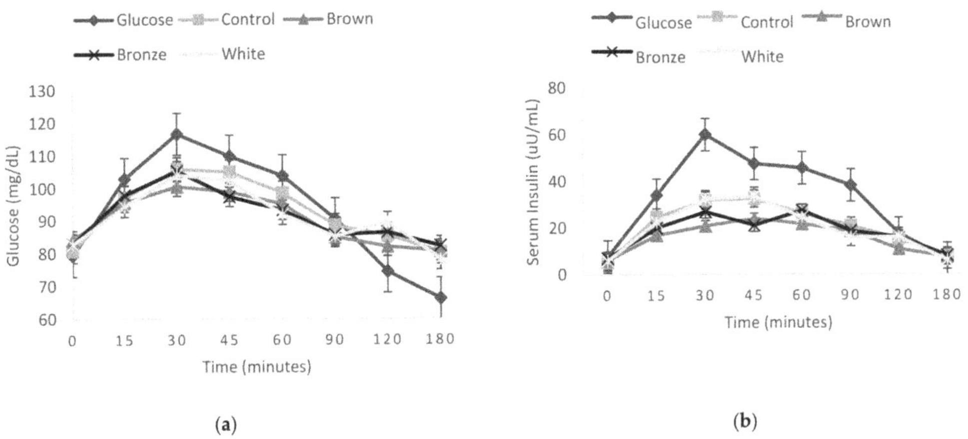

Figure 1. Fasting and postprandial glycemic (**a**) and insulinemic (**b**) responses to test meals bread intake containing three different sorghum genotypes and the control (rice bread). Values presented by means and their standard errors are represented by vertical bars.

4. Discussion

Based on the previous beneficial effects of sorghum bread intake on glucose metabolism [16], our findings support this hypothesis, showing improved glycemic and insulinemic responses after consuming sorghum bread in healthy adult men. All sorghum bread samples presented lower GI and II when compared to the control (rice bread). This is important because it was possible to diminish the interaction effects among the ingredients (flours, starch, eggs, and xanthan, etc.) since all the GF bread samples had precisely the same proportion of potato starch, cassava flour, and xanthan gum. Additionally, sorghum bread samples presented similar moisture contents, with no statistical difference. However, the control bread showed significantly less moisture ($p < 0.05$) compared to sorghum samples.

Several studies have found differences in chemical composition among sorghum genotypes such as tannins, phenolic compounds, resistant starch, and fiber content [7,8,14,17,19]. Regarding the sorghum bread genotypes, the brown sorghum bread presented lower carbohydrate amounts, lower lipids, and significantly higher fiber content and antioxidant activity (ORAC) than other sorghum bread genotypes. White sorghum bread showed low carbohydrate and lipids content and a high fiber amount. In contrast, bronze sorghum bread presented a higher resistant starch and protein content than other sorghum bread genotypes.

Brown sorghum bread showed a significantly lower 3 h AUC glucose response than the control bread and presented a low GI value. Since low GI is related to better glycemic response, brown sorghum bread can be considered a good alternative to improve glycemic response. According to Westman et al. [35], a diet with a low GI improves hemoglobin A1c, fasting glucose, and insulin compared to a normoglycemic diet in individuals with obesity and type 2 diabetes.

In our study, it was possible to determine that all sorghum bread samples presented a significantly higher fiber content than control bread ($p < 0.05$), contributing to the lower GI classification compared to the control. However, white sorghum bread presented the highest GI among sorghum bread samples. Its low resistant starch content can explain its low lipids and protein content, confirmed by Al Dhaheri et al. [36]. According to this study, which investigates the effect of nutritional composition on the glycemic index of different foods, samples with a high protein content reduce the glycemic response. Additionally, food's high fat and fiber content was related to a decreased postprandial glycemic response [36]. Therefore, brown sorghum bread has better nutritional effects since it shows significantly higher fiber and antioxidant activity, presenting the lowest GI.

Besides the fiber content, predominantly brown and white sorghum bread, all samples are GF, since gluten is not part of any ingredient and none presented cross-contamination industry disclosure. It is important since GFP tend to present less fiber content because they are usually prepared with corn starch, potato flour/starch, and low-fiber rice. Rice flour is one of the most used to produce GFP; however, it presents a low protein content and quickly digested carbohydrates [37].

According to Calvo-Lerma et al. [38], most GFP has a low nutritional profile, especially in bread and pasta. Similarly, Melini and Melini [39] who reviewed nutritional profile of GFP available on the market, showed that several GFP have a low protein and high fat and salt content compared to their equivalent gluten-containing products. Additionally, rice and corn that are the most frequently used in formulation of GFP are lacking in protein and fiber [40]. Other studies with GFP showed that white rice flour presented only 2.4 g of total fiber per 100 g, GF bread made with corn and potato starch had 3.34 g per 100 g, and commercial GF bread presented 1.2 to 5.6 g per 100 g of fiber [37,38,41,42]. On the other hand, the present study presented a better food composition for sorghum bread than rice flour bread (control). Similarly, Hariprasanna et al. [43] demonstrated that sorghum grain had a better nutritional profile than rice.

All sorghum bread samples presented more than 4.5 g of fiber per 100 g on a wet basis, as shown in Table 1. According to our analyses, sorghum bread samples presented 5.79 ± 0.03; 4.71 ± 0.13; and 5.48 ± 0.03 g per 100 g of serving for brown, bronze, and white sorghum, respectively. According to the FDA [44], food is a good source of fiber if it contains 10 to 19% of the dietary reference intake (DRI) per the amount customarily consumed, and is high or rich in fiber if it has more than 20%. So, food with at least 2.8 g of fiber is considered high or rich in fiber if it presents more than 5.5 g of dietary fiber per serving, since the fiber recommendation is 25 to 28 g per day. Thus, our brown sorghum bread can be classified as rich in fiber, and the other samples are a good fiber source.

Since fiber can be defined as any non-digestible carbohydrate and lignin not degraded in the upper gut, it has essential roles in decreasing postprandial glucose response [44]. Fiber is also associated with a reduced risk of cardiovascular diseases and diabetes mellitus type 2 (DM2), besides its relevance in DM2 treatment, lowering blood cholesterol, increasing satiety, and preventing constipation [45]. Therefore, the consumption of our sorghum bread samples, which are good fiber sources and rich in fiber, can improve human health due to several previously described benefits.

Another factor that influences postprandial glucose response is the presence of tannins. According to Poquette et al. [29], tannins in sorghum contribute to the poor digestibility of starch, which may lead to the slow absorption of carbohydrates. The lowest starch digestibility of sorghum among cereals is due to the strong association among starch, proteins, and tannins [7]. Additionally, sorghum proteins, especially after cooking, present a lower digestibility than cereals such as wheat and maize [46]. Furthermore, according to Espetia-Hernández [47], tannins form complexes with proteins and iron, reducing sorghum digestibility. Therefore, there are several mechanisms that may explain the low GI observed in sorghum bread, an added advantage for the glycemic control of type 2 diabetic people.

Furthermore, brown sorghum bread presented the highest antioxidant activity value than other samples (ORAC assay: 45.49 μmol TE/g). This result was probably because colored sorghum genotypes, as brown and red, had higher phenolic compounds concentration [16]. On the contrary, white sorghum presented the lowest antioxidant activity (ORAC assay: 22.41 μmol TE/g), similar to the study conducted by Awika and Rooney [8], in which white sorghum grain presented only 22 μmol TE/g. These findings can be explained because white sorghum is usually rich in tannins and has a reduced antioxidant activity. Moreover, according to Le Bourvellec and Renard [48], sorghum condensed tannins can bind starch and polysaccharides. Since phenolic compounds can form complexes with proteins and carbohydrates in foods leading to changes in structural properties that impact digestibility, this process is related to a decrease in glucose response.

Moraes et al. [15] found that the estimated glycemic index of sorghum flour was negatively correlated to phenolic compounds, specific flavonoids, antioxidant activity, and total, insoluble and soluble dietary fiber and b-glucan. However, RS did not correlate to the estimated GI. Basu et al. [49] studied the glucose metabolism with healthy volunteers after either rice (n = 8) or sorghum (n = 8) mixed meals consumption, and reported higher insulin sensitivity with sorghum than rice meals (identical calorie and macronutrient compositions).

Based on this result, we can speculate that sorghum bread's low postprandial glucose response may occur due to an improved insulin sensitivity without reducing insulin release. Galarregui et al. [50] presented that subjects with higher values of antioxidant capacity had a significantly lower insulin resistance (HOMA-IR) and correlation analyses showed inverse associations between GI and antioxidant capacity. In addition, a study conducted by Rosén et al. [51], with different varieties of rye bread, found that the content of phenolic compounds was negatively related to the early glucose response (T 0–60 min). The mechanism is probably multifactorial, including the effects of dietary fiber and a lowered rate of starch hydrolysis. Therefore, more studies are important to explore this complex mechanism.

Another study with sorghum, conducted by Park et al. [52], concluded that the administration of sorghum extract in mice significantly reduced serum glucose levels. However, only the treatment with a higher concentration significantly lowered the serum insulin level. Accordingly, Lakshmi et al. [53] demonstrated that the consumption of whole sorghum significantly lowered fasting and glucose 2 h AUC in type 2 diabetic individuals, likely due to fiber content.

Although dietary fiber lowers blood glucose levels by delaying gastric emptying, intestinal transit time, and carbohydrate absorption, Ray et al. [54] reported that consuming sorghum grain did not affect serum glucose or insulin levels compared with those in hyperlipidemic rats fed white rice [55]. These inconsistent results can be related to the type of sorghum consumed, sorghum grain versus extract, or the animal model used [52]. Therefore, the mechanisms of sorghum consumption and their effect on glucose, insulin, and antioxidant responses are still unclear and need more research.

This research had limitations due to the few sorghum genotypes (white, brown, and bronze) analyzed in gluten-free bread. Additionally, there was no direct gluten determination. Nevertheless, it is important to highlight that a study conducted with different sample profiles and other sorghum genotypes and products, besides bread, would be recommended since the interaction of ingredients may occur and present differences in GI.

5. Conclusions

Our results indicated that brown sorghum bread was the only sample classified as low GI and presented an improvement in postprandial glycemic responses in healthy adult men. This finding may occur due to a significantly higher fiber amount and the antioxidant activity of brown sorghum bread. Therefore, the consumption of brown sorghum should be encouraged as it produced better GFP due to its nutritional profile and health benefits than rice bread did. Finally, more research is required to explore the effects of different sorghum genotypes in food products on human health with different populations, including celiacs.

Author Contributions: L.R.d.R.G. conceived and designed the analysis, collected data, performed the analysis and wrote the paper; C.E.G.R. designed the analysis, performed the statistical analysis and reviewed the paper; M.A.M. designed and performed food analysis; V.S.N.d.S. and M.T.B.P. performed fiber analysis; R.B.A.B. conceived and designed the analysis and reviewed the paper. All authors have read and agreed to the published version of the manuscript.

Funding: This research was funded by FAP-DF (Brazilian Research Support Foundation in Federal District).

Institutional Review Board Statement: The study was conducted according to the guidelines of the Declaration of Helsinki, and approved by the Ethics Committee of University of Brasilia (protocol code 58257416.1.0000.0030 and date of approval 4 April 2017).

Informed Consent Statement: Informed consent was obtained from all subjects involved in the study.

Conflicts of Interest: The authors declare no conflict of interest.

References

1. Sapone, A.; Bai, J.C.; Ciacci, C.; Dolinsek, J.; Green, P.H.; Hadjivassiliou, M.; Kaukinen, K.; Rostami, K.; Sanders, D.S.; Schumann, M.; et al. Spectrum of gluten-related disorders: Consensus on new nomenclature and classification. *BMC Med.* **2012**, *10*, 13. [CrossRef]
2. Fasano, A.; Araya, M.; Bhatnagar, S.; Cameron, D.; Catassi, C.; Dirks, M.; Mearin, M.; Ortigosa, L.; Phillips, A. Federation of International Societies of Pediatric Gastroenterology, Hepatology, and nutrition consensus report on celiac disease. *J. Pediatr. Gastroenterol. Nutr.* **2008**, *47*, 214–219. [CrossRef]
3. Aguilar, N.; Albanell, E.; Miñarro, B.; Gallardo, J.; Capellas, M. Influence of final baking technologies in partially baked frozen gluten-free bread quality. *J. Food Sci.* **2015**, *80*, E619–E626. [CrossRef]
4. Taylor, J.R.; Schober, T.J.; Bean, S.R. Novel and non-food uses for sorghum and millets. *J. Cereal Sci.* **2006**, *44*, 252–271. [CrossRef]
5. Pontieri, P.; Mamone, G.; De Caro, S.; Tuinstra, M.R.; Roemer, E.; Okot, J.; De Vita, P.; Ficco, D.B.M.; Alifano, P.; Pignone, D.; et al. Sorghum, a healthy and gluten-free food for celiac patients as demonstrated by genome, biochemical, and immunochemical analyses. *J. Agric. Food Chem.* **2013**, *61*, 2565–2571. [CrossRef] [PubMed]
6. Khan, I.; Yousif, A.; Johnson, S.K.; Gamlath, S. Effect of sorghum flour addition on resistant starch content, phenolic profile and antioxidant capacity of durum wheat pasta. *Food Res. Int.* **2013**, *54*, 578–586. [CrossRef]
7. De Morais Cardoso, L.; Pinheiro, S.S.; Martino, H.S.D.; Pinheiro-Sant'ana, H.M. Sorghum (*Sorghum bicolor* L.): Nutrients, bioactive compounds and potential impact on human health. *Crit. Rev. Food Sci. Nutr.* **2015**, *57*, 372–390. [CrossRef]
8. Awika, J.M.; Rooney, L.W. Sorghum phytochemicals and their potential impact on human health. *Phytochemistry* **2004**, *65*, 1199–1221. [CrossRef] [PubMed]
9. Dayakar Rao, B.; Patil, J.V.; Hymavathi, T.V.; Nimal Reddy, K.; Rajendra Prasad, M.P. Creation of Demand for Millet Food through PCS Value Chain. In *Final Report of NAIP (ICAR)*; Directorate of Sorghum Research: Rajendranagar, India, 2014.
10. Rooney, L.W.; Waniska, R.D. Sorghum Food and Industrial Utilization. In *Sorghum: Origin, History, Technology and Production*, 1st ed.; Wayne Smith, C., Fredericksen, R.A., Eds.; John Wiley and Sons: New York, NY, USA, 2000; pp. 689–717.
11. Wolter, A.; Hager, A.S.; Zannini, E.; Arendt, E.K. Influence of sourdough on in vitro starch digestibility and predicted glycemic indices of gluten-free breads. *Food Funct.* **2014**, *5*, 564–572. [CrossRef] [PubMed]
12. Prasad, M.P.R. Glycaemic index and glycaemic load of sorghum products. *J. Sci. Food Agric.* **2014**, *95*, 1626–1630. [CrossRef] [PubMed]
13. Balota, M. Sorghum (*Sorghum vulgare* L.) marketability grain color and relationship to feed value. In *AREC-23NP*; Virginia Cooperative Extension Bulletin; Virginia Tech: Blacksburg, VA, USA, 2012; pp. 1–3.
14. Dykes, L.; Rooney, L.W. Review: Sorghum and millet phenols and antioxidants. *J. Cereal Sci.* **2006**, *44*, 236–251. [CrossRef]
15. Moraes, É.A.; Marineli, R.D.S.; Lenquiste, S.A.; Steel, C.J.; de Menezes, C.B.; Queiroz, V.A.V.; Júnior, M.R.M. Sorghum flour fractions: Correlations among polysaccharides, phenolic compounds, antioxidant activity and glycemic index. *Food Chem.* **2015**, *180*, 116–123. [CrossRef]
16. Rhodes, D.H.; Hoffmann, L., Jr.; Rooney, W.L.; Ramu, P.; Morris, G.P.; Kresovich, S. Genome-wide association study of grain polyphenol concentrations in global sorghum (*Sorghum bicolor* (L.) Moench) germplasm. *J. Agric. Food Chem.* **2014**, *62*, 10916–10927. [CrossRef] [PubMed]
17. Taylor, J.R.N.; Belton, P.S.; Beta, T.; Duodu, K.G. Increasing the utilisation of sorghum, millets and pseudocereals: Developments in the science of their phenolic phytochemicals, biofortification and protein functionality. *J. Cereal Sci.* **2014**, *59*, 257–275. [CrossRef]
18. Dlamini, N.R.; Taylor, J.R.; Rooney, L.W. The effect of sorghum type and processing on the antioxidant properties of African sorghum-based foods. *Food Chem.* **2007**, *105*, 1412–1419. [CrossRef]
19. Prior, R.L.; Wu, X. Anthocyanins: Structural characteristics that result in unique metabolic patterns and biological activities. *Free. Radic. Res.* **2006**, *40*, 1014–1028. [CrossRef] [PubMed]
20. Aruna, C.; Visarada, K.B.R.S. Sorghum Grain in Food and Brewing Industry. In *Breeding Sorghum for Diverse End Uses*; Woodhead Publishing: Hyderabad, India, 2019; pp. 209–228. [CrossRef]
21. De Aguiar, L.A.; Rodrigues, D.B.; Queiroz, V.A.V.; Melo, L.; de Oliveira Pineli, L.D.L. Comparison of two rapid descriptive sensory techniques for profiling and screening of drivers of liking of sorghum breads. *Food Res. Int.* **2020**, *131*, 108999. [CrossRef] [PubMed]
22. AOAC. *Official Methods of Analysis of the Association of Official Analytical Chemists*, 19th ed.; AOAC: Gaithersburg, MD, USA, 2012; Chapter 45 Met. 985.29; Volume 2, pp. 100–102. Available online: https://sbooks.best/downloads/Official-Methods-Of-Analysis-Of-Aoac-International-19th-Edition- (accessed on 10 May 2018).

23. AACC International. *AACC International Approved Methods*; AACC International: St. Paul, MN, USA, 2015; Available online: http://methods.aaccnet.org/toc.aspx (accessed on 10 May 2018).
24. AOCS. Approved Procedure Am 5-04 Rapid Determination of Oil/fat Utilizing High Temperature Solvent Extraction. 2005. Available online: http://www.academia.edu/30938058/AOCS (accessed on 10 May 2018).
25. FAO. Food energy-methods of analysis and conversion factors. In *Food and Agriculture Organization of the United Nations Technical Workshop Report*; FAO: Rome, Italy, 2003.
26. Prior, R.L.; Wu, X.; Schaich, K. Standardized Methods for the Determination of Antioxidant Capacity and Phenolics in Foods and Dietary Supplements. *J. Agric. Food Chem.* **2005**, *53*, 4290–4302. [CrossRef]
27. Cao, G.; Alessio, H.M.; Cutler, R.G. Oxygen-radical absorbance capacity assay for antioxidants. *Free Radic. Biol. Med.* **1993**, *14*, 303–311. [CrossRef]
28. Faul, F.; Erdfelder, E.; Lang, A.G.; Buchner, A. A flexible statistical power analysis program for the social, behavioral, and biomedical sciences. *Behav. Res. Methods* **2007**, *39*, 175–191. [CrossRef] [PubMed]
29. Poquette, N.M.; Gu, X.; Lee, S.O. Grain sorghum muffin reduces glucose and insulin responses in men. *Food Funct.* **2014**, *5*, 894–899. [CrossRef] [PubMed]
30. World Health Organization. *Obesity: Preventing and Managing the Global Epidemic: Report of a WHO Consultation*; Technical Report Series; WHO: Geneva, Switzerland, 2000; pp. 241–243.
31. Lukaski, H.C.; Bolonchuk, W.W.; Hall, C.B.; Siders, W.A. Validation of tetrapolar bioelectrical impedance method to assess human body composition. *J. Appl. Physiol.* **1986**, *60*, 1327–1332. [CrossRef]
32. Suresh, K. An overview of randomization techniques: An unbiased assessment of outcome in clinical research. *J. Hum. Reprod. Sci.* **2011**, *4*, 8–11. [CrossRef] [PubMed]
33. Benzie, I.F.F.; Strain, J.J. The Ferric Reducing Ability of Plasma (FRAP) as a Measure of "Antioxidant Power": The FRAP Assay. *Anal. Biochem.* **1996**, *239*, 70–76. [CrossRef]
34. FAO, Food and Agriculture Organization; World Health Organization, WHO. *Carbohydrates in Human Nutrition*; WHO: Geneva, Switzerland, 1998.
35. Westman, E.C.; Yancy, W.S., Jr.; Mavropoulus, J.C.; Marquart, M.; McDuffie, J.R. The effect of a low-carbohydrate, ketogenic diet versus a low-glycemic index diet on glycemic control in type 2 diabetes mellitus. *Nutr. Metab* **2008**, *5*, 36. [CrossRef]
36. Al Dhaheri, A.S.; Al, A.K.; Laleye, L.C.; Washi, S.A.; Jarrar, A.H.; Al Meqbaali, F.T.; Mohamad, M.N.; Masuadi, E.M. The effect of nutritional composition on the glycemic index and glycemic load values of selected Emirati foods. *BMC Nutr.* **2015**, *1*, 4. [CrossRef]
37. Rosell, C.M.; Marco, C. Rice. In *Gluten-Free Cereal Products and Beverages*; Arendt, E.K., Dal Bello, F., Eds.; Elsevier Science: London, UK, 2008; pp. 81–100.
38. Calvo-Lerma, J.; Crespo-Escobar, P.; Martínez-Barona, S.; Fornés-Ferrer, V.; Donat, E.; Ribes-Koninckx, C. Differences in the macronutrient and dietary fibre profile of gluten-free products as compared to their gluten-containing counterparts. *Eur. J. Clin. Nutr.* **2019**, *73*, 930–936. [CrossRef] [PubMed]
39. Melini, V.; Melini, F. Gluten-free diet: Gaps and needs for a healthier diet. *Nutrients* **2019**, *11*, 170. [CrossRef]
40. Hager, A.-S.; Wolter, A.; Jacob, F.; Zannini, E.; Arendt, E.K. Nutritional properties and ultra-structure of commercial gluten free flours from different botanical sources compared to wheat flours. *J. Cereal Sci.* **2012**, *56*, 239–247. [CrossRef]
41. Korus, J.; Witczak, M.; Ziobro, R.; Juszczak, L. The impact of resistant starch on characteristics of gluten-free dough and bread. *Food Hydrocoll.* **2009**, *23*, 988–995. [CrossRef]
42. Thompson, T. Folate, iron, and dietary fiber contents of the gluten-free diet. *J. Am. Diet. Assoc.* **2000**, *100*, 1389–1396. [CrossRef]
43. Hariprasanna, K.; Agte, V.; Elangovan, M.; Gite, S.; Kishore, A. Anti-Nutritional Factors and Antioxidant Capacity in Selected Genotypes of Sorghum [*Sorghum bicolor* (L.) Moench]. *Int. J. Agric. Sci.* **2015**, *7*, 620–625.
44. FDA. Food Labeling: Specific Requirements for Nutrient Content Claims, Nutrient Content Claims for "Good Source", "High", "More", and "High Potency". 2019. Available online: https://www.accessdata.fda.gov/scripts/cdrh/cfdocs/cfcfr/cfrsearch.cfm?fr=101.54 (accessed on 10 April 2020).
45. ADA. Standards of Medical Care for Patients with Diabetes Mellitus. *Diabetes Care* **2002**, *25* (Suppl. S1), S33–S49. [CrossRef]
46. Moraes, É.A.; Queiroz, V.A.V.; Shaffert, R.E.; Costa, N.M.B.; Nelson, J.D.; Ribeiro, S.M.R.; Martino, H.S.D. In Vivo protein quality of new sorghum genotypes for human consumption. *Food Chem.* **2012**, *134*, 1549–1555. [CrossRef] [PubMed]
47. Espitia-Hernández, P.; González, M.L.C.; Ascacio-Valdés, J.A.; Dávila-Medina, D.; Flores-Naveda, A.; Silva, T.; Chacón, X.R.; Sepúlveda, L. Sorghum (*Sorghum bicolor* L.) as a potencial source of bioactive substances and their biological properties. *Crit. Rev. Food Sci. Nutr.* **2020**, 1–12. [CrossRef]
48. Le Bourvellec, C.; Renard, C.M.G.C. Interactions between Polyphenols and Macromolecules: Quantification Methods and Mechanisms. *Crit. Rev. Food Sci. Nutr.* **2012**, *52*, 213–248. [CrossRef]
49. Basu, R.; Schiavon, M.; Petterson, X.-M.; Hinshaw, L.; Slama, M.; Carter, R.; Man, C.D.; Cobelli, C.; Basu, A. A novel natural tracer method to measure complex carbohydrate metabolism. *Am. J. Physiol. Endocrinol. Metab* **2019**, *317*, E483–E493. [CrossRef]
50. Galarregui, C.; Zulet, M.Á.; Cantero, I.; Marín-Alejandre, B.A.; Monreal, J.I.; Elorz, M.; Benito-Boillos, A.; Herrero, J.I.; Tur, J.A.; Abete, I.; et al. Interplay of Glycemic Index, Glycemic Load, and Dietary Antioxidant Capacity with Insulin Resistance in Subjects with a Cardiometabolic Risk Profile. *Int. J. Mol. Sci.* **2018**, *19*, 3662. [CrossRef]

51. Rosén, L.A.H.; Ostman, E.; Shewry, P.R.; Ward, J.L.; Andersson, A.A.M.; Piironen, V.; Lampi, A.-M.; Rakszegi, M.; Bedoe, Z.; Björck, I.M.E. Postprandial Glycemia, Insulinemia, and Satiety Responses in Healthy Subjects after Whole Grain Rye Bread Made from Different Rye Varieties. *J. Agric. Food Chem.* **2011**, *59*, 12139–12148. [CrossRef]
52. Park, J.H.; Lee, S.H.; Chung, I.M.; Park, Y. Sorghum extract exerts an anti-diabetic effect by improving insulin sensitivity via PPAR-γ in mice fed a high-fat diet. *Nutr. Res. Pract. (Nutr. Res. Pract.)* **2012**, *6*, 322–327. [CrossRef]
53. Lakshmi, K.B.; Vimala, V. Hypoglycemic effect of selected sorghum recipes. *Nutr. Res.* **1996**, *16*, 1651–1658. [CrossRef]
54. Ray, T.K.; Mansell, K.M.; Knight, L.C.; Malmud, L.S.; Owen, O.E.; Boden, G. Long-term effects of dietary fiber on glucose tolerance and gastric emptying in noninsulin-dependent diabetic patients. *Am. J. Clin. Nutr.* **1983**, *37*, 376–381. [CrossRef] [PubMed]
55. Lee, S.H.; Chung, I.M.; Cha, Y.S.; Park, Y. Millet consumption decreased serum concentration of triglyceride and C-reactive protein but not oxidative status in hyperlipidemic rats. *Nutr. Res.* **2010**, *30*, 290–296. [CrossRef] [PubMed]

foods

MDPI

Article

Synergistic Effect of Enzyme Hydrolysis and Microwave Reactor Pretreatment as an Efficient Procedure for Gluten Content Reduction

Ivana Gazikalović [1], Jelena Mijalković [2], Nataša Šekuljica [1], Sonja Jakovetić Tanasković [2], Aleksandra Đukić Vuković [2], Ljiljana Mojović [2] and Zorica Knežević-Jugović [2,*]

1 Innovation Center, Faculty of Technology and Metallurgy, Karnegijeva 4, 11000 Belgrade, Serbia; igazikalovic@tmf.bg.ac.rs (I.G.); nsekuljica@tmf.bg.ac.rs (N.Š.)
2 Department of Biotechnology and Biochemical Engineering, Faculty of Technology and Metallurgy, University of Belgrade, Karnegijeva 4, 11000 Belgrade, Serbia; jjovanovic@tmf.bg.ac.rs (J.M.); sjakovetic@tmf.bg.ac.rs (S.J.T.); adjukic@tmf.bg.ac.rs (A.Đ.V.); lmojovic@tmf.bg.ac.rs (L.M.)
* Correspondence: zknez@tmf.bg.ac.rs

Abstract: In this study, we assessed the effects of microwave irradiation of wheat gluten proteins as a pretreatment performed in a microwave reactor that could accurately control process parameters as a function of power and temperature, as well as comparing it with conventional heat treatment. The aim was to identify suitable combinations of partial enzymatic hydrolysis and microwave pretreatment parameters to produce gluten hydrolysates with reduced allergenicity and conserved techno-functional features for food application. FTIR analysis, and total and reactive SH group contents confirmed that the microwave-controlled heating can significantly change the secondary structure and conformation of gluten protein. The microwave treatment had the largest effect at 200 W and 100 °C, at which the content of gluten has been reduced by about 2.5-fold. The microwave pretreatment also accelerated the enzymatic hydrolysis of gluten, changing the kinetic profile. The apparent hydrolysis rate constants (k_2) were 1.00, 3.68, 3.48, 4.64 and 4.17 min^{-1} for untreated gluten, and those pretreated with microwave power of 200, 400, 600 and 800 W, respectively. Compared to the heat treatment, it appeared that microwave specific non-thermal effects had a significant influence on the gluten structure and allergenicity and, in combination with the enzymatic hydrolysis, ultimately yielded protein hydrolysates with enhanced antioxidant and functional properties.

Keywords: wheat gluten; microwave reactor; allergenicity; enzymatic hydrolysis; Alcalase; techno-functional properties; antioxidant activity

Citation: Gazikalović, I.; Mijalković, J.; Šekuljica, N.; Jakovetić Tanasković, S.; Đukić Vuković, A.; Mojović, L.; Knežević-Jugović, Z. Synergistic Effect of Enzyme Hydrolysis and Microwave Reactor Pretreatment as an Efficient Procedure for Gluten Content Reduction. *Foods* **2021**, *10*, 2214. https://doi.org/10.3390/foods10092214

Academic Editors: Maria Papageorgiou and Theodoros Varzakas

Received: 6 July 2021
Accepted: 13 September 2021
Published: 18 September 2021

1. Introduction

Apart from maize, wheat is the most globally cultivated cereal crop [1]. The unique properties of wheat flour, including specific viscoelastic properties, primarily reside in gluten-forming storage proteins of the wheat grain endosperm [2]. These unique features, together with the availability of wheat gluten as a by-product of the starch industry, have opened an enormous potential for a wide range of applications in the food bakery industry [3]. Although wheat gluten is the main determinant of dough properties, it also plays a vital role in the production and/or quality attributes of other foods by modification of the rheological, textural, and organoleptic properties of the final products. Moreover, it is possible to enlarge the field of gluten applications through its chemical or enzymatic modifications, with the aim of improving functional and nutritional properties for specific food and other applications [4–6].

Gluten makes up 80–85% of the total protein content in wheat, while 15–20% of the remaining proteins are non-glutenous [7]. Protein bodies found in the wheat endosperm contain prolamins, which represent the major gluten fraction, and other gluten constituents [8]. Wheat gluten is a rather complex protein composed of two seed storage proteins, gliadins and glutenins [9]. Glutenins, the major proteins of flour, are poorly soluble in alcohol because they are capable of forming large polymers that are stabilized by intermolecular disulfide bonds and hydrophobic interactions. On the other hand, gliadins are soluble in aqueous alcohol (for example 60–70% ethanol) and are mainly present in gluten as monomers interacting by non-covalent forces [10,11]. However, wheat storage protein, including gliadin and glutenin, are more complex proteins which are comprised of polymorphic polypeptides, showing more than 60 different molecular weights ranging from 30,000 to 90,000 Da. They can be closely divided into several groups according to their molecular weight (MW): (1) high molecular weight (HMW) group—HMW-glutenin subunits (HMW-GS), x- and y-type, with MW ranging 70–90 kDa; (2) medium molecular weight group (MMW)—ω5- and ω1,2-gliadin, with MW ranging 40–50 kDa; (3) low molecular weight (LMW) group—α/β- and γ- gliadin occurring as monomers, LMW-glutenin subunits (LMW-GS) occurring as aggregative proteins, with MW ranging 30–40 kDa [12]. Depending on the MW group, the proteins have different amino acid compositions [12,13].

Gluten proteins are generally rich in proline residues, making them indigestible and thus trigger an immune response in predisposed individuals [14]. Celiac Disease (CD) refers to chronic digestive problems and nutritional deficiencies. It is defined as an inflammatory disease of the upper small intestine and is caused by the consumption of gluten-containing foodstuff. The disease is chronic, and the only effective treatment is a strict, lifelong elimination of foods containing gluten from the diet, which is a big challenge for CD patients due to the frequent usage of gluten in the food industry [15,16]. Thus, food science research has paid considerable attention to the development of processing technology that reduces or eliminates "toxic" gluten and other protein sequences in raw materials and foods [17].

Enzymatic treatment of gluten seems to be a highly promising approach, aiming at the hydrolysis of toxic gluten sequences in vitro prior to ingestion; it has also been suggested as as an oral therapy for CD, in which dietary gluten is hydrolyzed by digestive peptidases which are already in the stomach, thus preventing CD-specific immune reactions in the small intestine (so-called medical approach) [18]. Even though oral therapy for celiac disease by digestive peptidases is an attractive approach, it creates several technological challenges such as ensuring rapid and complete enzymatic digestion of immunogenic gluten peptides in complex food matrices. Furthermore, some strongly allergenic sequences of the 33-mer peptide, from α/β-gliadins, seem to be resistant to gastrointestinal digestion [13]. Recent research has confirmed the potential of prolyl endopeptidases from different sources alone or in combination with cysteine endoprotease to detoxify gliadin peptides, but raise concerns regarding their possible efficacy in vivo, in the intestinal environment and in CD [13,19,20]. Namely, the ability of the enzyme to diffuse and access the epitopes is reduced by food matrix components, leading to incomplete allergen destruction. Additionally, the combination of low pH and the presence of the pepsin in vivo could enhance the inactivation of the enzyme, thus high concentrations and long reaction times are required to achieve complete detoxification and to prevent intestinal transport of toxic sequences [13,21]. Engineered synthetic and improved gluten-degrading enzymes, their combination and microencapsulation appear to be a future direction in enzyme therapy, but they are still in progress and in various phases of clinical testing [22–25]. On the other hand, enzymes with broader specificity such as Alcalase may hydrolyze in vitro more peptide bonds under selected reaction conditions and expose new sites that may not have been available to more specific enzymes [26]. The use of enzyme for in vitro gluten hydrolysis can be particularly advantageous since the enzymes are able not only to eliminate contaminant gluten, but also to enhance solubility and to obtain the required nutritional and functional properties [13,27]. Moreover, the hydrolysis may also produce bioactive peptides with various functionalities and potentially biological activities.

To the best of our knowledge, the enzymes obtained from various sources have been used for modifying the immunogenic fraction of gluten proteins. It appears to cause several changes in gluten's structure by cutting the large proteins to peptides of lower molecular weight, reducing the allergenic potential of wheat gluten since its allergenic epitopes contain between 5 and 20 amino acids [28]. However, the development of an efficient enzymatic processing technology for industrial application still requires progress. It is important to completely eliminate residual allergenic epitopes since they could have a deleterious effect on product quality and the health of coeliac patients during processing and subsequent peptic and tryptic digestion. Furthermore, due to the presence of large amounts of hydrophobic residues in hydrolyzed gluten fraction, the proteins typically have poor solubility and dispersibility. For example, the modified gluten network after endopeptidase treatment reduces the technological properties (viscoelasticity) of dough and baked products, which are supplemented by structuring agents such as pre-gelatinized starch, emulsifiers, and hydrocolloids [29,30]. This is the main disadvantage of the enzyme treatment. An effective treatment method must be able to cleave or mask the amino acid sequence present in toxic epitopes at a specific position or to alter the protein conformation of the allergen (protein denaturation, cross-linking, or aggregation). During the last decade, various approaches which imply the combination of partial and controlled enzyme hydrolysis with some physical pretreatment of gluten proteins and its fractions were applied and gave the possibility for obtaining final products with conserved techno-functional features [30–32].

The aim of this research was to utilize the advantages of a fully controlled and equipped microwave reactor system as a pretreatment step in order to facilitate the enzymatic hydrolysis reaction and to investigate the changes in wheat gluten proteins which occur during microwave treatment (MWT). This was conducted in order to investigate the MWT effects on gluten molecular structure and the possibility of certain toxic epitopes to become more easily available to the enzyme, which may favor their elimination. Conventional heating was also applied in order to compare its effects to the effects of microwave energy regarding gluten allergenicity. Changes in functional, antioxidant and metal-chelating properties were also investigated.

2. Materials and Methods

2.1. Materials

Gluten from wheat (min. 75% protein content, as declared by the supplier) and Alcalase 2.4L, a serine endopeptidase from *Bacillus licheniformis*, were obtained from Sigma-Aldrich (St. Louis, MO, USA). Bovine serum albumin (BSA), 2,2′-Azino-bis (3-ethylbenzothiazoline-6-sulfonic acid) diammonium salt (ABTS), 3-(2-Pyridyl)-5,6-diphenyl-1,2,4-triazine-4′,4′′-disulfonic acid sodium salt (ferrozine), 5,5′-(dithiobis-2-nitrobenzoate) (DTNB), $FeCl_2$ were also procured from Sigma-Aldrich (St. Louis, MO, USA). RIDASCREEN® Gliadin competitive (Art. No. R7021) ELISA test (R-Biopharm AG, Darmstadt, Germany) was used to quantify gliadin reactive epitopes. All of the other chemicals were of analytical grade.

2.2. Methods

2.2.1. Microwave Reactor Treatment

Microwave treatment was conducted using the microwave reactor system Anton Paar Monowave 300 (Anton Paar GmbH, Graz, Austria). A sample of 1 g of wheat gluten was weighed into designated reactor vials (G30) and distilled water was added, the vials were capped to avoid evaporation and then the samples were subjected to treatment. Such formed 10% (*w*/*v*) suspensions were treated by different constant microwave power, ranging from 200–800 W with magnetic stirring fixed to 600 rpm. Reaction temperature maximum was set at 100 °C and different microwave powers were applied. Treatment time at 100 °C was 1 min. Treated samples were then lyophilized (CHRIST Beta 2-8 LDPlus, Osterode am Harz, Germany) and stored in tightly secured plastic vials in a glass desiccator at room temperature for further analysis. The microwave power chosen for further experiments was based on the immunoassay results. Afterwards, the chosen microwave power was used in a set of experiments where the temperature was controlled and varied from 50–100 °C with an increment of 10 °C. Magnetic stirring speed was held constant and treatment time was 1 min. Total protein content ($N \times 5.7$) was determined by the Kjeldahl method [33]. The experiments were performed at room temperature (~25 °C).

2.2.2. Conventional Heat Treatment

Heat treatment was conducted using a heating unit with temperature control (Heidolph MR 3001, Heidolph Instruments GmbH & Co. KG, Schwabach, Germany) with an oil bath. The 10% (w/v) wheat gluten suspensions were heated to 50–100 °C range, with an increment of 10 °C, with magnetic stirring set to 600 rpm and samples were treated for 1 min at the designated temperatures. Samples were lyophilized (CHRIST Beta 2-8 LDPlus, Osterode am Harz, Germany) and stored in tightly secured plastic vials in a glass desiccator at room temperature for further analysis.

2.2.3. Enzymatic Hydrolysis

Immediately after the microwave reactor treatment, the sample was subjected to hydrolysis in a stirred tank reactor which consisted of pH electrode (Eutech instruments Pte Ltd, Singapore), overhead stirrer (Heidolph RZR 2020, Heidolph Instruments GmbH & Co. KG, Schwabach, Germany) and a heating unit (IKA® C-MAG HS 7, IKA®–Werke GmbH & Co. KG, Staufen, Germany), where the agitation speed was set to 200 rpm. The 2% (w/w) wheat gluten suspensions (total mass of 200 g in a 400 mL beaker) were stirred and allowed to equilibrate at 60 °C for 20 min. The reaction pH was adjusted to pH 8.0 using 0.8 N NaOH. The hydrolysis reaction was started by addition of Alcalase, where (E/S) ratio was 5.0%. Enzymatic hydrolysis was monitored by pH-stat assay, where DH was calculated according to Adler-Nissen [34], as shown in Equation (1):

$$DH(\%) = \frac{(h \times 100)}{h_{tot}} = \frac{N_b \times B \times 100}{\alpha \times m_p \times h_{tot}} \quad (1)$$

where h is the number of equivalent peptide bonds hydrolyzed at the time expressed in meq/g, h_{tot} is the total amount of peptide bonds per weight unit of a protein (for wheat gluten protein h_{tot} = 8.38 mmol/g of protein), N_b is the base normality, B is the amount of base consumed in mL, α is the degree of dissociation of α-amino groups (α = 0.926 at 60 °C and pH 8.0) and m_p is the protein mass in g. The hydrolysis was terminated by boiling the samples for 10 min and centrifuged at 7860× g at 4 °C (Eppendorf Centrifuge 5430 R, Hamburg, Germany). The samples were lyophilized (CHRIST Beta 2-8 LDPlus, Osterode am Harz, Germany) and stored in tightly secured plastic vials in a glass desiccator at room temperature for further analysis. Protein content in the hydrolysates was determined by the method of Lowry [35].

In order to interpret and better understand the obtained experimental results, a semi-empiric kinetic model that takes into account the enzyme deactivation and substrate inhibition was applied [36]. Afterwards, the kinetic parameters were calculated, and the initial rate of gluten proteins hydrolysis was interpreted via reaction rate constant (k_2), inhibition constant (K_I) and reaction rate constant of deactivation (k_d).

2.2.4. Quantification of Gliadin Reactive Epitopes

The gliadin content was determined in the untreated wheat gluten, microwave reactor treated gluten, conventionally heated gluten and in-wheat gluten hydrolysates with and without microwave reactor pre-treatment by using the ELISA RIDASCREEN® Gliadin Competitive kit (R-Biopharm, Darmstadt, Germany). The assay was performed according to the manufacturer's instructions.

2.2.5. Emulsifying Activity and Emulsion Stability Index

Emulsifying activity index (EAI) and emulsion stability index (ESI) were determined by the method of Pearce and Kinsella [37], with modifications. All of the previously lyophilized samples were dissolved in distilled water at 1.0% (w/v). The working solution consisted of 3/4 dissolved sample and 1/4 of sunflower oil and was mixed for 90 s using a laboratory homogenizer (Yellowline DI 25 basic, IKA® Works, Inc., Wilmington, NC, USA) at a speed of 9500 rpm. An aliquot of 50 μL was diluted 100 times with phosphate buffer (0.01 M, pH 7) containing 0.1% (w/v) SDS. The absorbance of the diluted emulsions was measured at 500 nm at 0 min (A_0) and at 10 min (A_{10}) at ambient temperature. The calculations are given in Equations (2)–(4):

$$T = \frac{2.303 \times A}{l} \quad (2)$$

$$EAI\left(\frac{m^2}{g}\right) = \frac{2 \times D \times T}{\varphi \times c \times 10{,}000} \quad (3)$$

$$ESI(h) = \frac{(A_0 \times \Delta t)}{\Delta A} \quad (4)$$

where T is turbidity, A is the absorbance measured at 500 nm (at 0 and 10 min), l is the optical path length of cuvette = 1 cm, D is the dilution factor = 100, c is the weight of protein per unit volume (g/mL), φ is the oil volume fraction in the emulsion, $\Delta A = A_0 - A_{10}$ and Δt = 10 min is the time interval.

2.2.6. Foam Capacity and Foam Stability

All of the samples were dissolved to form a 2.0% (w/v) solution. The method was slightly modified [38]. The initial volume of 50 mL of each sample solution was recorded and then the solution was whipped for 4 min with a laboratory homogenizer at 9500 rpm at ambient temperature. The same beaker was used for all samples. Immediately after whipping, the beaker was sealed with parafilm in order to evade air contact. Foam capacity (FC) was calculated as foam expansion at 0 min, as given in Equation (5):

$$FC(\%) = \frac{V_A - V_B}{V_A} \times 100 \quad (5)$$

where V_A is the volume after whipping (mL) and V_B is the volume recorded before 4 min of whipping (mL).

Foam stability (FS) was calculated as the percentage of liquid present in the foam after 30 min compared to the solution recorded at 4 min after whipping. The calculation is given in the Equation (6):

$$FS(\%) = \frac{V_A - V_B}{V_A} \times 100 \quad (6)$$

where V_A is the volume recorded after 30 min of rest (mL) and V_B is the volume recorded before 4 min of whipping (mL).

2.2.7. $ABTS^{\cdot+}$ Radical Scavenging Activity

The antioxidant properties of untreated wheat gluten, microwave reactor treated gluten, conventionally heated gluten, and wheat gluten hydrolysates with and without microwave reactor pretreatment were determined by a free-radical scavenging assay, ABTS, with modifications as described [39]. Determination of the ability of the aforementioned samples to scavenge the $ABTS^{\cdot+}$ radical was based on the $ABTS^{\cdot+}$ radical cation decolorization. The samples were prepared as 2 $mg_{protein}$/mL solutions in distilled water and vortexed. Then, 10 μL of prepared solutions were mixed with 1 mL of previously prepared $ABTS^{\cdot+}$ solution. The incubation time lasted for 5 min and absorbance was measured at 734 nm. ABTS radical scavenging activity (%) was calculated using Equation (7):

$$ABTS(\%) = \left(1 - \frac{A_S}{A_C}\right) \times 100 \quad (7)$$

where A_s represents the absorbance of the sample solution in the presence of $ABTS^{\cdot+}$ and A_c is the absorbance of the control solution with $ABTS^{\cdot+}$. Minimal inhibitory concentration of hydrolysates necessary to inhibit 50% of ABTS radical cation, at standard reaction conditions, was calculated and expressed in mg/mL.

2.2.8. Metal-Ion Chelating Activity

The metal-ion chelating activity (MICA) of the wheat gluten hydrolysates was determined using a ferrous ion chelating assay described by Decker and Welch [40], with modifications. A sample of 200 µL of each hydrolysate solution (4 $mg_{protein}$/mL in deionized water) was added to 800 µL of deionized water. Then, 100 µL of 2 mM $FeCl_2$ solution was added, vortexed and incubated for 3 min. Afterwards, 200 µL of 5 mM ferrozine solution was added, vortexed and incubated for another 10 min. The absorbance was recorded spectrophotometrically at 562 nm, with deionized water as blank. Metal-ion chelating activity was then calculated as given in Equation (8):

$$MICA(\%) = \left(1 - \frac{A_1}{A_0}\right) \quad (8)$$

where A_0 is the absorbance of the control and A_1 is the absorbance of the sample, both measured at 562 nm. Minimal inhibitory concentration of hydrolysates necessary to chelate 50% of ferrous ions, at standard reaction conditions, was calculated and expressed in mg/mL.

2.2.9. Quantification of Total and Reactive SH Groups

For the purpose of verifying the influence of microwave pretreatment on the potential structural changes of gluten proteins, the effect of microwave heating and conventional heat pretreatments on the changes in content of total sulfhydryl and reactive (SH) groups was determined spectrophotometrically by Ellman's procedure using 5,5′-(dithiobis-2-nitrobenzoate), DTNB, which reacts with exposed SH groups to yield a product with a maximum absorbance at 412 nm. Analysis was conducted as previously described [41,42].

2.2.10. FTIR Analysis

Fourier transformation infrared spectroscopy (FTIR) absorbance spectra of lyophilized samples were acquired using Nicolet iS10 FTIR Spectrometer (Thermo Scientific™). Absorbance spectra at 4 cm^{-1} resolution were collected over the scanning range of 400 to 4000 cm^{-1}. The background of spectra was corrected by spectrum of air. All of the analyses were performed at room temperature.

2.2.11. Sodium Dodecyl-Sulfate Polyacrylamide Gel Electrophoresis (SDS-PAGE) Electrophoresis

SDS-PAGE electrophoresis was performed on hydrolysate samples using a 12% Precise™ Protein Gel on a Hoefer™ Mighty Small™ II Mini Vertical Electrophoresis System. Hydrolysate samples and sample buffer were mixed in a 1:1 ratio and boiled for 5 min. Afterwards, 20 µL of the reduced protein sample was used to load on to the separating gel. The separation was performed under 40 mA current for 100 min. Spectra Multicolor Broad Range Protein Ladder (Thermo Scientific), a protein standard containing 10 pre-stained proteins with molecular weights ranging 10–260 kDa, was used. The gel was stained with Coomassie Brilliant Blue R-250. Molecular weights (Mw) were then estimated on the basis of the protein standard.

2.2.12. Statistical Analysis

Two independent experiments were performed for each of the experimental sets (microwave or heat treatment; enzymatic hydrolysis), and the results are presented as the mean values with standard deviations (SD). One-way ANOVA with repeated measures (within subjects), followed by Tukey's test, was used to determine the statistical significance of mean values differences for comparison of gluten treatments at level less than 0.05. On the contrary, one-way ANOVA followed by Tukey's test was used to examine the relationship between the studied parameters, i.e., dependent variables of two and three technical repeats per allergenicity detection and antioxidant and functional analyses, respectively (significance level was $p < 0.05$). Statistical analyses were performed using OriginPro 9.0 software (OriginLab Corporation, Northampton, MA, USA).

3. Results and Discussion

3.1. Microwave-Enhanced Heating Process Induced Changes in Gluten Allergenicity Detection

The effects of microwave irradiation on natural gluten proteins as a pretreatment performed in a microwave reactor were studied as a function of reaction parameters, power and temperature, and compared with conventional heat treatment. Since wheat gluten as a raw natural material contains impurities, including carbohydrates and proteins which are very sensitive to temperatures above 90 °C due to denaturation and degradation of natural compounds as well as adverse reactions such as Maillard reaction and others, the tendency to find alternative less aggressive physical treatments is increasingly pronounced [43]. The aim is to obtain modified gluten proteins in high yield with a lower value of toxic immunogenic epitopes and preserved techno-functional properties, which further implies that special conditions are necessary to perform the reaction process. Thus, the first step was focused on the investigation of the possibility to apply a physical treatment on the gluten proteins by using a microwave reactor with temperature control in order to modify gluten molecules in a way which will be able to influence the reduction in gluten allergenicity. The influence of the microwave irradiation power (200–800 W) and the temperature (50–100 °C) under controlled conditions was examined, and the obtained results were presented in Figure 1.

The results revealed that the microwave treatment of wheat gluten resulted in gluten content reduction in all of the treated samples (Figure 1a) in comparison with untreated gluten. The greatest reduction in detected relative gluten content was recorded for the sample treated at power of 200 W, retaining 39.65 ± 1.69% of its initial gluten content. At low power, an initial decrease in content of toxic gluten epitopes appeared, then the lowest reduction in toxic gluten epitopes was attained at 400 W (63.16 ± 1.98% of initial content) and, at that point, the decrease in relative gluten content at higher applied power was observed. The mechanism of the microwave influence on gluten allergenicity at the molecular level appeared to be rather complex and based on the alteration of the protein conformation of epitopes. It appeared that the denaturation and unfolding of the gluten molecule were higher at 400 W than at 200 W, thus rendering hidden epitopes more accessible by the antibody used in ELISA, resulting in higher antigenicity. However, further increasing of microwave power to 800 W could have a different effect on gluten structure including disulfide bonded cross-linking between gliadin and glutenin, which could cause destruction and/or masking of some epitopes, or additional denaturation of epitope resulting in reduced antigenicity. Furthermore, the aggregation and loss of protein solubility, rather than the epitope destruction, may be responsible for the observed decrease in gluten immunoreactivity at higher microwave power. Thus, future studies on immunoreactivity of different soluble and insoluble gluten fractions are required to additional understand the mechanisms of the inactivation of gluten toxic epitopes by microwave treatment [44]. Similar studies have demonstrated that at lower dose of applied energy, the gliadin immunoreactivity increased, reached the maximum value; at higher applied doses (500 W for 2 min), a decrease in gliadin immune response was observed [45]. Based on the results obtained, the microwave power of 200 W was selected to examine the effect of temperature on the detectable value of gluten content.

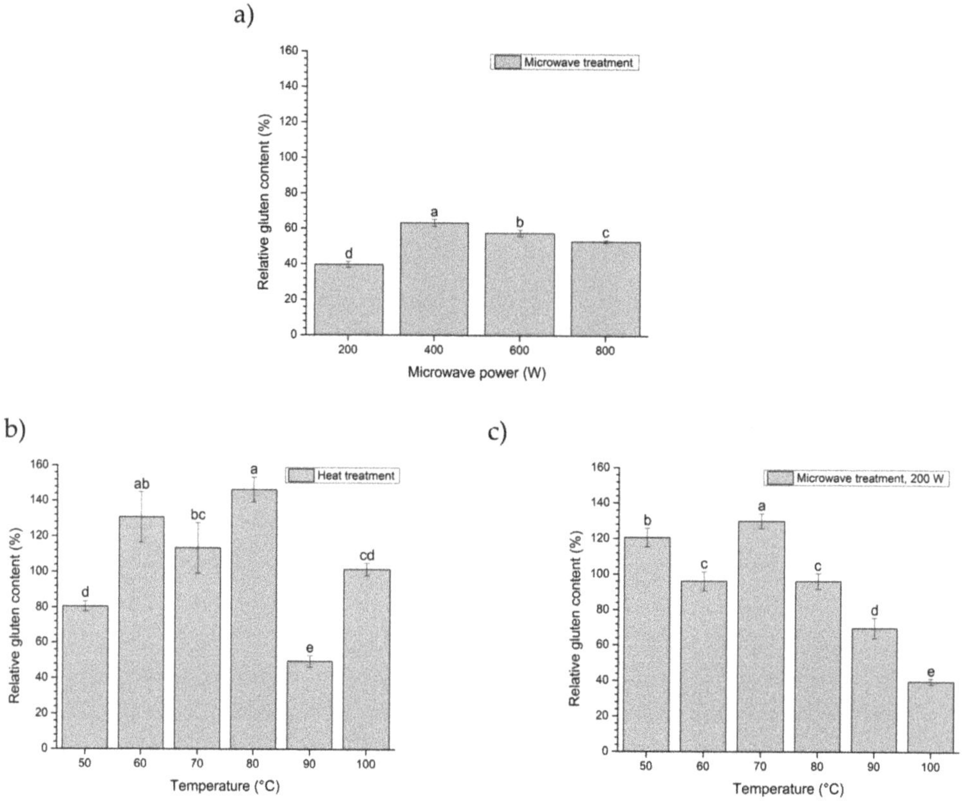

Figure 1. Relative gluten content (%) detected after (**a**) microwave treatment of wheat gluten at different microwave powers (200–800 W); (**b**) heat treatment of wheat gluten at different temperatures (50–100 °C); and (**c**) microwave treatment of wheat gluten at 200 W at different controlled temperatures (50–100 °C). All measurements were compared to an untreated gluten sample, considered as control (100%). Results are expressed as mean ± standard deviation (n = 2). Means with different letters in the same figure are significantly different ($p < 0.05$).

In order to compare microwave treatment to the effect of conventional heat treatment, gluten samples were treated on a heating unit to 50–100 °C with an increment of 10 °C, under the same conditions. The same temperatures were set on the microwave reactor and power of 200 W was applied. The results (Figure 1b,c) showed that gluten samples treated with microwave power, exhibited somewhat of a declining trend in relative gluten content value for the selected temperatures. The lowest relative gluten content value detected for conventionally heated gluten samples was at 90 °C and was 49.56 ± 3.25%, while the microwave treatment was most effective for the detoxification of gluten at the highest temperature (100 °C), at which the relative gluten content was 39.65 ± 1.69%. These results suggested that the gluten allergenicity detection was greatly affected by microwave treatment. Namely, an approximate 2.5-fold reduction in detected gluten content was achieved through the simple application of 200 W of microwave power. As microwave heating and conventional heating at the same temperature showed rather different effects on gluten content, it can be concluded that microwave specific effects (non-thermal effects) had influence on the gluten structure and on the gluten allergenicity. Thus, it can be emphasized that the microwave treatment under controlled conditions destroys the network of hydrogen bonds, causing changes in dipole rotation of gluten protein molecules and migration of ions in the working aqueous environment [46].

Compared to the heat treatment, the application of microwave treatment is a promising alternative since the reaction time is drastically reduced, which increases the economy of the treatment process and the exposure of gluten proteins to temperatures for a shorter period of time, which affects their characteristics. The time required for the microwave treatment of 200 W to reach the 50–100 °C temperatures was a few seconds, compared to heat treatment where 50–100 °C were reached in 2.75–25 min. Furthermore, the heat treatment appears to be undesirable because of its known negative effects on protein nutritional and functional properties as a consequence of the cross-linking of disulfide peptide bonds that take place by the mechanism of acylation of free amino groups [47]. The amino acid lysine is the most susceptible to this type of reaction, but also the amino acids serine, cysteine and cystine, as well as tryptophan-giving reaction products that significantly reduce the nutritional value of proteins. Generally speaking, the process involves the application of high pressures, shear force, as well as high temperatures for a long period of time result in irreversible denaturation of gluten proteins caused by deamidation of aspartic acid and glycine residues, disruption of peptide bond on aspartic acid, and destruction of amino acid residues [48]. Similarly, Lamacchia et al. [46] reported that the gliadins from treated flours (microwave oven: 1000 W, 2 min of treatment to reach a temperature of 110–120 °C) showed significantly reduced cross-reactivity with the R5 antibody. However, although the microwave pretreatment has been found to lead to gluten secondary structure alterations related to the polymer's disaggregation phenomenon, the pretreatment inefficiency to detoxify the gluten for celiac disease patients was observed [44].

There are many contradictory reports on the influence of microwaves on gluten celiac-related toxicity and there is scientific proof indicating inefficiency of this approach [44,49]. For example, although the microwave treatment of soaked wheat kernels has been documented to decrease the toxic epitope content in gluten with the R5-ELISA assay and also, after deamidation, with in vitro assay on gut-derived T-cell lines of celiac patients [48,50], other works reported that it neither destroyed the gluten nor chemically modified the toxic epitopes [44,49]. Namely, a recent paper by Gianfrani et al. [49] demonstrated that, by LC-MS/MS and by in vitro assay with T-cells of celiac patients, despite the early encouraging results about the drastic reduction in R5-immunoreactivity of the undigested soluble microwave treated gluten fraction, the MWT did not affect immune toxicity of gluten [49]. They confirmed that the treatment altered the wheat kernel protein solubility and apparently caused a drastic reduction in gluten, up to 70 ppm, which is in accordance with a previous report [48], but also established that the immunoreactive R5 gliadin components of microwave treated kernels were not simply extracted, most likely remaining attached to the substrate due to protein denaturation. Thus, the differences in the literature may be explained by different gluten extraction procedures and analytical tools for gluten detection, as well as different operating systems of the microwaves which mostly have not been specified. Moreover, in order to reproduce the same temperature and energy profile in the sample, time and power control are very important in this process. The present results confirm the potential of the designed MWT to detoxify wheat gluten but raise concerns regarding its possible efficacy in vivo, in the intestinal environment. Thus, the aim of the present research is to overcome the disadvantages of prior investigations by proposing a method of detoxification whereby gluten is based on the combination of microwave treatment and enzymatic hydrolysis to achieve better effects.

3.2. Combined Effect of Microwave Pretreatment (MWT) and Biocatalyst Alcalase on Gluten Content Reduction

The next series of experiments were performed in order to investigate the combined effect of the MWT and biocatalyst Alcalase on the detectable gluten content in the resulting hydrolysates (MWGHs). The starting point for this study was the finding that the microwave treatment alone resulted in the retention of toxicity and, on the other hand, the high proline content in gluten makes it highly resistant to complete enzymatic hydrolysis.

In order to investigate the combined effect of microwave pretreatment and a biocatalytic process on gluten content detection, microwave reactor powers of 200, 400, 600 and

800 W were applied. The pre-treatment step was followed by immediate gluten hydrolysis with Alcalase (Figure 2) and the aim of this experimental setup was to investigate whether microwave energy can facilitate the hydrolysis of toxic epitopes. By using a microwave reactor Anton Paar Monowave 300, it is possible to completely control the treatment conditions: temperature, treatment time and power, which is of great importance when it comes to protein treatment, where the pretreatment process itself can directly affect the final protein characteristics.

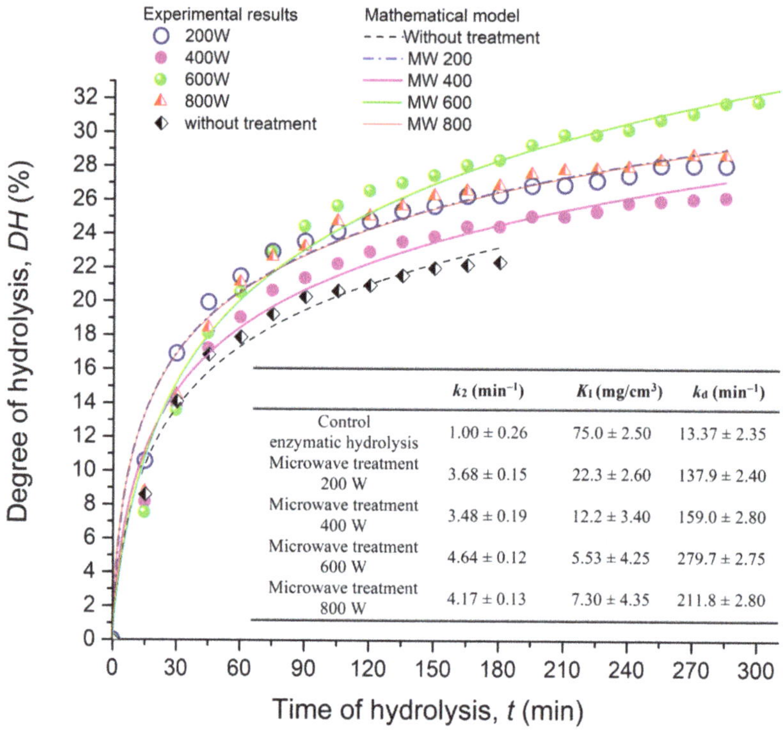

	k_2 (min^{-1})	K_1 (mg/cm^3)	k_d (min^{-1})
Control enzymatic hydrolysis	1.00 ± 0.26	75.0 ± 2.50	13.37 ± 2.35
Microwave treatment 200 W	3.68 ± 0.15	22.3 ± 2.60	137.9 ± 2.40
Microwave treatment 400 W	3.48 ± 0.19	12.2 ± 3.40	159.0 ± 2.80
Microwave treatment 600 W	4.64 ± 0.12	5.53 ± 4.25	279.7 ± 2.75
Microwave treatment 800 W	4.17 ± 0.13	7.30 ± 4.35	211.8 ± 2.80

Figure 2. Comparison of the susceptibility of differently microwave treated gluten proteins to enzymatic hydrolysis conducted with commercial food-grade protease, Alcalase. The experimental results were fitted by using the empirical kinetic model with substrate inhibition and enzyme deactivation. Reaction conditions for hydrolysis of presented curves: gluten concentration 2% (*w*/*w*), *E*/*S* ratio 5%, temperature 60 °C and pH 8). Inserted table—Values of the kinetic constants for enzymatic hydrolysis of microwave pretreated gluten proteins.

It is obvious from Figure 2 that the microwave-enhanced heat treatment significantly changed the hydrolysis pattern of gluten proteins ($p < 0.05$) in all of the examined irradiation powers (200–800 W). The initial reaction rate increased with the increase in the microwave power, while the achieved degree of hydrolysis increased with the increase in power up to 600 W, followed by a decrease when 800 W was applied. The achieved degree of hydrolysis varied from 20% to 32% over the 300 min time period. The hydrolysis proceeded at a rapid rate during the initial 45 min of the reaction and the recorded *DH* was about 16% for CGH, and more than 17% for MWGHs. Afterward, the enzymatic hydrolysis proceeded with a slow increase in hydrolysis rate, for the next 150 to 200 min, and then entered a steady state. In this way, it was confirmed that the microwave-treated gluten proteins as substrate had a considerable susceptibility to Alcalase; thus, the prepared hydrolysates could be used to identify their allergenicity characteristics. The represented *DH* profile with time is similar to the typical hydrolysis curve reported by Elmalimadi et al., (2017) [27] and Kong et al.,

(2007) [26]. The evident initial increase in susceptibility to enzymatic hydrolysis may be attributed to microwave-heat-induced conformational changes of gluten molecules, which can cause full or partial unfolding of polypeptides and resulted in exposure of buried peptide bonds, making them more accessible to the enzyme attack. However, treatment at 800 W caused a significant decrease ($p < 0.05$) in the susceptibility to enzymatic hydrolysis compared to 200 W, suggesting the previously mentioned gluten aggregation which in turn protected the internal bonds of the proteins.

To quantitatively compare the effects of microwave treatment power, the experimental data were fitted to a semi-empiric kinetic model that took into account the enzyme deactivation and substrate inhibition. The predicted mathematical kinetic model, considering the hydrolysis reaction as a zero-order reaction, aligned well with the obtained results, $R^2 \geq 0.98$. Based on the calculated values for the apparent reaction rate constant (k_2), the microwave pretreatment enhanced the reaction rate of enzymatic hydrolysis of gluten when compared to the gluten without treatment (Table inserted in Figure 2). For microwave treatment at power of 200 W, a 3.6-fold increase in the reaction rate constant appeared, but the overall level of gluten proteins hydrolysis, *DH* increased only 3%. However, at more intensive microwave treatment at a power of 600 W, the 4.6-fold increase in reaction rate constant led to a significant enhancement of *DH*, even ~9%. Both inhibition and reaction rate constants of deactivation also varied significantly along the microwave power, ranging from 5.53 to 22.3 mg/cm^3 and 137.9 to 211.8 min^{-1}, respectively. It is plausible that the protein microwave pretreatment caused the structural changes of gluten molecules to a form having an increased susceptibility to Alcalase but decreased binding to enzyme that may affect both enzyme inhibition by substrate (K_i) and enzyme stability (k_d).

The application of 600 and 800 W did not result in a significant gluten reduction ($p > 0.05$, Figure 3a). However, the relative gluten content of CGH and MWGH samples pretreated at 200 W and 400 W significantly improved compared to untreated gluten; samples showed an approximate 10-fold reduction. Nevertheless, the pretreatment at 200 W has proven to be the most effective; therefore, a closer comparison with the control hydrolysis was investigated (Figure 3b). In order to investigate whether a complete enzymatic hydrolysis needed to be performed in order to achieve the greatest gluten content reduction, samples were taken at different time of hydrolysis, lyophilized and gluten content was determined. It was apparent that the treatment at 200 W slightly contributed to reducing gluten content of the hydrolysate compared to CGH. The treatment had the highest contribution after 90 min when the content of gluten was reduced by an estimated 11.86%. It was apparent that the prolonged reaction time above 90 min did not support significantly towards further detoxification of gluten.

3.3. Influence of Microwave Pretreatment on Structural Changes of Gluten Proteins

In an effort to better understand the previous results and the influence of microwave pretreatment on the enzymatic hydrolysis, its impact on the structure of gluten main protein fractions in terms of total and reactive SH group contents and FTIR analysis were performed and compared with conventional heat treatment. The obtained results are presented in Figure 4.

It seemed clear that both content of reactive and total SH groups were strongly dependent on microwave power and temperature, but also the type of treatment. Regarding the content of SH groups presented in Figure 4a, the microwave treatment of gluten protein resulted in a statistically significant ($p < 0.05$) increase in the content of reactive SH groups, at the power of 800 W, which showed an advantage over the lower microwave powers. Control gluten sample demonstrated significant differences between the amount of total and reactive SH groups, 1.68 $\pm$ 0.103 and 0.58 $\pm$ 0.103 µmol/g, respectively. The increase in the content of total and reactive SH groups at 800 W could be explained by heat-induced conformational changes, which can cause partial unfolding of polypeptides, resulting in exposure of buried SH groups (reactive SH = 0.36 to 1.40 µmol/g and total SH = 0.63 to 1.50 µmol/g). Namely, during microwave treatment, most SH residues in gluten that existed in the interior of protein molecules were exposed with controlled-heat denaturation. As the susceptibility of gluten to hydrolysis after MWT 600 W was rather high (*DH* ~ 32%), the decrease in total SH content might be explained by the formation of predominantly intramolecular disulfide bonds due to sulfhydryl oxidation. On the other side, a MWT at a power greater than 600 W, the increase in total SH groups may attribute to disruption of intramolecular disulfide bonds induced by a temperature higher than 100 °C. Besides, decrease in protein susceptibility to hydrolysis and increase in substrate inhibition could be associated with the presence of intermolecular disulfide bond, leading to the protein aggregates formation. The protein aggregates formation is in accordance with the measurement of detectable amount of gluten contents (Figure 1a,c).

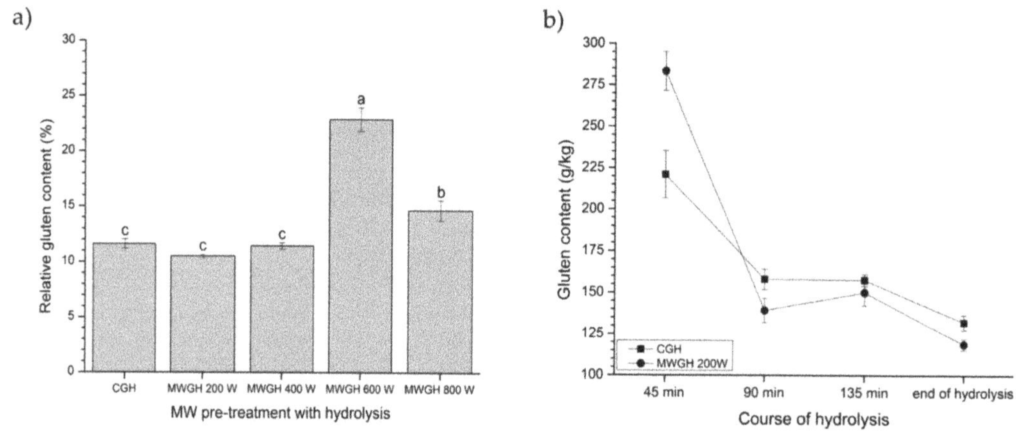

Figure 3. (**a**) Relative gluten content (%) of the control gluten hydrolysis (CGH) and microwave pretreated gluten hydrolysates (200–800 W) (Alcalase, pH 8, 60 °C) compared to untreated gluten (100%) and (**b**) gluten content reduction (g/kg) during hydrolysis, CGH and MWGH 200 W. Results are expressed as mean ± standard deviation ($n = 2$). Means with different letters in the same figure are significantly different ($p < 0.05$).

When the conventional heat treatment was applied (Figure 4b), low significant differences among the content of reactive SH groups of heated and untreated gluten were noticed ($p < 0.05$). Besides, the content of total SH groups decreased after the heat treatment at 50 and 100 °C from 1.2 to 0.3 μmol/g, respectively ($p < 0.05$). Microwave heating of gluten proteins at 50 °C with input power 200 W (Figure 4c) appeared to lead to an increase in the reactive SH groups due to unfolding of proteins but with accompanying aggregation, causing a decrease in the total SH content, and probably an increase in disulfide bonds. During the experiment performed at fixed input power of 200 W, the different time treatment caused various temperatures. It can be emphasized that longer exposure time caused higher temperature (100 °C) due to which gluten proteins aggregates were formed, i.e., gliadin molecules were polymerized and cross-linked with glutenins. At the same time, the oxidation of free SH groups occurred, S-S bonds were formed and consequently cross-linking of gliadin aggregates with glutenins were evident. Obviously, MWT caused a higher effect in the range from 60 to 90 °C on both reactive and total SH groups' content in comparison to heat treatment, particularly at 90 °C, indicating different mechanism of gluten protein denaturation by heat and microwave. It is evident that microwave power and temperatures caused certain changes on protein molecules that seemed to have some relation to the gluten protein unfolding and aggregation. The stated assumptions are in full accordance with the literature data, which analyzed effect of thermal treatment on the gluten structure [44,48]. However, it is very important to state that there is no data in the literature on the use of temperature-controlled microwave reactor, which is the main purpose and benefit of this research.

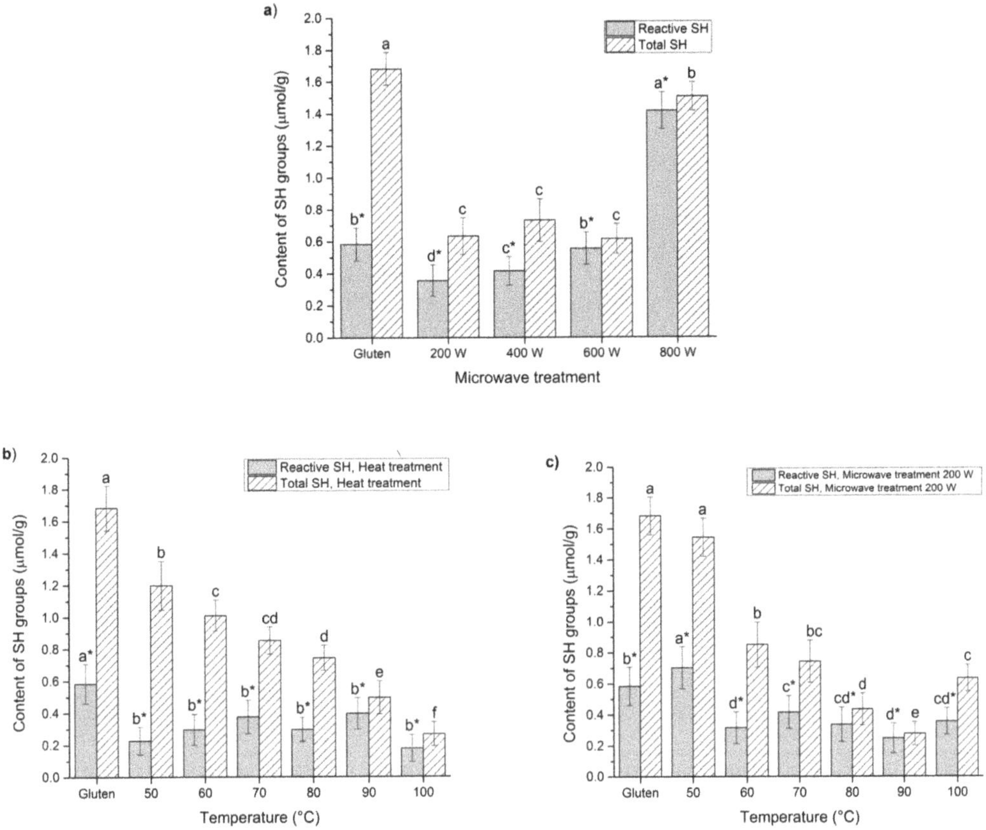

Figure 4. Content of total and reactive SH groups detected after: (**a**) microwave treatment of wheat gluten at different powers (200–800 W); (**b**) heat treatment of wheat gluten at different temperatures (50–100 °C); and (**c**) microwave treatment of wheat gluten at 200 W and controlled different temperatures (50–100 °C). All measurements were compared at the same protein concentration 2 mg/mL. Results are expressed as mean ± standard deviation (n = 3). Means with different letters in the same figure are significantly different ($p < 0.05$).

Theoretically, the major role of SH groups lies in the determination and stabilization of the three-dimensional structure of proteins, thus, the change in SH content by heating represents a first indication that some of the fundamental structural changes in gluten protein functionality exists. It is universally known that glutenin subunits develop ordered fibrous macromolecular polymers with intermolecular disulfide linkages, while gliadins only form intramolecular disulfide linkages [51]. Thus, it can be concluded that the influence of microwave-controlled heat treatment is reflected in terms of the protein native state unfolding and making the gluten protein chains more accessible to Alcalase penetration which is reflected in an enhanced enzymatic hydrolysis reaction. Additionally, these results suggest that MWT at 200 W input power might induce enzymatic hydrolysis cleavage of an existing protein aggregate and the SH groups might take a part in this phenomenon by reducing S-S bonds, which are responsible for maintaining aggregates' structure.

The effects of microwave-controlled treatment on the gluten secondary structure of the solid microwave-treated and hydrolyzed samples were analyzed by FTIR (Figure S1). The deconvolution of the Amide I band (Figures 5 and 6) of control untreated and microwave treated glutens, as well as gluten hydrolysates, enabled the analysis of the main secondary structure elements, namely the relative intensities of the extended β-sheet

(1623–1641 cm^{-1}), the intermolecular β-sheet (1612 cm^{-1}), the α-helix (1648–1657 cm^{-1}) and the coils (2 to 7 α-helices coiled together; 1662–1686 cm^{-1}). The relative intensities of the gluten main secondary structure elements are summarized in Table 1.

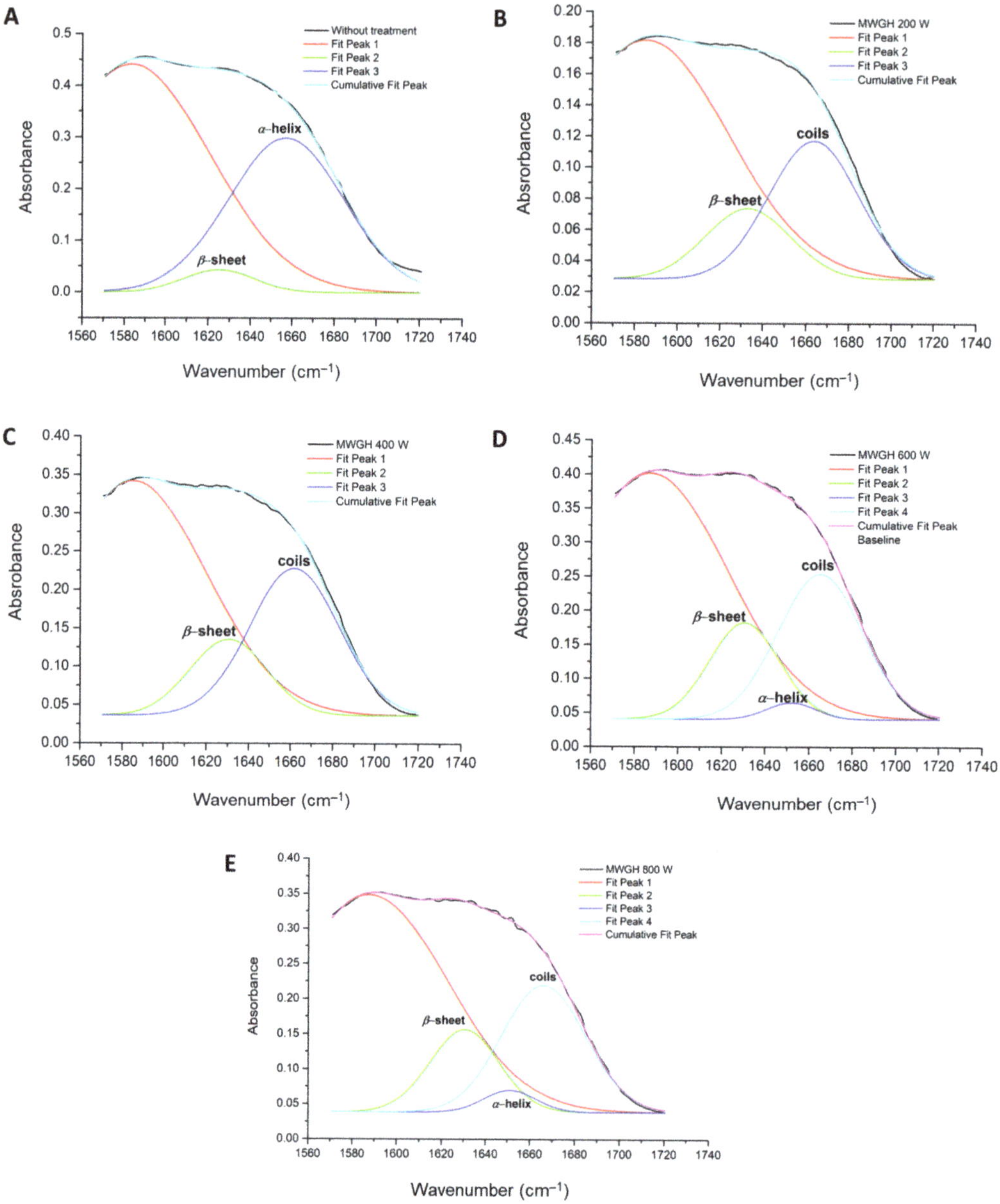

Figure 5. Peak deconvolution of Amide I band in the FTIR spectra of hydrolysates of gluten proteins (**A**) without pretreatment and microwave pretreated at powers of (**B**) 200 W, (**C**) 400 W, (**D**) 600 W and (**E**) 800 W using Peak and Baseline functions (baseline subtracting, deconvolution, second derivate and Gaussian fitting mode) in the OriginPro Lab 9.0.

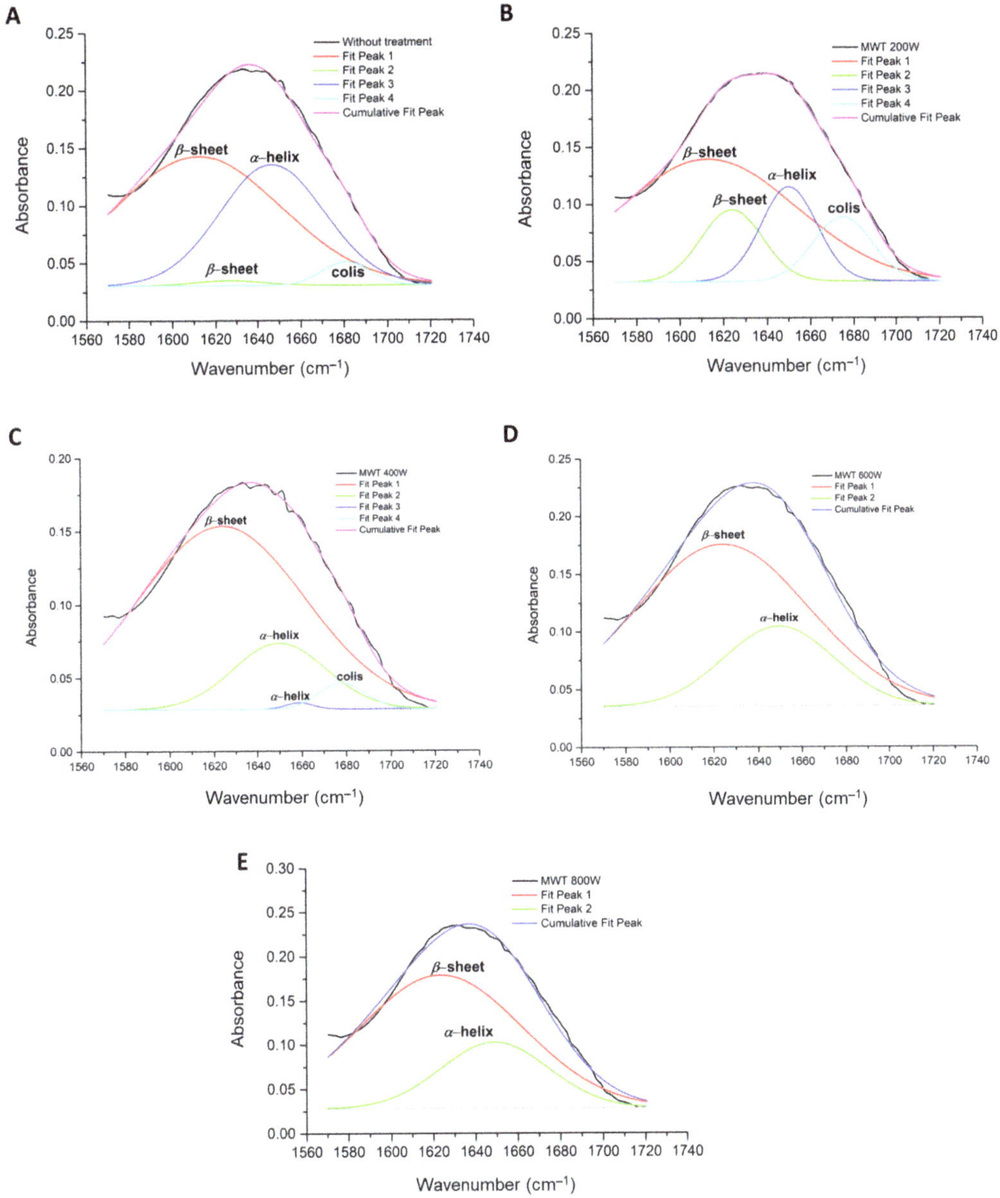

Figure 6. Peak deconvolution of Amide I band in the FTIR spectra of gluten proteins (**A**) without pretreatment and microwave pretreated at powers of (**B**) 200 W, (**C**) 400 W, (**D**) 600 W and (**E**) 800 W using Peak and Baseline functions (baseline subtracting, deconvolution, second derivate and Gaussian fitting mode) in the OriginPro Lab 9.0.

Table 1. Secondary structure band assignments in the gluten protein samples pretreated with microwaves and in their hydrolysates.

Secondary Structure	Band Assignment in Gluten Microwave Treated Samples				
	Without Treatment	200 W	400 W	600 W	800 W
α-helix	38.9%	16.6%	17.3%	25.6%	26.2%
random	n.d.	n.d.	0.49%	n.d.	n.d.
β-sheet (extended)	1.03%	13.7%	78.8%	74.4%	73.8%
β-sheet (intermolecular)	56.8%	57.7%	n.d.	n.d.	n.d.
coils	3.36%	11.9%	3.04%	n.d.	n.d.
Secondary Structure	**Band Assignment in Hydrolysates of Microwave Pretreated Gluten**				
	Without Treatment	**200 W**	**400 W**	**600 W**	**800 W**
α-helix	41.9%	n.d.	n.d.	1.77%	2.80%
β-sheet (extended)	3.77%	13.7%	13.8%	14.8%	14.1%
coils	n.d.	28.0%	31.3%	26.6%	26.0%
side chain	54.3%	58.3%	54.9%	56.8%	57.2%

n.d.—not determined.

The secondary structure of gluten samples and hydrolysate samples significantly changed when the sample was pretreated with microwave-controlled irradiation at different powers of 200–800 W, in comparison to samples without treatment. When the microwave power of 200 W was applied, the total α-helix decreased, whereas the β-sheet and coils significantly increased with the time. With the increase in the applied power from 200 to 800 W, the contribution of α-helix was significantly augmented. It can be emphasized that the β-sheet (intermolecular) and α-helix of the gluten protein molecules, without treatment and after microwave pre-treatment, were found in the interior of polypeptide chains, and β-sheet (extended, i.e., β-turn) were formed because of the reversal of polypeptide chains just after microwave pre-treatment with 400, 600 and 800 W. The promotion of β-helix and extended β-sheet existed due to contribution of the microwave irradiation in reactor system to the formation of disulfide bonds, which is in accordance with the previously discussed results in the Figure 3a, or iso-peptide bonds [44,52].

On the other hand, the intensity of bands assignment in hydrolysates of microwave pretreated gluten clearly demonstrated that microwave irradiation had induced facilitated hydrolysis by Alcalase. The decrease in the relative abundance of α-helix secondary structures suggested a different arrangement of the polypeptide bonds in the hydrolysates of microwave treated glutens, especially after application of microwave powers at 200 and 400 W. Considering this observation, it can be stated that combination of microwave pretreatment and Alcalase hydrolysis promoted intermolecular reactions, so that to increase the structural reorganization of α-helix to coils (28.0% and 31.3% at 200 and 400 W, respectively), and augmented the relative abundance of side chains (58.3% at 200 W). Extended β-sheet structure was significantly changed after synergistic effect of microwave treatment and Alcalase hydrolysis, suggested that partial hydrolyzed polypeptide chains formed more stable structures, which may be crucial for further techno-functional properties and allergenicity.

Comprehensively, the main absorption bands of proteins appear due to the presence of Amide I band stretching vibrations of C=O group (≈1645 cm^{-1}) and Amide II band in plane N-H bending vibrations (≈1545 cm^{-1}) [37,38]. Noticeable change in the spectra was observed in the Amide I and II bands region (≈1700–1200 cm^{-1}) of the hydrolysates. Untreated wheat gluten showed a sharp band at ≈1633 cm^{-1} (Supplementary Figure S2, which changes its shape as shown in Figure S3), where all hydrolysates showed a band at ≈1590 cm^{-1}, which was supposedly a result of structural changes in the secondary structure of the protein.

The absorption peaks in the wavelength range of 3100–3500 cm^{-1} (Amide A), which is attributed to N-H and O-H vibrations of the hydroxyl groups, have changed the shape and the intensity after microwave pretreatment due to the interaction of OH vibrations of water molecules and N-H stretching caused by enzymatic hydrolysis of gluten proteins. The applied microwave treatment led to changes in the intensity of tensile vibrations of C–H_2, C–H and =C–H bonds at wavelength 1450, 1456, 2933 and 3060 cm^{-1}, which can be related to the hydrophobic interactions and conformational changes of molecules caused by the action of electromagnetic waves. Compared to the control hydrolysates, the more pronounced absorption bands of all obtained gluten hydrolysates were observed at the wavelength of 1300–1700 cm^{-1} corresponding to the Amide I and Amide II regions, which means that there were structural changes within the amide bond. In particular, the peak losses at 1516 cm^{-1} were observed compared to the control gluten hydrolysis without pre-treatment. Additionally, an increase in peak intensity at 1651 cm^{-1} was observed in all gluten hydrolysates obtained from microwave-pretreated proteins, compared to the control. This increase in peak intensity was a confirmation of the existence of a certain proportion of the α-coil within the hydrolysis structure. It is necessary to point out that reduced peak intensity was observed within the Amid VI region at 560 cm^{-1} for all gluten hydrolysates, and the changes or/modifications for gluten microwave pre-treated molecules, compared to the control untreated gluten. Vibrations of the S–S bonds are detected at this wavelength, thus it can be confirmed that there was a change in the number of disulfide connections after microwave treatments at various power. This observation confirmed the previously stated claim that microwave-controlled pre-treatment had an effect on increasing/decreasing the content of sulfhydryl groups in the gluten protein molecules.

Additionally, the denaturation and aggregation of protein molecules after microwave treatment were analyzed by using the denatured SDS-PAGE electrophoresis (Figure 7). The SDS-PAGE results (Figure 7a) clearly demonstrated that both glutenin and gliadin fractions from the MWT (200–800 W) gluten proteins did not present any significant difference in the presented electrophoretic profiles and also in comparison with the control gluten, including any increase or decline in the number of protein bands. The demonstrated migration pattern observed for all the MWT gluten samples pointed out that the primary structure of the gluten proteins was not modified, so microwave treatment with controlled heat effect did not have influence on the gluten protein hydrolysis even at high temperature. Results are in a good accordance with the available literature data, which generally stated that electromagnetic irradiation caused by microwave oven cannot modify the primary structure of proteins, since the energy of the chemical bonds is superior to the quantum energy of the microwave [44,53].

SDS-PAGE electrophoresis showed a different profile between two tested hydrolysates (Figure 7b). The complete absence of protein fractions with molecular weights greater than 45 kDa after enzymatic hydrolysis was observed in both samples. However, difference in detected protein fractions occurred for the molecular weights lesser than 50 kDa. For the 200 W microwave pretreated gluten hydrolysate, a complete absence of gliadin and glutenin fractions in the range from 25–50 kDa was observed. This absence of fractions corresponding to LMW GS and ω1,2-gliadins (32–39 and 39–44 kDa, respectively), γ- and α/β-gliadins (31–35 and 28–35 kDa, respectively) can be ascribed to the effect of microwave treatment applied on gliadin and glutenin fractions. Since CGH did not undergo any heat treatment prior to enzymatic hydrolysis temperature equilibration at 60 °C, it was clear that fast and short application of microwave energy of 200 W was sufficient to induce changes in the gluten protein fractions, making them more available to the enzyme. Singh and MacRitchie [54] found that the polymerization of glutenins occurred at temperatures below 100 °C, while gliadins polymerized at higher temperatures. The competitive ELISA test indicated that microwave treatment of wheat gluten reduced detectable gliadin content to a significant level, which is similar with findings of Lamacchia et al. [48].

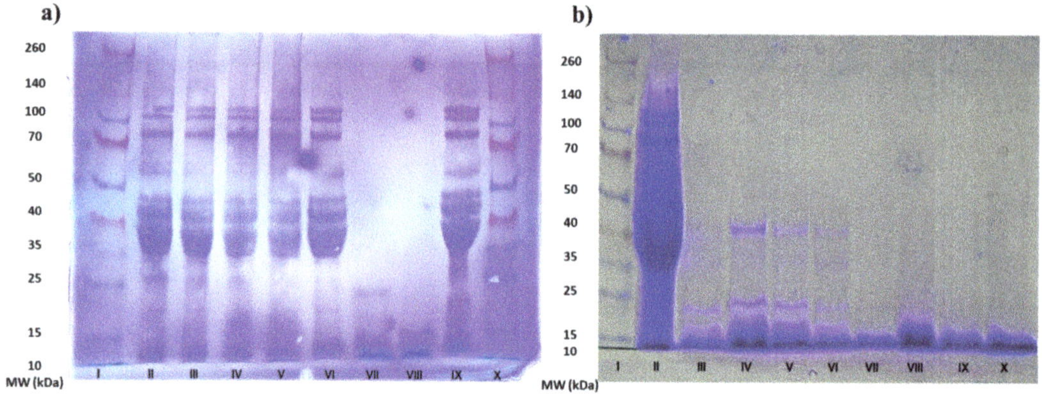

Figure 7. SDS-PAGE profiles of Gluten, MWT gluten at different microwave power (200–800 W), CGH and MWGH. (**a**) Band: I—protein standard, II—MWT 200 W, III—MWT 400 W, IV—MWT 600 W, V—MWT 800 W, VI—Gluten, VII—CHG, VIII—MWGH, IX—Gluten, X—protein standard; (**b**) Band: I—protein standard, II—Gluten, III—CGH, IV—CGH (at 45 min), V—CGH (at 90 min), VI—CGH (at 135 min), VII—MWGH, VIII—MWGH (at 45 min), IX—MWGH (at 90 min), X—MWGH (at 135 min).

3.4. Beneficial Effect of the Enzymatic Hydrolysis on Antioxidant and Functional Properties

Given the fact that gluten proteins are responsible for the formation of the network and structure of many food products, after examining the influence of controlled microwave pretreatment on the structural changes, the availability of peptide bonds to the biocatalyst and the amount of allergenic toxic epitopes, it was important to examine whether the synergistic effect of enzyme hydrolysis and microwave pretreatment affected techno-functional properties and antioxidant activities of the hydrolysates. Thus, the emulsifying activity index (*EAI*) and emulsion stability (*ESI*) have been tested for untreated gluten, CGH and MWGH, and results are presented in Table 2.

The results indicated that depending on the processing and microwave pretreatment conditions, enzymatic hydrolysis can improve both antioxidant and functional properties of gluten. The most significant improvement ($p < 0.05$) in *EAI* was achieved for the MWGH after 135 min of hydrolysis (50.66 ± 3.10 m^2/g) compared to raw wheat gluten (11.63 ± 1.10 m^2/g) or CGH after 90 min (36.19 ± 0.08). However, extensive hydrolysis reduced the *EAI* value, which is in accordance with the literature data for completely hydrolyzed protein samples [26,55]. Smaller peptides appeared to form weaker coats around the oil droplets, thus yielding lower *EAI* values. The extensive hydrolysis of the wheat gluten samples, which generated smaller peptides, resulted in 25.63 ± 2.85 m^2/g and 22.05 ± 3.60 m^2/g for the CGH and MWGH, respectively. On the other hand, *ESI* increased with the degree of hydrolysis compared to untreated gluten and the greatest improvement in stability was achieved for the completely hydrolyzed CGH of 3.30 ± 0.31 h. It can be assumed that Alcalase has hydrolyzed the gluten proteins in a way that showed great specificity towards peptide bonds rich in hydrophobic amino acids. Due to the greater exposure of hydrophobic amino acid residues on the surface of molecules, greater ability of gluten hydrolysates to form stable emulsion systems also manifested.

Table 2. ABTS (%), MICA (%), ABTS IC_{50} ($mg_{protein}/mL$), MICA IC_{50} ($mg_{protein}/mL$), *EAI* (m^2/g) and *ESI* (h), *FC* (%) and *FS* (%) values for untreated gluten, control gluten hydrolysate (CGH) and microwave pretreated gluten hydrolysate (MWGH).

Sample	ABTS, %	MICA, %	ABTS IC_{50}, (mg/mL)	MICA IC_{50}, (mg/mL)	*EAI*, m^2/g	*ESI*, h	*FC*, %	*FS*, %
Gluten	7.52 ± 0.22^{b}	nd*	63.12 ± 20.85^{b}	nd*	11.63 ± 1.10^{f}	0.33 ± 0.04^{e}	50.93 ± 6.55^{ab}	48.91 ± 5.13^{a}
CGH (at 45 min)	66.86 ± 0.88^{a}	91.24 ± 0.51^{c}	1.10 ± 0.17^{a}	1.07 ± 0.00^{ab}	41.02 ± 2.96^{b}	1.64 ± 0.85^{bc}	60.67 ± 1.41^{a}	55.05 ± 4.43^{a}
CGH (at 90 min)	69.27 ± 0.66^{a}	93.44 ± 0.71^{b}	1.22 ± 0.02^{a}	0.84 ± 0.05^{cd}	36.19 ± 0.08^{b}	1.22 ± 0.47^{cde}	54.04 ± 0.48^{ab}	41.27 ± 4.49^{a}
CGH (at 135 min)	67.45 ± 0.15^{a}	94.24 ± 0.05^{ab}	1.11 ± 0.19^{a}	0.81 ± 0.01^{cd}	29.29 ± 0.80^{c}	2.51 ± 0.46^{ab}	53.09 ± 0.51^{ab}	38.37 ± 1.64^{a}
CGH	69.93 ± 1.02^{a}	94.57 ± 0.09^{b}	1.08 ± 0.15^{a}	0.74 ± 0.03^{d}	25.63 ± 2.85^{ce}	3.30 ± 0.31^{a}	47.06 ± 9.31^{ab}	11.11 ± 7.86^{b}
MWGH (at 45 min)	66.50 ± 0.51^{a}	90.12 ± 0.40^{c}	1.06 ± 0.11^{a}	1.20 ± 0.09^{a}	27.59 ± 1.12^{cd}	3.06 ± 0.42^{a}	60.93 ± 3.26^{a}	53.32 ± 0.54^{a}
MWGH (at 90 min)	69.56 ± 0.51^{a}	94.75 ± 0.05^{ab}	1.04 ± 0.14^{a}	0.84 ± 0.01^{cd}	27.50 ± 0.95^{cd}	1.32 ± 0.11^{be}	56.47 ± 1.29^{ab}	42.76 ± 8.73^{a}
MWGH (at 135 min)	68.69 ± 0.51^{a}	94.80 ± 0.47^{ab}	0.96 ± 0.06^{a}	0.92 ± 0.01^{bc}	50.66 ± 3.10^{a}	1.10 ± 0.27^{cde}	59.50 ± 5.05^{a}	53.7 ± 2.62^{a}
MWGH	70.29 ± 1.09^{a}	96.00 ± 0.11^{a}	1.03 ± 0.17^{a}	0.75 ± 0.02^{d}	22.05 ± 3.60^{de}	1.64 ± 0.30^{bd}	41.65 ± 1.70^{b}	10.56 ± 2.74^{b}

Results are expressed as mean ± standard deviation ($n = 3$). Means with different letters in the same column are significantly different ($p < 0.05$). *nd—could not be determined.

The basic prerequisites for a food protein to be a good foam agent is the ability to be rapidly adsorbed during bubbling process at the air-water interface and undergo a rapid conformational change and rearrangement of functional groups at the interface, and to have the ability to form cohesive viscoelastic film using intermolecular interactions. Thus, prepared hydrolysates, which originated from the microwave-modified and non-modified gluten proteins, were also analyzed on the ability to form cohesive and stable foam (Table 2). It is evident that hydrolysates prepared using the non-extensive action of an endo-protease, Alcalase, i.e., with hydrolysates with lower *DH*, especially after ~45 min of proteolysis where *DH* of ~17% was achieved, possessed the highest foam capacity. This applied to both the hydrolysates obtained from untreated gluten (60.67 ± 1.41%) and the hydrolysates obtained from microwave pre-treated proteins (60.93 ± 3.26%). Further continuation of hydrolysis led to a statistically significant ($p < 0.05$) reduction in the ability to form foam in both cases. The microwave treatment led to different Alcalase activity and substrate specificity, causing a slightly better foam capacity of the hydrolysates. Therefore, the microwave pretreatment had greater effect on the partial aggregation of gluten proteins that were subsequently hydrolyzed by Alcalase, and during hydrolysis, lower molecular weight peptides were formed, with low net charge and high surface hydrophobicity and as such, were ideal to form foam of higher capacity and stability than gluten proteins without treatment. In other words, due to changes in the structure of gluten proteins (*SH* groups content and structural changes referred by the FTIR analysis) caused by microwave pretreatment and subsequently under the influence of Alcalase, they take such a conformation that they allowed the reduction in the boundary stress between the aqueous and air phases. The formed foams showed high stability during 30 min: for CGH stability was measured as 55.05 ± 4.43% and for MWGH was 53.32 ± 0.54%. Yalcin et al. [56] reported that the emulsifying and foaming ability and stability values of microwave-heated gluten samples were slightly greater than those of the control sample, which is in accordance with the results of this research. An identical conclusion was reached in an earlier study examining the effect of Alcalase on the hydrolysis of traditional treated gluten proteins on functional properties [55]. However, it was difficult to discuss microwave-induced effects because microwave devices with precise control of the microwave power and temperature were not used in most cases.

In terms of the potential health benefits of gluten hydrolysates, the antioxidant ability was measured. Therefore, the free-radical scavenging activity determined by using the $ABTS^{\cdot+}$ radical showed significant improvement in the gluten hydrolysates compared to untreated wheat gluten proteins (Table 2). However, no significant difference was achieved between MWGH and CGH without pretreatment. MWGH had 70.29 ± 1.09% of ABTS activity. All of the samples taken after 45, 90 and 135 min of hydrolysis showed no significant difference in ABTS activity among the tested samples. It can be concluded that after 45 min of hydrolysis, a plateau of free-radical scavenging activity was achieved. In comparison to raw wheat gluten, ABTS IC_{50} values for all of the hydrolysate samples were significantly reduced. Since no significant difference was achieved by microwave pretreatment, it can be concluded that it did not improve the production of more active free-radical scavenging peptides. Metal-ion chelating activities of hydrolysates were also improved by enzymatic hydrolysis in comparison to raw wheat gluten. MICA activity of raw wheat gluten could not be determined. However, in this case, the hydrolysates showed significant difference between MWGH and CGH. The highest MICA of 96.00 ± 0.11% was achieved for the MWGH, revealing that the microwave treatment improved the metal-ion chelating activity of gluten hydrolysates.

4. Conclusions

It appeared that microwave specific non-thermal effects had significant influence on the gluten structure and gluten allergenicity, enhancing its hydrolysis and, in combination with the enzymatic hydrolysis, ultimately yielded protein hydrolysates with enhanced free-radical scavenging and metal-ion chelating activity. The combination of enzyme hydrolysis and microwave reactor pretreatment (200 W, 100 °C, 1 min) seemed to be an efficient procedure for gluten content reduction, resulting in almost 10-fold reduction in immunoreactive epitopes with R5 competitive ELISA. Future studies on immunoreactivity of different soluble and insoluble gluten fractions, particularly clinical trials, are required to additionally understand mechanism of inactivation of gluten toxic epitopes by this combined procedure.

Supplementary Materials: The following are available online at https://www.mdpi.com/article/10.3390/foods10092214/s1, Figure S1: FTIR spectra of the Amide I region (1700–1600 cm^{-1}) of the control hydrolysis and microwave pretreated (200–800 W) gluten hydrolysates (Alcalase, pH 8, T 60 °C), Figure S2: FTIR spectra of untreated wheat gluten. Figure S3: FTIR of (a) untreated gluten and MW treated gluten (200–800 W) and (b) gluten, control gluten hydrolysate and MWGH (200–800 W).

Author Contributions: Conceptualization, I.G., J.M. and Z.K.-J.; methodology, J.M. and A.Đ.V.; software, I.G. and A.Đ.V.; validation, S.J.T., N.Š. and J.M.; formal analysis, I.G.; investigation, I.G. and J.M.; resources, Z.K.-J. and L.M.; data curation, J.M., Z.K.-J. and L.M.; writing—original draft preparation, I.G. and J.M.; writing—review and editing Z.K.-J.; visualization, I.G. and N.Š.; supervision, Z.K.-J.; project administration, Z.K.-J.; funding acquisition, Z.K.-J. and L.M. All authors have read and agreed to the published version of the manuscript.

Funding: This research was funded by Ministry of Education, Science and Technological Development of the Republic of Serbia (Contract No. 451-03-9/2021-14/200135 and Contract No. 451-03-9/2021-14/200287) and EUREKA project E! 13082 BIOFLOSBAKE-LAVGLU.

Institutional Review Board Statement: Not applicable.

Informed Consent Statement: Not applicable.

Data Availability Statement: Not applicable.

Acknowledgments: The authors are grateful to Dušan Mijin (Faculty of Technology and Metallurgy) for the kind contribution to the realization of this work in materials and microwave esuipment, and for performing FTIR analysis.

Conflicts of Interest: The authors declare no conflict of interest. The funders had no role in the design of the study; in the collection, analyses, or interpretation of data; in the writing of the manuscript, or in the decision to publish the results.

References

1. Ooms, N.; Delcour, J.A. How to impact gluten protein network formation during wheat flour dough making. *Curr. Opin. Food Sci.* **2019**, *25*, 88–97. [CrossRef]
2. Delcour, J.A.; Joye, I.J.; Pareyt, B.; Wilderjans, E.; Brijs, K.; Lagrain, B. Wheat Gluten Functionality as a Quality Determinant in Cereal-Based Food Products. *Annu. Rev. Food Sci. Technol.* **2012**, *3*, 469–492. [CrossRef] [PubMed]
3. Day, L. Wheat gluten: Production, properties and application. In *Handbook of Food Proteins*; Phillips, G.O., Williams, P.A., Eds.; Woodhead Publishing: Sawston, UK, 2011; pp. 267–288. [CrossRef]
4. Cappelli, A.; Oliva, N.; Cini, E. A Systematic Review of Gluten-Free Dough and Bread: Dough Rheology, Bread Characteristics, and Improvement Strategies. *Appl. Sci.* **2020**, *10*, 6559. [CrossRef]
5. Pourmohammadi, K.; Abedi, E. Enzymatic modifications of gluten protein: Oxidative enzymes. *Food Chem.* **2021**, *356*, 129679. [CrossRef]
6. Fucinos, C.; Estevez, N.; Míguez, M.; Fajardo, P.; Chapela, M.J.; Gondar, D.; Rua, M.L. Effectiveness of proteolytic enzymes to remove gluten residues and feasibility of incorporating them into cleaning products for industrial purposes. *Food Res. Int.* **2019**, *120*, 167–177. [CrossRef] [PubMed]
7. Arendt, E.K.; Zannini, E. Wheat and other Triticum grains. In *Cereal Grains for the Food and Beverage Industries*; Woodhead Publishing Limited: Sawston, UK, 2013; pp. 1–67.
8. Field, J.M.; Shewry, P.R.; Burgess, S.R.; Forde, J.; Parmar, S.; Miflin, B.J. The presence of high molecular weight aggregates in the protein bodies of developing endosperms of wheat and other cereals. *J. Cereal Sci.* **1983**, *1*, 33–41. [CrossRef]
9. Biesiekierski, J.R. What is gluten? *J. Gastroenterol. Hepatol.* **2017**, *32*, 78–81. [CrossRef] [PubMed]
10. Wieser, H.; Koehler, P. The Biochemical Basis of Celiac Disease. *Cereal Chem.* **2008**, *85*, 1–13. [CrossRef]
11. Rallabhandi, P.; Sharma, G.M.; Pereira, M.; Williams, K.M. Immunological Characterization of the Gluten Fractions and Their Hydrolysates from Wheat, Rye and Barley. *J. Agric. Food Chem.* **2015**, *63*, 1825–1832. [CrossRef]

12. Wieser, H. Detection of gluten. In *Gluten-Free Cereal Products and Beverages*; Arendt, E.K., Dal Bello, F., Eds.; Academic Press: Cambridge, MA, USA, 2008; pp. 47–80.
13. Wieser, H.; Koehler, P. Detoxification of Gluten by Means of Enzymatic Treatment. *J. AOAC Int.* **2012**, *95*, 356–363. [CrossRef]
14. Hausch, F.; Shan, L.; Santiago, N.A.; Gray, G.M.; Khosla, C. Intestinal digestive resistance of immunodominant gliadin peptides. *Am. J. Physiol. Gastrointest.* **2002**, *283*, G996–G1003. [CrossRef]
15. Scherf, K.A.; Koehler, P.; Wieser, H. Gluten and wheat sensitivities—An overview. *J. Cereal Sci.* **2016**, *67*, 2–11. [CrossRef]
16. Allred, L.K.; Ritter, B.W. Recognition of Gliadin and Glutenin Fractions in Four Commercial Gluten Assays. *J. AOAC Int.* **2010**, *93*, 190–196. [CrossRef]
17. Huang, H.-W.; Hsu, C.-P.; Yang, B.B.; Wang, C.-Y. Potential Utility of High-Pressure Processing to Address the Risk of Food Allergen Concerns. *Compr. Rev. Food Sci. Food Saf.* **2014**, *13*, 78–90. [CrossRef]
18. Janssen, G.; Christis, C.; Kooy-Winkelaar, Y.; Edens, L.; Smith, D.; van Veelen, P.; Koning, F. Ineffective Degradation of Immunogenic Gluten Epitopes by Currently Available Digestive Enzyme Supplements. *PLoS ONE* **2015**, *10*, e0128065. [CrossRef]
19. Shan, L.; Marti, T.; Sollid, L.M.; Gray, G.M.; Khosla, C. Comparative biochemical analysis of three bacterial prolyl endopeptidases: Implications for coeliac sprue. *Biochem. J.* **2004**, *383*, 311–318. [CrossRef]
20. Martinez, M.; Gómez-Cabellos, S.; Giménez, M.J.; Barro, F.; Diaz, I.; Diaz-Mendozam, M. Plant Proteases: From key enzymes in germination to allies for fighting human gluten-related disorders. *Front. Plant Sci.* **2019**, *10*, 721. [CrossRef] [PubMed]
21. Matysiak-Budnik, T.; Candalh, C.; Cellier, C.; Dugave, C.; Namane, A.; Vidal-Martinez, T.; Cerf-Bensussan, N.; Heyman, M. Limited Efficiency of Prolyl-Endopeptidase in the Detoxification of Gliadin Peptides in Celiac Disease. *Gastroenterology* **2005**, *129*, 786–796. [CrossRef] [PubMed]
22. Segura, V.; Ruiz-Carnicer, Á.; Sousa, C.; de Lourdes Moreno, M. New Insights into Non-Dietary Treatment in Celiac Disease: Emerging Therapeutic Options. *Nutrients* **2021**, *13*, 2146. [CrossRef] [PubMed]
23. Ehren, J.; Morón, B.; Martin, E.; Bethune, M.T.; Gray, G.M.; Khosla, C. A foodgrade enzyme preparation with modest gluten detoxification properties. *PLoS ONE* **2009**, *4*, e6313. [CrossRef]
24. Osorio, C.E.; Wen, N.; Mejías, J.H.; Mitchell, S.; von Wettstein, D.; Rustgi, S. Directed-Mutagenesis of Flavobacterium meningosepticum prolyl-oligopeptidase and a glutamine-specific endopeptidase from barley. *Front. Nutr.* **2020**, *7*, 11. [CrossRef]
25. Lähdeaho, M.L.; Kaukinen, K.; Laurila, K.; Vuotikka, P.; Koivurova, O.P.; Kärjä-Lahdensuu, T.; Marcantonio, A.; Adelman, D.C.; Mäki, M. Glutenase ALV003 attenuates gluten-induced mucosal injury in patients with celiac disease. *Gastroenterology* **2014**, *146*, 1649–1658. [CrossRef]
26. Kong, X.; Zhou, H.; Qian, H. Enzymatic hydrolysis of wheat gluten by proteases and properties of the resulting hydrolysates. *Food Chem.* **2007**, *102*, 759–763. [CrossRef]
27. Elmalimadi, M.B.; Stefanović, A.B.; Šekuljica, N.Ž.; Žuža, M.G.; Luković, N.D.; Jovanović, J.R.; Knežević-Jugović, Z.D. The synergistic effect of heat treatment on alcalase-assisted hydrolysis of wheat gluten: Functional and antioxidant properties. *J. Food Process. Preserv.* **2017**, *41*, e13207. [CrossRef]
28. Merz, M.; Appel, D.; Berends, P.; Rabe, S.; Blank, I.; Stressler, T.; Fischer, L. Batch-to-batch variation and storage stability of the commercial peptidase preparation Flavourzyme in respect of key enzyme activities and its influence on process reproducibility. *Eur. Food Res. Technol.* **2016**, *242*, 1005–1012. [CrossRef]
29. Walter, T.; Wieser, H.; Koehler, P. Degradation of gluten in rye sourdough products by means of a proline-specific peptidase. *Eur. Food Res. Technol.* **2015**, *240*, 517–524. [CrossRef]
30. Rai, S.; Kaur, A.; Chopra, C.S. Gluten-free products for celiac susceptible people. *Front. Nutr.* **2018**, *5*, 116. [CrossRef]
31. Yang, X.; Li, Y.; Li, S.; Oladejo, A.O.; Wang, Y.; Huang, S.; Zhou, C.; Wang, Y.; Mao, L.; Zhang, Y.; et al. Effects of multi-frequency ultrasound pretreatment under low power density on the enzymolysis and the structure characterization of defatted wheat germ protein. *Ultrason. Sonochem.* **2017**, *38*, 410–420. [CrossRef] [PubMed]
32. Zhang, Y.; Li, J.; Li, S.; Ma, H.; Zhang, H. Mechanism study of multimode ultrasound pretreatment on the enzymolysis of wheat gluten. *J. Sci. Food Agric.* **2018**, *98*, 1530–1538. [CrossRef]
33. Kjeldahl Method (ISO 5983-2005). *Animal Feeding Stuffs: Determination of Nitrogen Content and Calculation of Crude Protein Content. Part 1: Kjeldahl Method*; International Organization for Standardization: Geneva, Switzerland, 2005.
34. Adler-Nissen, J. *Enzymic Hydrolysis of Food Proteins*; Elsevier Applied Science Publishers Ltd.: Amsterdam, The Netherlands, 1986; p. 427.
35. Lowry, O.H.; Rosebrough, N.J.; Farr, A.L.; Randall, R.J. Protein measurement with the Folin phenol reagent. *J. Biol. Chem.* **1951**, *193*, 265–275. [CrossRef]
36. Jovanović, J.R.; Stefanović, A.B.; Šekuljica, N.Ž.; Jakovetić Tanasković, S.M.; Dojčinović, M.B.; Bugarski, B.M.; Knežević-Jugović, Z.D. Ultrasound Pretreatment as an Useful Tool to Enhance Egg White Protein Hydrolysis: Kinetics, Reaction Model, and Thermodinamics. *J. Food Sci.* **2016**, *81*, C2664–C2675. [CrossRef]
37. Pearce, K.N.; Kinsella, J.E. Emulsifying properties of Proteins: Evaluation of a Turbidimetric Technique. *J. Agric. Food Chem.* **1979**, *26*, 716–723. [CrossRef]
38. Stefanović, A.B.; Jovanović, J.R.; Balanč, B.D.; Šekuljica, N.Ž.; Jakovetić Tanasković, S.M.; Dojčinović, M.B.; Knežević-Jugović, Z.D. Influence of ultrasound probe treatment time and protease type on functional and physicochemical characteristics of egg white protein hydrolysates. *Poult. Sci.* **2018**, *97*, 2218–2229. [CrossRef]

39. Jovanović, J.R.; Stefanović, A.B.; Žuža, M.G.; Jakovetić, S.M.; Šekuljica, N.; Bugarski, B.M.; Knežević-Jugović, Z.D. Improvement of antioxidant properties of egg white protein enyzmatic hydrolysates by membrane ultrafiltration. *Hem. Ind.* **2016**, *70*, 419–428. [CrossRef]
40. Decker, E.A.; Welch, B. Role of ferritin as a lipid oxidation catalyst in muscle food. *J. Agric. Food Chem.* **1990**, *38*, 674–677. [CrossRef]
41. Shimada, K.; Cheftel, J.C. Determination of sulfhydryl groups and disulfide bonds in heat-induced gels of soy protein isolate. *J. Agric. Food Chem.* **1988**, *36*, 147–153. [CrossRef]
42. Stefanović, A.B.; Jovanović, J.R.; Grbavčić, S.Ž.; Šekuljica, N.Ž.; Manojlović, V.B.; Bugarski, B.M.; Knežević-Jugović, Z.D. Impact of ultrasound on egg white proteins as a pretreatment for functional hydrolysates production. *Eur. Food Res. Technol.* **2014**, *239*, 979–993. [CrossRef]
43. Woo, K.S.; Kim, H.Y.; Hwang, I.G.; Lee, S.H.; Jeong, H.S. Characteristics of the thermal degradation of glucose and maltose solutions. *Prev. Nutr. Food Sci.* **2015**, *20*, 102–109. [CrossRef] [PubMed]
44. Mahroug, H.; Ribeiro, M.; Rhazi, L.; Bentallah, L.; Zidoune, M.N.; Nunes, F.M.; Igrejas, G. How microwave treatment of gluten affects its toxicity for celiac patients? A study on the effect of microwaves on the structure, conformation, functionality and immunogenicity of gluten. *Food Chem.* **2019**, *297*, 124986. [CrossRef] [PubMed]
45. Leszczynska, J.; Lacka, A.; Szemraj, J.; Lukamowicz, J.; Zegota, H. The effect of microwave treatment on the immunoreactivity of gliadin and wheat flour. *Eur. Food Res. Technol.* **2003**, *217*, 387–391. [CrossRef]
46. Lamacchia, C.; Landriscina, L.; D'Agnello, P. Changes in wheat kernel proteins induces by microwave treatment. *Food Chem.* **2016**, *197*, 634–640. [CrossRef] [PubMed]
47. Cheftel, J.C. Chemical and nutritional modification of food proteins due to processing and storage. In *Food Proteins*; Whitaker, J.R., Tannenbaum, S.R., Eds.; Avi Publishing Co.: Westport, CT, USA, 1977; pp. 401–445.
48. Wang, C.H.; Damodaran, S. Thermal destruction of cysteine and cystine residues of egg protein under conditions of gelation. *J. Food Sci.* **1990**, *55*, 1077–1080. [CrossRef]
49. Gianfrani, C.; Mamone, G.; la Gatta, B.; Camarca, A.; Di Stasio, L.; Maurano, F.; Picascia, S.; Capozzi, V.; Perna, G.; Picariello, G.; et al. Microwave-based treatments of wheat kernels do not abolish gluten epitopes implicated in celiac disease. *Food Chem. Toxicol.* **2017**, *101*, 105–113. [CrossRef]
50. Lamacchia, C.; Di Luccia, A.; Gianfrani, C. Method for the Detoxification of Gluten Proteins from Grains of Cereals. Patent Cooperation Treaty N. PCT/IB2013/000797, 29 April 2013.
51. Ma, S.; Han, W.; Li, L.; Zheng, X.; Wang, X. The thermal stability, structural changeability, and aggregability of glutenin and gliadin proteins induced by wheat bran dietary fiber. *Food Funct.* **2019**, *10*, 172–179. [CrossRef]
52. Xiang, S.; Zou, H.; Liu, Y.; Ruan, R. Effects of microwave heating on the protein structure, digestion properties and Maillard products of gluten. *J. Food Sci. Technol.* **2020**, *57*, 2139–2149. [CrossRef] [PubMed]
53. Han, Z.; Cai, M.-J.; Cheng, J.-H.; Sun, D.-W. Effects of electric fields and electromagnetic wave on food protein structure and functionality: A review. *Trends Food Sci. Technol.* **2018**, *75*, 1–9. [CrossRef]
54. Singh, H.; MacRitchie, F. Changes in proteins induced by heating gluten dispersions at high temperature. *J. Cereal Sci.* **2004**, *39*, 297–301. [CrossRef]
55. Elmalimadi, M.B.; Jovanović, J.R.; Stefanović, A.B.; Tanasković, S.J.; Đurović, S.B.; Bugarski, B.M.; Knežević-Jugović, Z.D. Controlled enzymatic hydrolysis for improved exploitation of the antioxidant potential of wheat gluten. *Ind. Crops Prod.* **2017**, *109*, 548–557. [CrossRef]
56. Yalcin, E.; Sakiyan, O.; Sumnu, G.; Celik, S.; Koksel, H. Functional properties of microwave treated gluten. *Eur. Food Res. Technol.* **2008**, *227*, 1411–1417. [CrossRef]

foods

Review

High Protein Substitutes for Gluten in Gluten-Free Bread

Adriana Skendi [1], Maria Papageorgiou [1,*] and Theodoros Varzakas [2]

1 Department of Food Science and Technology, International Hellenic University, P.O. Box 141, 57400 Thessaloniki, Greece; andrianaskendi@hotmail.com
2 Department of Food Science and Technology, University of the Peloponnese, 24100 Kalamata, Greece; t.varzakas@uop.gr
* Correspondence: mariapapage@ihu.gr; Tel.: +30-2310013775

Abstract: Gluten-free products have come into the market in order to alleviate health problems such as celiac disease. In this review, recent advances in gluten-free bread are described along with plant-based gluten-free proteins. A comparison with animal-based gluten-free proteins is made reporting on different high protein sources of animal origin. Sea microorganisms- and insect-based proteins are also mentioned, and the optimization of the structure of gluten-free bread with added high protein sources is highlighted along with protein digestibility issues. The latter is an issue for consideration that can be manipulated by a careful design of the mixture in terms of phenolic compounds, soluble carbohydrates and fibres, but also the baking process itself. Additionally, the presence of enzymes and different hydrocolloids are key factors controlling quality features of the final product.

Keywords: gluten-free bread; plant proteins; animal proteins; microalgae; optimized bread structure; protein substitutes

Citation: Skendi, A.; Papageorgiou, M.; Varzakas, T. High Protein Substitutes for Gluten in Gluten-Free Bread. *Foods* **2021**, *10*, 1997. https://doi.org/10.3390/foods10091997

Academic Editor: Ke-Xue Zhu

Received: 16 July 2021
Accepted: 21 August 2021
Published: 25 August 2021

1. Introduction

There is an increasing number of people that sufferhealth problems from the consumption of wheat and other cereals that contain gluten as well as all their derived products. They can experience different health conditions such as celiac disease, non-celiac gluten sensitivity (NCGS), wheat allergy or irritable bowel syndrome [1].

Celiac disease is a specific immune response that is caused after consumption of gluten present in wheat, rye, barley and related grains from genetically predisposed patients. On the other hand, wheat allergy is triggered by the consumption of insoluble gliadins of wheat that react with immunoglobulin E (IgE) causing allergy symptoms that could be life-threatening. Contrary to celiac disease, wheat allergy is not reported to cause permanent damage to the gastrointestinal system [1,2]. For celiac patients and those diagnosed with wheat allergy, the only treatment is to adhere to a very strict diet that does not contain gluten or a wheat-free diet, respectively.

According to Catassi et al. [3], NCGS is a syndrome observed in patients that are not affected by either celiac disease or wheat allergy and are characterized by intestinal and extra-intestinal symptoms due to the ingestion of foods containing gluten. In their review, Barbaro et al. [4] and Biesiekierski and Iven [1] report that is still not clear if is the gluten or other wheat components that cause NCGS, but at the same time, they suspected that its prevalence to be higher than that of celiac disease. Thus, patients diagnosed with NCGS have to eliminate gluten from their diet to some extent. The difference with other celiac and allergic patients lies in the extent to which each one can tolerate the gluten depending on the severity of the symptoms he experiences.

The population affected by the celiac disease, that by necessity has to consume only gluten-free products, was estimated to be approximately 1% of the total Western population with some Western European populations showing higher prevalence [2]. On the other hand, nowadays, gluten-free foods are not consumed only by those consumers experiencing digestive problems with gluten or wheat. Specifically, another group of consumers exists that deliberately avoids or limits gluten as a part of a diet regime or other expected health benefits. Thus, the demand for gluten-free products is high and besides this, there is foreseen a continuous increase in the trade of gluten-free products reaching 8.3 billion US dollars in 2025 [5].

The food industry is focusing on the production of gluten-free products in order to fulfil this increasing need. Among the gluten-free products that are produced, bread occupies a special place. Conventional wheat bread, or bread made with other cereals such as rye, barley and oat, represents a staple food present on a daily basis in the table of the consumers. Eliminating bread constitutes a great deprivation for those following a gluten-free diet. Therefore, the production of good quality and tasty gluten-free bread represents a major challenge for the bakery industry.

The recipe for gluten-free bread varies depending on the gluten-free ingredients used. In the market, there are circulating different gluten-free mixtures. The most common wheat flour substitutes for the production of gluten-free bread are rice and/or maize flours combined with starch of different origins (e.g. potato, corn, cassava). These ingredients are the most abundant and the cheapest. Gluten-free mixtures are mainly composed of carbohydrates and lack in protein content. The latter not only affects the required daily dietary amount of proteins but also largely affects the bread structure and quality. In conventional wheat bread, the open cellular structure is due to the elasticity of the gluten that after mixing with water is able to entrap carbon dioxide (CO_2) produced by yeasts during fermentation in the leavened dough, causing the dough to rise. Although being a minor component of gluten, the glutenin macro polymer (GMP) fraction is considered the main contributor of the elastic properties observed in the wheat dough, playing an important role in breadmaking [6].

Besides, the improvement of the nutritional profile of gluten-free products has become an important target due to a low protein and fibre content and a high fat and salt content raising health issues to consumers suffering from other diseases such as cardiovascular diseases or diabetes. However, during the last years, efforts have been made to improve this, e.g., the level of dietary fibre content [7,8]. A gluten-free bread that is made of maize flour instead of wheat flour is also considered low in the amino acids lysine and tryptophan and high in other large amino acids such as leucine and valine [9]. Therefore, the amino acid profile must also be considered.

Since gluten, which is responsible for obtaining raised bread loaves, is missing in gluten-free breads, its structure deteriorates from that of conventional bread. It is difficult to mimic the properties of gluten with other proteins. A great number of studies have tried to improve both the quality and the nutritional profile of gluten-free bread by increasing protein content using appropriate protein concentrates or isolates obtained from microorganisms, animals and plants. According to Akharume et al. [10], protein ingredients in commerce fall into three categories; protein flours, protein concentrates and protein isolates that contain 10–20%, 55–60% and more than 80% protein content, respectively. Gorissen et al. [11] have monitored protein content and amino acid composition of some of the commercially available plant-based protein isolates such as oat, lupin, wheat, hemp, microalgae, soy, brown rice, pea, corn, potato, milk, whey, caseinate, casein, egg and human skeletal muscle protein. They observed that the content of essential amino acids of plant-based protein isolates was lower than that of animal-based proteins. In addition, the profile of amino acids differs, with methionine and lysine being higher in animal-based proteins. They suggest the use of a balanced combination of different plant-based proteins in order to increase the quality of protein in the blend.

The literature reports the development of a range of gluten-free breads enriched with alternative sources for protein trying to improve the quality and taste of the final product. In a search with keywords "gluten-free AND protein AND isolate" "gluten-free AND protein AND concentrate" present in the article title, keywords and abstract, the scopus database returned 74 and 53 total results, of which 45 and 32, respectively, were published from 2016 until now (June 2021). Thus, this subject represents an issue that attracts researchers' interest during these last five years. In this review, we gathered published data and report recent advances made in the development of gluten-free products with high protein substitutes for gluten, with the main focus on bread.

2. Plant-Based Gluten-Free Protein

Many studies found in the recent literature report the use in a gluten-free bread recipe of highly concentrated protein sources of plant origin (Table 1). The use of these sources intends to improve not only bread quality but also its nutritive values. Using protein-rich sources of plant origin has gained interest due to the limitation of animal origin counterparts due to their allergenic character. Generally, the gluten-free flours of plant origin differ in the content and quality of protein. Wu et al. [12] prepared breads from a range of gluten (white wheat, wholemeal wheat, spelt and rye) and gluten-free (lupin, buckwheat, chickpea, amaranth) flours standardized at 10% protein with maize starch. They observed differences in the proportions of essential amino acids (0.31–0.35 and 0.34–0.41 for gluten and gluten-free flours, respectively) with lysine and arginine showing much higher levels in gluten-free flours compared to glutenous flours, whereas the opposite was observed for proline. Low levels of proline in gluten-free flours were considered responsible for the lower rising capacity during proofing.

Table 1. Different high protein sources of vegetable origin reported in recent published literature to improve gluten-free bread.

Source	Concentration (% in the Starchy Flour Mixture)	Control Bread	Literature
Gluten-free cereals			
rice protein	30%	100% maize starch	[13]
rice protein	5%, 10%	50% rice flour: 50% maize starch	[14]
rice bran protein concentrate	2%, 4%	100% rice flour	[15]
Zein	15%	15% vital wheat gluten: 85% rice starch	[16]
Zein	15%	100% wheat flour, 100% starch from: rice, maize, potato	[17]
Zein	2.5%, 5%, 10%	100% wheat flour, ~88% maize starch: ~12% potato starch	[18]
Legumes			
Pea protein	30%	100% maize starch	[13]
Pea protein	5%, 10%	50% rice flour: 50% maize starch	[14]
Pea protein	~10%	80% maize starch: 20% potato starch	[19]
Pea protein	2%	100% potato starch	[20]
Lupin protein	~10%	80% maize starch: 20% potato starch	[19]
Lupin protein	2%	100% potato starch	[20]

Table 1. *Cont.*

Source	Concentration (% in the Starchy Flour Mixture)	Control Bread	Literature
Soy protein	~10%	80% maize starch: 20% potato starch	[19]
Soy protein	2%, 4%, 6%	100% rice flour	[21]
Soy protein	4%	100% rice flour	[22]
Soy protein	2%	100% potato starch	[20]
Oil seeds			
Rapeseed protein	6%, 9%, 12%, 15%	80% corn starch: 20% potato starch	[23]
Rapeseed protein	6%, 9%, 12%, 15%	80% corn starch: 20% potato starch	[24]
Canola protein extract	3%, 6%, 9%	100% wheat flour, 100% rice flour	[25]
Sunflower protein	5%, 10%, 20%	70% rice flour: 30% maize starch	[26]
Tubers			
Potato protein	2%, 6%, 10%	80% maize starch: 20% potato starch	[27]
Potato protein	2%	100% potato starch	[20]

In recent years, different plant-based high protein isolates have been used for gluten-free bread preparation (Table 1). Rice flour is considered as the most suitable among the cereal flours for the production of gluten-free bread because it is considered hypoallergenic and has high digestibility besides a bland taste and white colour that do not affect the final bread quality. Derived from rice flour, rice protein isolate is used as a safe source of protein in gluten-free bread production [13]. On the other hand, rice bran protein concentrate obtained from rice bran that is a by-product of rice production is also considered a non-allergenic protein. It is utilized for increasing the protein content of gluten-free bread [15]. It is isolated from rice bran via an alkaline-acid extraction technique resulting in 68% protein on a dry basis. Zein, a protein from another non-gluten cereal, maize, was used to provide extensibility to starch-based doughs [16,17] and firmness to the bread crumb comparable to wheat breads [18].

Besides non-gluten cereals, proteins from legumes have attained the interest of researchers due to their adequate protein profile. Although being high in lysine, they are deficient in amino acids methionine, cystine and cysteine [28].

Soy protein isolate is obtained by extraction from the soy bean, and is of high biological value due to high amounts of the essential amino acids lysine and methionine [29]. It has the ability to alter water absorption of the dough mixture and thereby impacts its rheology. Moreover, soy protein isolates have high foaming capacities, as well as high foam stability [20]. Contradictory results are reported in the literature about the resulted bread volume and crumb structure [20,22]. Horstmann, Foschia and Arendt [20] report that soy-protein-enriched bread (2% on potato starch) produces bread with a low volume and a dense crumb structure and a low consumer acceptance score, whereas Masure, Wouters, Fierens and Delcour [22] showed that the addition of soy (4% on rice flour) produced bread with a similar volume of that of the control bread made of rice flour and in a homogeneous crumb structure with very large gas cells.

Lupin protein represents another protein-rich source gaining interest in gluten-free bread production [19]. It was noted that the extraction method used to obtain lupin protein concentrates/isolates plays a significant role in the functional qualities of lupin proteins, including their chemical composition, emulsification, rheology and thermal properties [30].

Pea protein, an extract from pea seeds, has become a popular product in the food industry due to its well-balanced amino acid profile rich in the essential amino acid lysine [31]. According to Horstmann, Foschia and Arendt [20], the emulsifying capacity of legume isolates decreases in the following order: soy > lupin > pea, whereas emulsifying stability decreases in the order: lupin protein > soy protein > pea protein.

A highly concentrated protein of legume origin is being used recently not only to equilibrate the amino acid profile of gluten-free bread but also to strengthen the protein matrix in these breads. Micrographs in Figure 1 reveal the matrix created in the gluten-free breads with different protein sources. Compared to the control bread made of rice flour and maize starch (Figure 1f), cross-sectional photographs of gluten-free bread with added rice and pea protein (Figure 1g and h, respectively) show a higher number of small filaments connected to starch granules. These filaments are more evident for rice protein and represent protein molecules that link starch granules [14]. Incorporation of rice protein until 10% increases the specific volume of the gluten-free bread whereas when added 5% of pea proteins, there is not observed any difference with the control. Any further increase (10%) decreased the specific volume of the bread [14]. In their study, Ziobro, Juszczak, Witczak and Korus [19] observed that the pea and soy protein addition at 10% level has a negative effect on bread volume, whereas lupine protein showed similar values with control bread (maize and potato starch). The crumb structure of pea, soy and lupine breads was more porous and has more large pores than control bread. Nevertheless, lupine is considered a better option due to the lower hardness of crumb than pea and soy.

The incorporation of legume proteins produces gluten-free breads with darker crumb and crust [13]. Among proteins of vegetal origin added to gluten-free bread, pea protein decreases the luminosity of the bread crust to a greater extent than rice, due to the higher lysine content that reacts with carbohydrates producing more coloured products during the Maillard reaction [13]. Although different flavours are produced during the baking of breads there are not all considered attractive to the consumers, it was observed that the acceptability of panellist decreases in the following range: lupine protein, pea protein, soy protein.

Different regulatory authorities such as European [32], Australian and New Zealand [33] legislation consider rapeseed protein isolate (containing 96% on dry basis protein) as a novel food ingredient for use in food products, whereas the US FDA gives GRAS status [34]. Its inclusion in the gluten-free bread recipe provides not only valuable amino acids but also affects the pasting characteristics of starch and modifies rheological characteristics of dough [24] as well as improves quality characteristics, sensory attributes and storage of gluten-free bread [23]. Levels of addition higher than 9% resulted in higher bread volume and lower hardness of the breads during storage compared to control bread. In another study, when canola proteins were added to a white rice flour bread recipe up to 6% resulted in an improvement of the technological properties of the dough and the resulted bread [25].

Sunflower protein concentrate represents another source of protein. Containing more than 75% proteins, it is obtained by extraction from the cake that remains after oil extraction from the seeds. Its high water-holding capacity decreased bread hardness during the storage, whereas the dark colour decreased the brightness of the bread making them more attractive to the panellist due to the similarity with whole flour [26].

Potato protein is extracted from the remains after the removal of starch [35]. The nutritional quality of potato protein is considered high in quality because it contains a high proportion of lysine, approaching the quality of proteins in eggs [36]. This isolate (>90% protein dry base) was employed by Witczak, Juszczak, Ziobro and Korus [27] for the production of gluten-free bread.

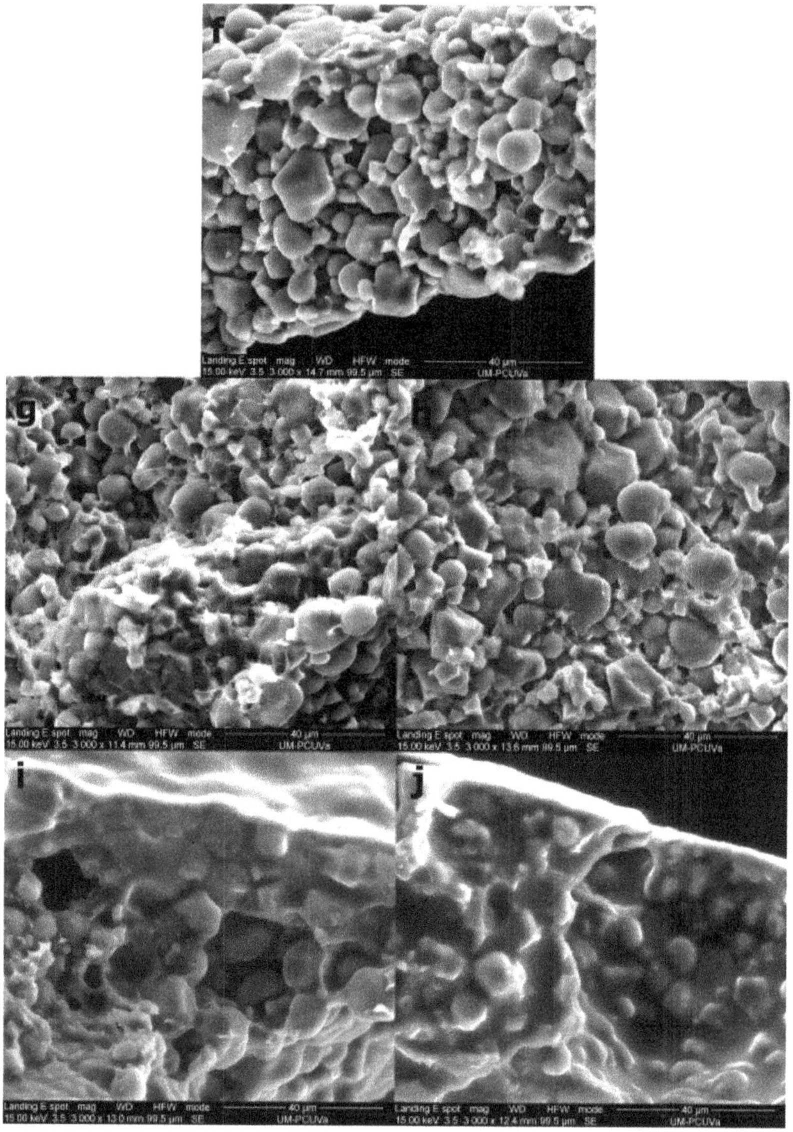

Figure 1. Micrographs of crust cross-section at 3000× magnification. Images correspond to breads supplemented with 10% protein and control sample: Control (50% rice flour: 50% maize starch) (**f**), Rice protein 10% (**g**), Pea protein 10% (**h**), Egg white powder 10% (**i**), Whey protein 10% (**j**) [14].

3. Animal-Based Gluten-Free Protein

Among the high protein sources of animal origin, those based on milk and eggs are widely used to increase the protein content and improve amino acids' profile in gluten-free breads. These characteristics made these sources highly popular for gluten-free breads in the recent scientific literature (Table 2). Proteins of animal origin have good solubility, high emulsifying and foaming capacity and high stability. This fact is observed in surface photographs (Figure 1) where a film is visible covering the starch granules in the case when egg white powder and whey protein were used. There is observed a very distinct difference when compared to the control bread or to breads with rice and pea protein isolates [14]. Animal proteins such as whey protein and egg white powder were found to decrease the crispness in the crust of gluten-free breads [14]. The type of protein added affects the crust colour since it reacts with the carbohydrates triggering Maillard reactions. In general, when protein sources of animal-origin are added to a gluten-free recipe, they decrease the luminosity of crust, producing breads with a darker and more reddish crust than the control bread [13,19,37]. There are differences in the decrease in luminosity among the different protein sources, with whey protein yielding a darker crust than egg white powder due to the high lysine content that this protein contains and the importance it has in the Maillard reaction [14]. In the literature, there are reported contradictory results for the crust and crumb colour, possibly due to the differences in the level of proteins added [13,15,19,37]. Generally, the higher the protein addition level, the darker the colour [37].

Table 2. Different high protein sources of animal origin in recent published literature to improve gluten-free bread.

Source	Concentration (% of Starchy Flour Mixture)	Control Bread	Literature
Dairy			
whey protein	10%, 20%, 30%	100% wheat flour, 50% cassava starch: 50% chickpea flour	[37]
whey protein	30%	100% maize starch	[13]
whey protein	5%, 10%	50% rice flour: 50% maize starch	[14]
whey protein	2%, 4%, 6%	100% rice flour	[21]
whey protein	12% *	50% quinoaflour: 50% (maize starch, potato starch, modified maize starch, modified potato starch)	[38]
Eggs			
egg white powder	30%	100% maize starch	[13]
egg white powder	5%, 10%	50% rice flour: 50% maize starch	[14]
egg white powder	2%, 4%	100% rice flour	[15]
egg white powder	~10%	80% maize starch: 20% potato starch	[19]
egg white powder	5%, 10%, 15%	Commercial gluten-free flour (mixture of garbanzo bean flour, potato starch, tapioca flour, whole grain sorghum flour and fava bean flour)	[39]
egg white powder	4%	100% rice flour	[22]
Other animal sources			
collagen	~10%	80% maize starch: 20% potato starch	[19]

* the added protein amount is calculated based on the total flour mixture, and its addition is associated with removal of only the same amount of starch fraction.

The volume and the texture parameters of the breads with added animal protein content vary depending on the protein type and the level of addition [13,15,19,38,39]. For example, Ziobro, Juszczak, Witczak and Korus [19] report an increase in the specific volume of breads when 10% albumin was added. Contrarily, Phongthai, D'Amico, Schoenlechner and Rawdkuen [15] report an increase in the specific volume of bread when the level increases to 2% and then a decrease for a further increase to 4%, whereas Sahagún and Gómez [13] report a decrease at the 30% addition level. A comparison is difficult since in the above studies, the egg protein source was added in different flour mixtures, and different levels of hydration were applied resulting in different dough systems. The same is valid for the texture and the structure of these breads. Moreover, the composition of the high protein source can affect the quality of the bread. Han, Romero, Nishijima, Ichimura, Handa, Xu and Zhang [39] compared two egg white sources of similar composition but differing in water solubility and reported that the powder with more water-soluble protein aggregates was associated with larger improvement in bread quality. On the other hand, Masure, Wouters, Fierens and Delcour [22] reported that regular egg white powder and dry heated egg white powder produced similar bread loaf volume, but regular egg white showed lower firmness during storage than the dry heated counterpart. Another factor that must be taken into consideration is the basic flour mixture used to prepare the gluten-free bread. In their study, Aprodu and Banu [38] observed that the type of starch also affects the efficiency of whey protein on thermo-mechanical properties, specific volume and firmness of the bread crumb.

Collagen is reported as more effective than albumin to reduce the hardness and prevents staling of gluten-free breads [19]. Generally, the hardness of the crumb is increased with increasing protein concentration [37], suggesting the need for optimization of the level of protein enrichment.

4. Sea Microorganisms- and Insect-Based Proteins

Another protein source used for the preparation of gluten-free breads can be obtained from algae, seaweed and insects (Table 3). Seaweeds or macroalgae are complex multicellular organisms that grow in salt and marine environments, and most of them can be used for direct human nutrition.

Table 3. Different high protein sources from algae and insects reported in recent published literature to enrich gluten-free breads.

Source	Concentration (% of Starchy Flour Mixture)	Control Bread	Literature
Algae			
Chlorella powder (*Chlorella sorokiniana*)	2.1%, 4.2%	25% rice flour: 58.3% maize starch: 16.7% pea flour	[40]
Microalgae powder (*Nannochloropsis gaditana* L2; *Chlamydomonas* sp. EL5)	1%, 3%	31% rice flour: 46% buckwheat: 23% potato starch	[41]
Spirulina (strain LEB -18)	1–4%	100% rice flour	[42]
Brown algae powder (*Ascophyllum nodosum*)	2%, 4%, 6%, 8%, 10%	45% white rice flour: 45% maize flour: 10% millet flour	[43]
Insects			
Cricket powder (*Acheta domesticus*)	2%, 6%, 10%	80% maize starch: 20% potato starch	[44]
Cricket powder (*Acheta domesticus*)	5.5%	80% maize flour: 20% rice flour	[45]
Cricket powder (*Gryllus assimilis*)	10%, 20%	70% rice flour: 30% maize starch	[46]

Microalgae are considered as a rich source of protein of high quality (rich profile of essential amino acids), besides other bioactive compounds (e.g., polyunsaturated fatty acids, carotenoids, vitamins) [47]. Becker [47] in his review reports that the protein content of different algae that can be used in the food industry varies 6–71% of the dry matter, but most of them contain more than 28%. In addition, the amino-acid profile of microalgae protein is considered well-balanced and comparable with that recommended by WHO/FAO and that of eggs and soybean [47]. Some varieties of macroalgal species have been used to obtain bioactive peptides that exert beneficial health effects beyond those benefits associated with basic nutritional values [48].

Among microalgae, Chlorella species were reported to have a high amount of protein in combination with an adequate amount of essential amino acids, especially higher levels of the essential amino acids lysine and tryptophane, in order to provide adequate nutrition [49]. Besides protein, they contain in high amounts carbohydrates (8–64%) and lipids (2–22%) [47]. In order to minimize production cost, instead of refined protein, the industry tries to promote the use of algal biomass as a whole in powder form and not as a protein isolate. The use of these powders was reported to alter sensory characteristics of the bread such as colour, aroma and flavour as well as texture [40,41]. Breads enriched with C. sorokiniana were characterised by low luminosity and a deep green colour [40]. Similarly, other authors observed a decrease in bread luminosity and an increase in green and yellow colour when two other microalgae were used as the species, observing differences that are due to both species and level of addition [41]. Changes in dough and bread colour are due to the presence of pigments (mainly chlorophyll) in microalgal biomass.

In their study, Khemiri, Khelifi, Nunes, Ferreira, Gouveia, Smaali and Raymundo [41] evaluated the effect of the incorporation of two different microalgae (*Nannochloropsis gaditana L2* and *Chlamydomonas* sp. EL5) on the dough properties. They suggest that for levels 1% and 3%, there was no need to adjust the water content and observed that dough mixing curves of microalgae-added doughs were similar to that of the control. The incorporation of microalgae increased dough development time and stability of the doughs, but differences were noticed based on microalgae strains and the level of addition. On the other hand, texture parameters (firmness, adhesiveness and cohesiveness) of the doughs were not affected by microalgae addition. Contrarily, the addition of microalgae significantly increased the firmness and adhesiveness of the gluten-free bread crumb as microalgal biomass incorporation increased, but without producing significant changes in bread cohesiveness. These changes were considered positive by the authors since it helped to strengthen the texture of the gluten-free bread by reinforcing the protein structure of the bread and reduce the brittleness that characterizes gluten-free breads [41]. In another study, breads with microalgae *C. sorokiniana* at a higher level of addition (4.2%) showed increased crumb porosity in comparison with the control bread. This behaviour was attributed to the high protein and lipid contents present in the powder [40]. Generally, besides the odour, taste and texture, panellists appreciated the intense green colour of the 3% supplemented bread scoring higher than the control [41].

Dough and bread pH is a parameter of great importance since it determines the growth kinetic of yeasts/microorganisms during fermentation, and the final bread pH is linked with bread taste and storage but often is not reported. The addition of microalgal biomass increased the pH of the gluten-free dough with values varying from 5.77 in control to 6.05 and 6.01 for *Chlamydomonas* sp. EL5 and *N. gaditana* L2, respectively, when the microalgae were added at 3% [41]. Similarly, Różyło, Hameed Hassoon, Gawlik-Dziki, Siastała and Dziki [43] observed an increase in the pH of dough that ranged from 5.05 in control dough to 5.2 in the dough with 10% brown algae added. These values differ due to the different initial mixture used to prepare the gluten-free bread as well as the composition and level of the microalgae added. The addition of brown algae increased the bread volume [43]. The authors suggested that algae components are hydrated, swelled and gelatinized at a slower rate compared to the control flour mixture. Moreover, they report that it was due to the algal protein enrichment that improved the rheological properties of dough and increased the gas retention capacity of the dough that resulted in higher bread volume. Since algae contain pigments, their addition decreased the lightness of bread crumb. The brown algae addition decreased bread firmness and the staling rate, whereas the bread elasticity was increased with algae addition. Nevertheless, sensory tests demonstrated that the addition of much less amount (2%) of brown algae can lead to an acceptable gluten-free bread.

Spirulina was another microalgae used to enhance the protein content by 20% in rice bread when its level was increased from 1 to 4% [42]. Besides the decrease in the brightness of bread with increasing spirulina level, an increase in the specific volume of the bread was observed. Dough elasticity influences gas retention capacity and the specific volume of breads. Panellists preferred gluten-free bread with 1% content of spirulina than that with 4% because of the lighter crumb colour, although the general score was similar for breads with 1 and 4% spirulina.

Red seaweed (*Palmaria palmata)* contains 9–26% protein, and the amount of vital amino acid lysine can reach 5.9 g/100 g of total amino acids [50]. Hydrolyzation of a protein extract from *P. palmata* with food-grade enzymes released renin inhibitory peptides [51,52]. The protein hydrolysates from *Palmaria palmata* rich in inhibitory peptides that were used to increase the nutritional profile of wheat bread [51] could be used as an alternative protein source.

Besides microalgae, insects have attracted researchers as a new source of protein since they can contain more than 40–75% of their dry base. In addition, their protein is 77–98% digestible from the human organisms and rich in (46–96%) essential amino acids [53]. Among the insects, the Orthoptera (Grasshoppers and locusts)were found to have the higher amount of proteins (61–77) [53].

One study where cricket powder (*Acheta domesticus*) was used to replace starch (from rice and corn) at levels 2%, 6% and 10% resulted in exceeding a two-, four- and seven-fold increase in nutritional value, in terms of the protein content, compared to control bread [44]. In addition, a significant decrease in crumb lightness was observed. This replacement induced changes at the molecular level, stabilizing water transport, that delayed the process of bread staling and resulted in reduced bread hardness [54]. In another recent study [45], where the cricket powder was used for the fortification of gluten-free mixture made of rice and maize flour with protein at about 5.5%, there was observed that the bread developed a unique bouquet of volatile organic compounds. The aroma compounds developed were considered similar to those of the reference dough regarding acetoin and acetate, slightly higher ethanol and lactate and a little lower 1,4-butanediol content. In addition, cricket-powder-enriched samples developed a typical flavouring profile made by nuances of hexanoic and nonanoic acid, 2,4-nonadienal, 1-hexanol, 1-heptanol, and 1-octen-3-ol, 2,4-butanedione, 2-heptanone and 3-octen-2-one. Besides this, cricket powder inclusion in the gluten-free recipe increased the amount of soluble proteins in cricket dough compared to the reference dough.

Another type of cricket powder (*Gryllus assimilis*) was used at the 10 and 20% level of addition in a mixture of rice flour and maize starch to improve the protein content of the gluten-free bread [46]. The authors observed that enrichment with cricket powder increased the hardness of the bread due to the formation of a more stable structure compared to the control. Nevertheless, when added at the same levels with lentil and buckwheat flour, the increase in the hardness derived from the cricket powder was less. Moreover, when compared to lentil and buckwheat flour, as a protein source, cricket flour was more efficient to improve the cohesiveness and the springiness of the breads. Bread with high cohesiveness keeps its integrity during slicing and mastication whereas that with high springiness has the ability to return to its original shape after compressing. Moreover, due to higher protein content, breads with cricket powder have higher porosity compared to respective breads made with lentil and buckwheat flour. The authors reported that in order to obtain the best in terms of bread quality, no oil should be used in the case of cricket powder since it is rich in lipids.

5. Optimizing the Structure of Gluten-Free Bread with Added High Protein Sources

It is advisable to use transglutaminase (TGase) in gluten-free bread containing added protein sources because it helps the creation of a protein network in the gluten-free dough [27,42,55–57]. The simultaneous addition of TGase with protein was reported to be more effective than adding them separately [57]. Transglutaminase is an enzyme that forms ε-(γ-glutamyl) lysine cross-links when acting with proteins [58], affecting protein properties such as water-holding capacity, gelation capability, thermal stability, etc. This behaviour is utilized to promote a protein network that in its turn improves viscoelastic properties of the gluten-free dough. The amount and the enzyme, its origin as well as the protein substrate largely affect the efficiency of the structure created [59]. Selmo and Salas-Mellado [42] reported that when the concentration of the protein source (spirulina) is high, the amount of transglutaminase should be less in order to increase the specific volume and decrease the firmness of the gluten-free bread. In a previous study, Dłużewska, Marciniak-Lukasiak and Kurek [56] reported that higher levels of TGase (10U/g protein) negatively affected the texture, staling and sensory characteristics of gluten-free breads with soy and whey protein isolate. Moreover, they observed that microbiological TGase was more efficient in increasing the specific volume and crumb porosity of gluten-free bread with soy protein isolate compared to whey protein isolate [56]. In another study, when the enrichment of rice flour with three different protein sources (soy, casein and whey protein isolate)was studied, the addition of TGase further increased the specific volume of bread, reduced the second proofing time and decreased the hardness of the crumb [57]. On the contrary, the findings of Moore, Heinbockel, Dockery, Ulmer and Arendt [59] reported skim milk powder and egg powder but not soya flour as a good substrate for TGase, which allowed the creation of substantial protein networking that improved loaf volume, crumb characteristics and the appearance of gluten-free breads.

In order to obtain gluten-free bread with a high specific volume, it is of great importance to adjust the hydration of the doughs based on the final selected recipe [60]. Especially when a high protein source is added, it will alter the water distribution among the recipe components. This variation will be dependent on the difference in water-binding capacities of proteins in the recipe. The ability of proteins to absorb water affects dough rheology and bread volume. Contrary to the wheat-based dough, the amount of water in the case of gluten-free breads cannot be calculated based on farinograph consistency. Baking tests of gluten-free breads prepared with the stepwise increase/decrease of water are generally needed. Nevertheless, it was observed that more water is needed to be used in a bread recipe with vegetal proteins than the breads with animal proteins to obtain maximum values of specific volume [13]. Water levels of 150% and 115% are needed when pea and rice protein are added at the 30% level, compared to 90% needed for the control (100% maize starch). For the same level of addition, the animal proteins egg white and whey protein needed 85% and 40% water, respectively. Although, Sahagún and Gómez [13] and Bravo-Núñez et al. [61] optimized the hydration level of bread, they use a very high level of protein source (30%) that can be blamed for the negative effect observed in bread specific volume and texture. Thus, besides the moisture content, optimization of the addition level depending on the basic flour mixture is also needed.

Besides enzymes, the addition of hydrocolloids was found to be very effective in order to improve the gluten-free bread structure when a rich source of protein is added. Hydrocolloids have the ability to increase the viscosity and water holding ability of the dough system due to high molecular weight and help to create a more stable structure.

Xanthan gum is reported to be used together with rich protein for better results [56,59]. In these studies, the optimal recommended amounts of xanthan vary from 0.75 to 1.5 to be added no matter what type of protein was used (skim milk powder, soy flour, egg powder, soy protein isolate, whey protein isolate). Moreover, methylcellulose is another hydrocolloid used in gluten-free mixtures together with rich protein sources [42]. Selmo and Salas-Mellado [42] reported that optimal levels to be used are above 1.5% but less than 2.2%. The simultaneous addition of proteins and hydroxypropyl methylcellulose (HPMC) improves the porosity of gluten-free bread, being similar to that of wheat bread [21]. According to Manik and Nur [62], HPMC and xanthan gum are considered the most suitable hydrocolloids for a gluten-free bread with the first being more effective in increasing the volume of the bread and producing a softer crumb.

In addition, the extraction procedure, as well as a treatment performed on the protein source, could affect the properties of proteins that are of great importance for the dough behaviour and final bread quality. Zein's limited ability to improve dough extensibility was surpassed by applying thermal treatment [16] and extrusion [17]. These treatments improved the gluten-free dough structure. In Figure 2, one can visualize the microstructure of the dough during extension [16]. Dough samples with rice starch and untreated zein showed a higher number of broken fibrous linkages during stretching that was not observed in thermally treated zein at 160 °C (Z-V-160) or in samples with added wheat gluten, as indicated by white arrows. This suggests that thermally treated zein was able to create a more similar structure to gluten compared to the untreated one, and the thermal treatment could further improve the quality of bread.

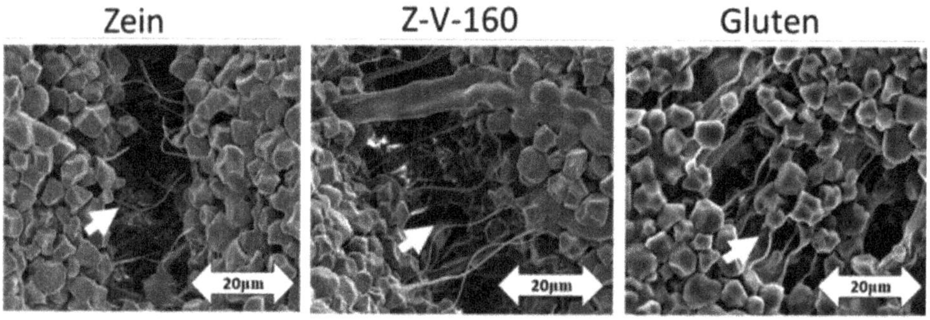

Figure 2. Images of dough samples containing starch with zein, thermally-treated-zein (Z-V-160) and gluten [16].

From the abovementioned studies, one can conclude that the right combination of the flour source rich in carbohydrates, protein source and hydrocolloids/enzymes in addition to the right hydration level and possible protein isolate treatments could play an important role in the optimal structure-forming of breads.

6. Protein Digestibility

Simonato et al. [63] reported that protein-specific immune responses in people with wheat "allergy" were due to reactivity caused primarily to baked crumb and crust and not to dough. Thus, enhancing protein digestion techniques can lower the allergenic effects of gluten proteins. Wu, Taylor, Nebl, Ng and Bennett [12] observed differences between the gluten and gluten-free flours in terms of digestibility and size distribution of undigested peptides. Gluten-free products (pre-mix, doughs and breads) differ in their counts of total undigested peptides from the respective gluten products; in gluten containing products, they remained constant for pre-mix to baked breads while in non-gluten ones, digestibility was increased significantly during proofing but remained unchanged during baking.

Digestibility of bread proteins is affected by the interaction of macro-nutrients and micro-nutrients during all stages of bread preparation such as mixing, proofing and baking. During baking, digestibility of proteins is decreased due to the denaturation and reaction between neighbouring proteins and carbohydrates (reducing sugars) [64]. Phenolic compounds were considered a significant factor that affected protein digestibility [12,65–67] and in the second place were listed the carbohydrates [12]. Phenolics affect recognition sites of digestive enzymes during mixing and mediate protein-protein cross-linking and thus decrease digestibility. On the other hand, carbohydrates (reducing sugars) responsible to trigger Maillard reactions (mostly at the surface or the bread crust) with proteins contribute to a lesser extent, especially during moderate baking conditions [12]. Besides the aforementioned factors, the presence of some micro-compounds such as enzyme inhibitors or phytic acid present in some of the gluten-free flours affect protein digestibility of breads since proteins are bound by phytic acid in insoluble binary and ternary structures, making them unavailable for digestion [68].

In their study, Wu, Taylor, Nebl, Ng and Bennett [12] suggest an appropriate design of formulation in order to improve protein digestion and increase protein availability. The increased ratio of soluble carbohydrate to protein, limiting the amount of fibre and phenolics, could improve protein digestion, but these changes affect negatively other dietary components of bread (antioxidants and fibre content). On the other hand, maximizing gas production and retention during fermentation could limit the opportunity for protein cross-linking since it helps to maximize the separation of parallel protein strands. In addition, modification of baking conditions (moisture heating) could modify protein digestibility. Thus, the availability of proteins can be increased by applying a careful design of the mixture as well as the baking process.

7. Conclusions

In this review, plant-based gluten-free proteins are compared to animal-based gluten-free proteins, with the main focus on gluten-free bread. Protein digestibility issues are very important in this context and should be taken seriously into account. Moreover, the optimization of the structure of gluten-free bread with added high protein sources should be considered along with factors such as hydration, the protein extraction procedure, viscosity and water holding ability of the dough system. The presence of enzymes and different hydrocolloids are key factors controlling the above parameters, while the digestibility of the added protein is also an issue for consideration that can be manipulated by a careful design of the mixture in terms of phenolic compounds, soluble carbohydrates and fibres but also the baking process itself.

Author Contributions: Conceptualization: M.P., A.S. and T.V.; Writing—original draft preparation, A.S., M.P., T.V.; Writing—review and editing: M.P., A.S., T.V. All authors have read and agreed to the published version of the manuscript.

Funding: This research received no external funding.

Acknowledgments: The authors gratefully acknowledge COST Action 18101 SOURDOMICS—Sourdough biotechnology network towards novel, healthier and sustainable food and bioprocesses (https://sourdomics.com/; https://www.cost.eu/actions/CA18101/), accessed on 1st June 2021, where the author M.P. is the vice-leader of the working group entitled "Recovery, characterization and selection of autochthonous conventional & nonconventional (pseudo)cereal seeds "; the author A.S. is a beneficiary of the Short-Term Scientific Mission "Assessing the technological role of sourdoughs or enzymes in cereals based products" and is supported by COST (European Cooperation in Science and Technology) (https://www.cost.eu/). COST is a funding agency for research and innovation networks.

Conflicts of Interest: The authors declare no conflict of interest.

Abbreviations

Non-celiac gluten sensitivity (NCGS); Carbon dioxide (CO_2); glutenin macro polymer (GMP); hydroxypropyl methylcellulose (HPMC); immunoglobulin E (IgE); Total Nitrogen (TN), Transglutaminase (TGase); World Health Organization (WHO).

References

1. Biesiekierski, J.R.; Iven, J. Non-coeliac gluten sensitivity: Piecing the puzzle together. *Unit. Eur. Gastroenterol. J.* **2015**, *3*, 160–165. [CrossRef]
2. Dubé, C.; Rostom, A.; Sy, R.; Cranney, A.; Saloojee, N.; Garritty, C.; Sampson, M.; Zhang, L.; Yazdi, F.; Mamaladze, V.; et al. The prevalence of celiac disease in average-risk and at-risk Western European populations: A systematic review. *Gastroenterology* **2005**, *128*, S57–S67. [CrossRef] [PubMed]
3. Catassi, C.; Elli, L.; Bonaz, B.; Bouma, G.; Carroccio, A.; Castillejo, G.; Cellier, C.; Cristofori, F.; De Magistris, L.; Dolinsek, J.; et al. Diagnosis of Non-Celiac Gluten Sensitivity (NCGS): The Salerno Experts' Criteria. *Nutrients* **2015**, *7*, 4966–4977. [CrossRef] [PubMed]
4. Barbaro, M.R.; Cremon, C.; Stanghellini, V.; Barbara, G. Recent advances in understanding non-celiac gluten sensitivity. *F1000Res* **2018**, *7*, F1000 Faculty Rev-1631. [CrossRef]
5. Wunsch, N.-G. Gluten-Free Food Market Value Worldwide 2020–2025. Available online: https://www.statista.com/statistics/248467/global-gluten-free-food-market-size/ (accessed on 1 June 2021).
6. Skendi, A.; Papageorgiou, M. Chapter 1-Introduction in wheat and breadmaking. In *Trends in Wheat and Bread Making*; Galanakis, C.M., Ed.; Academic Press: Piscataway, NJ, USA, 2021; pp. 1–27. [CrossRef]
7. Melini, V.; Melini, F. Gluten-Free Diet: Gaps and Needs for a Healthier Diet. *Nutrients* **2019**, *11*, 170. [CrossRef] [PubMed]
8. Allen, B.; Orfila, C. The Availability and Nutritional Adequacy of Gluten-Free Bread and Pasta. *Nutrients* **2018**, *10*, 1370. [CrossRef]
9. Van Hees, N.J.M.; Giltay, E.J.; Tielemans, S.M.A.J.; Geleijnse, J.M.; Puvill, T.; Janssen, N.; van der Does, W. Essential amino acids in the gluten-free diet and serum in relation to depression in patients with celiac disease. *PLoS ONE* **2015**, *10*, e0122619. [CrossRef]
10. Akharume, F.U.; Aluko, R.E.; Adedeji, A.A. Modification of plant proteins for improved functionality: A review. *Compr. Rev. Food Sci. Food Saf.* **2021**, *20*, 198–224. [CrossRef]
11. Gorissen, S.H.M.; Crombag, J.J.R.; Senden, J.M.G.; Waterval, W.A.H.; Bierau, J.; Verdijk, L.B.; van Loon, L.J.C. Protein content and amino acid composition of commercially available plant-based protein isolates. *Amino Acids* **2018**, *50*, 1685–1695. [CrossRef]
12. Wu, T.; Taylor, C.; Nebl, T.; Ng, K.; Bennett, L.E. Effects of chemical composition and baking on in vitro digestibility of proteins in breads made from selected gluten-containing and gluten-free flours. *Food Chem.* **2017**, *233*, 514–524. [CrossRef]
13. Sahagún, M.; Gómez, M. Assessing Influence of Protein Source on Characteristics of Gluten-Free Breads Optimising their Hydration Level. *Food Bioproc. Technol.* **2018**, *11*, 1686–1694. [CrossRef]
14. Pico, J.; Reguilón, M.P.; Bernal, J.; Gómez, M. Effect of rice, pea, egg white and whey proteins on crust quality of rice flour-corn starch based gluten-free breads. *J. Cereal Sci.* **2019**, *86*, 92–101. [CrossRef]
15. Phongthai, S.; D'Amico, S.; Schoenlechner, R.; Rawdkuen, S. Comparative study of rice bran protein concentrate and egg albumin on gluten-free bread properties. *J. Cereal Sci.* **2016**, *72*, 38–45. [CrossRef]
16. Federici, E.; Selling, G.W.; Campanella, O.H.; Jones, O.G. Thermal treatment of dry zein to improve rheological properties in gluten-free dough. *Food Hydrocol.* **2021**, *115*, 106629. [CrossRef]
17. Federici, E.; Jones, O.G.; Selling, G.W.; Tagliasco, M.; Campanella, O.H. Effect of zein extrusion and starch type on the rheological behavior of gluten-free dough. *J. Cereal Sci.* **2020**, *91*, 102866. [CrossRef]
18. Berta, M.; Koelewijn, I.; Öhgren, C.; Stading, M. Effect of zein protein and hydroxypropyl methylcellulose on the texture of model gluten-free bread. *J. Texture Stud.* **2019**, *50*, 341–349. [CrossRef] [PubMed]
19. Ziobro, R.; Juszczak, L.; Witczak, M.; Korus, J. Non-gluten proteins as structure forming agents in gluten free bread. *J. Food Sci. Technol.* **2016**, *53*, 571–580. [CrossRef]
20. Horstmann, S.W.; Foschia, M.; Arendt, E.K. Correlation analysis of protein quality characteristics with gluten-free bread properties. *Food Funct.* **2017**, *8*, 2465–2474. [CrossRef]
21. Srikanlaya, C.; Therdthai, N.; Ritthiruangdej, P.; Zhou, W. Effect of hydroxypropyl methylcellulose, whey protein concentrate and soy protein isolate enrichment on characteristics of gluten-free rice dough and bread. *Int. J. Food Sci. Technol.* **2018**, *53*, 1760–1770. [CrossRef]
22. Masure, H.G.; Wouters, A.G.B.; Fierens, E.; Delcour, J.A. Impact of egg white and soy proteins on structure formation and crumb firming in gluten-free breads. *Food Hydrocol.* **2019**, *95*, 406–417. [CrossRef]
23. Korus, J.; Chmielewska, A.; Witczak, M.; Ziobro, R.; Juszczak, L. Rapeseed protein as a novel ingredient of gluten-free bread. *Eur. Food Res. Technol.* **2021**, *247*, 2015–2025. [CrossRef]
24. Witczak, M.; Chmielewska, A.; Ziobro, R.; Korus, J.; Juszczak, L. Rapeseed protein as a novel ingredient of gluten-free dough: Rheological and thermal properties. *Food Hydrocol.* **2021**, *118*, 106813. [CrossRef]
25. Salah, K.; Olkhovatov, E.A.; Aïder, M. Effect of canola proteins on rice flour bread and mathematical modelling of the baking process. *J. Food Sci. Technol.* **2019**, *56*, 3744–3753. [CrossRef]
26. Zorzi, C.Z.; Garske, R.P.; Flôres, S.H.; Thys, R.C.S. Sunflower protein concentrate: A possible and beneficial ingredient for gluten-free bread. *Innovat. Food Sci. Emerg. Technol.* **2020**, *66*, 102539. [CrossRef]
27. Witczak, T.; Juszczak, L.; Ziobro, R.; Korus, J. Rheology of gluten-free dough and physical characteristics of bread with potato protein. *J. Food Proc. Eng.* **2017**, *40*, e12491. [CrossRef]
28. Farzana, W.; Khalil, I.A. Protein quality of tropical food legumes. *J. Sci. Technol.* **1999**, *23*, 13–19.
29. Iqbal, A.; Khalil, I.A.; Ateeq, N.; Sayyar Khan, M. Nutritional quality of important food legumes. *Food Chem.* **2006**, *97*, 331–335. [CrossRef]

30. Lo, B.; Kasapis, S.; Farahnaky, A. Lupin protein: Isolation and techno-functional properties, a review. *Food Hydrocoll.* **2021**, *112*, 106318. [CrossRef]
31. Nunes, M.C.; Raymundo, A.; Sousa, I. Rheological behaviour and microstructure of pea protein/kappa-carrageenan/starch gels with different setting conditions. *Food Hydrocol.* **2006**, *20*, 106–113. [CrossRef]
32. E.U. Commission Implementing Decision of 1 July 2014 authorising the placing on the market of rapeseed protein as a novel food ingredient under Regulation (EC) No 258/97 of the European Parliament and of the Council. *Off. J. Eur. Union* **2014**, *L196*, 27.
33. FSANZ—Food Standards Australia New Zealand. Approval Report–Application A1175. Rapeseed Protein Isolate as a Novel Food. 15 December 2020. p. 34. Available online: https://www.foodstandards.gov.au/code/applications/Documents/A1175%20Approval%20Report.pdf (accessed on 1 June 2021).
34. FDA. GRAS Notice No. GRN 000683 Canola Protein Isolate. Available online: https://www.cfsanappsexternal.fda.gov/scripts/fdcc/index.cfm?set=GRASNotices&id=683&sort=GRN_No&order=DESC&startrow=1&type=advanced&search=%C2%A4Canola%20protein%C2%A4%C2%A4 (accessed on 1 June 2021).
35. Waglay, A.; Karboune, S.; Alli, I. Potato protein isolates: Recovery and characterization of their properties. *Food Chem.* **2014**, *142*, 373–382. [CrossRef]
36. Santos, J.; Calero, N.; Guerrero, A.; Muñoz, J. Relationship of rheological and microstructural properties with physical stability of potato protein-based emulsions stabilized by guar gum. *Food Hydrocol.* **2015**, *44*, 109–114. [CrossRef]
37. Komeroski, M.R.; Homem, R.V.; Schmidt, H.d.O.; Rockett, F.C.; de Lira, L.; Vitória da Farias, D.; Kist, T.L.; Doneda, D.; Rios, A.d.O.; Ruffo de Oliveira, V. Effect of whey protein and mixed flours on the quality parameters of gluten-free breads. *Int. J. Gastron. Food Sci.* **2021**, *24*, 100361. [CrossRef]
38. Han, A.; Romero, H.M.; Nishijima, N.; Ichimura, T.; Handa, A.; Xu, C.; Zhang, Y. Effect of egg white solids on the rheological properties and bread making performance of gluten-free batter. *Food Hydrocol.* **2019**, *87*, 287–296. [CrossRef]
39. Aprodu, I.; Banu, I. Effect of starch and dairy proteins on the gluten free bread formulation based on quinoa. *J. Food Meas. Charact.* **2021**, *15*, 2264–2274. [CrossRef]
40. Diprat, A.B.; Silveira Thys, R.C.; Rodrigues, E.; Rech, R. Chlorella sorokiniana: A new alternative source of carotenoids and proteins for gluten-free bread. *LWT* **2020**, *134*, 109974. [CrossRef]
41. Khemiri, S.; Khelifi, N.; Nunes, M.C.; Ferreira, A.; Gouveia, L.; Smaali, I.; Raymundo, A. Microalgae biomass as an additional ingredient of gluten-free bread: Dough rheology, texture quality and nutritional properties. *Algal Res.* **2020**, *50*, 101998. [CrossRef]
42. Selmo, M.S.; Salas-Mellado, M.M. Technological quality of bread from rice flour with Spirulina. *Int. Food Res. J.* **2014**, *21*, 1523–1528.
43. Różyło, R.; Hameed Hassoon, W.; Gawlik-Dziki, U.; Siastała, M.; Dziki, D. Study on the physical and antioxidant properties of gluten-free bread with brown algae. *CyTA J. Food* **2017**, *15*, 196–203. [CrossRef]
44. Kowalczewski, P.Ł.; Gumienna, M.; Rybicka, I.; Górna, B.; Sarbak, P.; Dziedzic, K.; Kmiecik, D. Nutritional value and biological activity of gluten-free bread enriched with cricket powder. *Molecules* **2021**, *26*, 1184. [CrossRef]
45. Nissen, L.; Samaei, S.P.; Babini, E.; Gianotti, A. Gluten free sourdough bread enriched with cricket flour for protein fortification: Antioxidant improvement and Volatilome characterization. *Food Chem.* **2020**, *333*, 127410. [CrossRef]
46. Da Rosa Machado, C.; Thys, R.C.S. Cricket powder (Gryllus assimilis) as a new alternative protein source for gluten-free breads. *Innov. Food Sci. Emerg. Technol.* **2019**, *56*, 102180. [CrossRef]
47. Becker, E.W. Micro-algae as a source of protein. *Biotechnol. Adv.* **2007**, *25*, 207–210. [CrossRef] [PubMed]
48. Aneiros, A.; Garateix, A. Bioactive peptides from marine sources: Pharmacological properties and isolation procedures. *J. Chromatogr. B* **2004**, *803*, 41–53. [CrossRef] [PubMed]
49. Viegas, C.V.; Hachemi, I.; Mäki-Arvela, P.; Smeds, A.; Aho, A.; Freitas, S.P.; da Silva Gorgônio, C.M.; Carbonetti, G.; Peurla, M.; Paranko, J.; et al. Algal products beyond lipids: Comprehensive characterization of different products in direct saponification of green alga Chlorella sp. *Algal Res.* **2015**, *11*, 156–164. [CrossRef]
50. Galland-Irmouli, A.-V.; Fleurence, J.; Lamghari, R.; Luçon, M.; Rouxel, C.; Barbaroux, O.; Bronowicki, J.-P.; Villaume, C.; Guéant, J.-L. Nutritional value of proteins from edible seaweed Palmaria palmata (dulse). *J. Nutr. Biochem.* **1999**, *10*, 353–359. [CrossRef]
51. Fitzgerald, C.; Gallagher, E.; Doran, L.; Auty, M.; Prieto, J.; Hayes, M. Increasing the health benefits of bread: Assessment of the physical and sensory qualities of bread formulated using a renin inhibitory Palmaria palmata protein hydrolysate. *LWT Food Sci. Technol.* **2014**, *56*, 398–405. [CrossRef]
52. Fitzgerald, C.; Mora-Soler, L.; Gallagher, E.; O'Connor, P.; Prieto, J.; Soler-Vila, A.; Hayes, M. Isolation and Characterization of Bioactive Pro-Peptides with in Vitro Renin Inhibitory Activities from the Macroalga Palmaria palmata. *J. Agric. Food Chem.* **2012**, *60*, 7421–7427. [CrossRef]
53. Verkerk, M.C.; Tramper, J.; van Trijp, J.C.M.; Martens, D.E. Insect cells for human food. *Biotechnol. Adv.* **2007**, *25*, 198–202. [CrossRef]
54. Kowalczewski, P.Ł.; Walkowiak, K.; Masewicz, Ł.; Smarzyński, K.; Thanh-Blicharz, J.L.; Kačániová, M.; Baranowska, H.M. LF NMR spectroscopy analysis of water dynamics and texture of Gluten-Free bread with cricket powder during storage. *Food Sci. Technol. Int.* **2021**. [CrossRef]
55. Felipe da Silva, F.; Tainara de Morais, C.; Camila Rubira, S.; Myriam de las Mercedes, S.-M. Pão sem gluten enriquecido com a microalga Spirulina platensis/ Elaboration of gluten-free bread enriched with the microalgae Spirulina platensis. *Brazil. J. Food Technol.* **2011**, *14*, 308–316. [CrossRef]

56. Dłużewska, E.; Marciniak-Lukasiak, K.; Kurek, N. Effect of transglutaminase additive on the quality of gluten-free bread. *CyTA J. Food* **2015**, *13*, 80–86. [CrossRef]
57. Shin, M.; Gang, D.-O.; Song, J.-Y. Effects of protein and transglutaminase on the preparation of gluten-free rice bread. *Food Sci. Biotechnol.* **2010**, *19*, 951–956. [CrossRef]
58. Kuraishi, C.; Yamazaki, K.; Susa, Y. Transglutaminase: Its utilization in the food industry. *Food Rev. Int.* **2001**, *17*, 221–246. [CrossRef]
59. Moore, M.M.; Heinbockel, M.; Dockery, P.; Ulmer, H.M.; Arendt, E.K. Network Formation in Gluten-Free Bread with Application of Transglutaminase. *Cereal Chem.* **2006**, *83*, 28–36. [CrossRef]
60. Skendi, A.; Mouselemidou, P.; Papageorgiou, M.; Papastergiadis, E. Effect of acorn meal-water combinations on technological properties and fine structure of gluten-free bread. *Food Chem.* **2018**, *253*, 119–126. [CrossRef]
61. Bravo-Núñez, Á.; Sahagún, M.; Gómez, M. Assessing the Importance of Protein Interactions and Hydration Level on Protein-Enriched Gluten-Free Breads: A Novel Approach. *Food Bioproc. Technol.* **2019**, *12*, 820–828. [CrossRef]
62. Manik, L.C.M.; Nur, M. The recent development of gluten-free bread quality using hydrocolloids. *IOP Conf. Ser. Earth Environ. Sci.* **2021**, *733*, 012101. [CrossRef]
63. Simonato, B.; Pasini, G.; Giannattasio, M.; Peruffo, A.D.B.; De Lazzari, F.; Curioni, A. Food Allergy to Wheat Products: The Effect of Bread Baking and in Vitro Digestion on Wheat Allergenic Proteins. A Study with Bread Dough, Crumb, and Crust. *J. Agric. Food Chem.* **2001**, *49*, 5668–5673. [CrossRef]
64. Smith, F.; Pan, X.; Bellido, V.; Toole, G.A.; Gates, F.K.; Wickham, M.S.J.; Shewry, P.R.; Bakalis, S.; Padfield, P.; Mills, E.N.C. Digestibility of gluten proteins is reduced by baking and enhanced by starch digestion. *Mol. Nutr. Food Res.* **2015**, *59*, 2034–2043. [CrossRef] [PubMed]
65. Villarino, C.B.J.; Jayasena, V.; Coorey, R.; Chakrabarti-Bell, S.; Foley, R.; Fanning, K.; Johnson, S.K. The effects of lupin (Lupinus angustifolius) addition to wheat bread on its nutritional, phytochemical and bioactive composition and protein quality. *Food Res. Int.* **2015**, *76*, 58–65. [CrossRef]
66. Sun-Waterhouse, D.; Sivam, A.S.; Cooney, J.; Zhou, J.; Perera, C.O.; Waterhouse, G.I.N. Effects of added fruit polyphenols and pectin on the properties of finished breads revealed by HPLC/LC-MS and Size-Exclusion HPLC. *Food Res. Int.* **2011**, *44*, 3047–3056. [CrossRef]
67. Świeca, M.; Gawlik-Dziki, U.; Dziki, D.; Baraniak, B.; Czyz, J. The influence of protein-flavonoid interactions on protein digestibility in vitro and the antioxidant quality of breads enriched with onion skin. *Food Chem.* **2013**, *141*, 451–458. [CrossRef] [PubMed]
68. Kies, A.K.; De Jonge, L.H.; Kemme, P.A.; Jongbloed, A.W. Interaction between Protein, Phytate, and Microbial Phytase. In Vitro Studies. *J. Agric. Food Chem.* **2006**, *54*, 1753–1758. [CrossRef]

foods

MDPI

Review

Sourdough Biotechnology Applied to Gluten-Free Baked Goods: Rescuing the Tradition

Laura Ramos [1,2], Alicia Alonso-Hernando [1,*], Miriam Martínez-Castro [1], Jose Alejandro Morán-Pérez [1], Patricia Cabrero-Lobato [1], Ana Pascual-Maté [1], Eduardo Téllez-Jiménez [1] and Jorge R. Mujico [1]

[1] Facultad de Ciencias de la Salud, Universidad Isabel I, 09003 Burgos, Spain; laura.ramos@cajal.csic.es (L.R.); miriam.martinez@ui1.es (M.M.-C.); jaleomp@gmail.com (J.A.M.-P.); patricia.cabrero@ui1.es (P.C.-L.); ana.pascual@ui1.es (A.P.-M.); eduardo.tellez@ui1.es (E.T.-J.); jrmujico@gmail.com (J.R.M.)

[2] Unidad de Citometría de Flujo y Separación Celular, Instituto Cajal, CSIC, 28002 Madrid, Spain

* Correspondence: alicia.alonso@ui1.es; Tel.: +34-947-671-731

Abstract: Recent studies suggest that the beneficial properties provided by sourdough fermentation may be translated to the development of new GF products that could improve their technological and nutritional properties. The main objective of this manuscript is to review the current evidence regarding the elaboration of GF baked goods, and to present the latest knowledge about the so-called sourdough biotechnology. A bibliographic search of articles published in the last 12 years has been carried out. It is common to use additives, such as hydrocolloids, proteins, enzymes, and emulsifiers, to technologically improve GF products. Sourdough is a mixture of flour and water fermented by an ecosystem of lactic acid bacteria (LAB) and yeasts that provide technological and nutritional improvements to the bakery products. LAB-synthesized biopolymers can mimic gluten molecules. Sourdough biotechnology is an ecological and cost-effective technology with great potential in the field of GF products. Further research is necessary to optimize the process and select species of microorganisms robust enough to be competitive in any circumstance.

Keywords: celiac disease; gluten-free; food additives; sourdough; microbiota; lactic acid bacteria

Citation: Ramos, L.; Alonso-Hernando, A.; Martínez-Castro, M.; Morán-Pérez, J.A.; Cabrero-Lobato, P.; Pascual-Maté, A.; Téllez-Jiménez, E.; Mujico, J.R. Sourdough Biotechnology Applied to Gluten-Free Baked Goods: Rescuing the Tradition. *Foods* **2021**, *10*, 1498. https://doi.org/10.3390/foods10071498

Academic Editors: Maria Papageorgiou and Theodoros Varzakas

Received: 11 May 2021
Accepted: 16 June 2021
Published: 28 June 2021

Publisher's Note: MDPI stays neutral with regard to jurisdictional claims in published maps and institutional affiliations.

1. Introduction

Celiac disease (CD) is an immune-mediated systemic disease, caused by gluten and related prolamins intake in genetically susceptible individuals. CD can only be treated by a lifetime adherence to a gluten-free (GF) diet, by removing wheat, barley, rye, oats, and their hybrids from the daily food intake [1–3].

When CD patients continuously ingest gluten, the small intestine mucosa is damaged by an increased number of lymphocytes and can evolve into villus atrophy and crypt hyperplasia [3]. The sustained consumption of gluten in these patients, even at trace levels, maintains the pathology and the intestinal damage, although there are no apparent clinical symptoms. The damage is accompanied by a malabsorption of nutrients that can lead to chronic diarrhea, abdominal distension, and reduced physical growth (the classic CD triad). Although CD has been traditionally considered as a gastrointestinal disease, nowadays, it is classified as an autoimmune-mediated systemic disease, affecting several organs and tissues [4].

The worldwide prevalence of CD is around 1.4% [5], with a heterogeneous distribution, that mainly affects Caucasians, and is more frequent in women than in men (in a ratio of approximately 2.8:1) [5,6]. The major problem related to this disease are the undiagnosed cases, since they can present atypical or no symptoms at all. It is estimated that 83% of celiac patients are not conscious of their disease [7], a percentage that increases up to 90% in pediatric patients [8], a phenomenon known as the "celiac iceberg" [9].

Commercialized GF products usually present technologically associated drawbacks related to the elasticity and cohesion of the dough, two properties provided by gluten proteins. As gases produced during fermentation are difficult to retain, they also show less volume and fluffy texture. These GF products are clearly inferior compared to their gluten-containing (GC) counterparts, since they are worse at a sensorial level, present low nutritional quality, and are more expensive [10,11]. The development of high-quality GF bakery products is a challenge for the food science and technology community, which is going through two different approaches: (i) from the technological perspective, using aeration by high pressure, flour pretreatment with ultrasounds, partial baking with freezing cycles, hydrothermal and extrusion treatments, etc.; and (ii) from the scientific perspective, with modified formulations, such as using additives–adjuvants, and/or the sourdough-based biotechnology [12–15].

Sourdough is a mixture of flour and water that is fermented by the action of microorganisms. The fermentation process can be spontaneous or directed by the addition of commercial starter cultures. Sourdough microbiota is composed by different lactic acid bacteria (LAB) and yeasts, in a ratio of approximately 100:1; both types of microorganisms can be naturally found in the cereal grains (and, consequently, in their flours), or provided by the "house microbiota" present in the physical environment where sourdough is made [16]. The main function of LAB is the acidification of the dough, producing chemical, metabolic and enzymatic modifications, whereas the main function of yeast is carbon dioxide (CO_2) production.

In the elaboration of baked goods, there is a tendency towards the recuperation of sourdough fermentation due to its numerous beneficial properties caused by the fermentation and acidification of dough by the native microbiota. Table 1 presents some sourdough properties that improve the quality of bakery products. These beneficial properties include organoleptic [17], nutritional [18,19], and functional [20] improvements, as well as an extension of the shelf life of baked goods [21]. Recent studies suggest that these positive effects may be translated to the development of new GF products, solving their low-quality properties.

Table 1. Properties of sourdough responsible of improving the quality of bakery products.

Sourdough Property/Function	References
Sensory improvements	[17]
Nutritional improvements	[18,19]
Functional improvements	[20]
Shelf-life extension	[21]

The general objective of this paper is to analyze the scientific evidence regarding the production of GF baked goods (mainly bread), and to present the latest knowledge about sourdough biotechnology. The use of additives or adjuvants in GF bakery products, alone or in combination with sourdough biotechnology, the autochthonous LAB and yeast naturally present in GF sourdoughs, and the microorganisms that synthesize gluten-like molecules that thereby improve the bakery products, will be also described.

A bibliographic search was performed between September and December of 2020 in Scopus, ScienceDirect, PubMed/Medline, and FSTA (Food Science & Technology Abstracts) databases. The following keywords and Boolean operators, both in Spanish and in English, were used: (adjuvant OR additive OR hydrocolloid OR protein OR enzymes OR emulsifiers) AND gluten-free bread; (lactic acid bacteria OR LAB OR sourdough OR yeast OR microbiota OR microbiome OR ecology OR biota) AND gluten-free NOT human; (lactic acid bacteria OR LAB OR exopolysaccharides OR EPS OR sourdough AND gluten-free). The search was restricted to studies containing the terms of reference in both the title and the abstract, using the query [TIAB] (TITLE AND ABSTRACT). The search was limited to studies published during the last 12 years, including research papers, meta-analysis, reviews and/or systematic reviews, books, or thesis. Then, a total of 92 studies that met these criteria which analyze ingredients, or final products mainly based on GF cereals (rice, corn/maize, millet, sorghum) and/or pseudo-cereals (buckwheat, quinoa, amaranth, teff) were finally selected for this review.

2. Technological Aspects of Using Additives and Adjuvants in Gluten-Free Baked Goods and Joint Contributions with Sourdoughs

Hydrocolloids, proteins, enzymes, and emulsifiers are the most-used additives and adjuvants in the preparation of GF bakery products. Its widespread use implies that they are also common ingredients in GF bread formulations that include sourdoughs, both in already developed products or products under research. The technological advantages provided by these compounds are briefly described in the following subsections.

2.1. Hydrocolloids

Hydrocolloids are a group of water-soluble polymers that are used in the formulation of GF doughs because they improve the properties of the final product in terms of structure, volume, texture, and palatability, as well as shelf-life extension. With very different chemical structures, they can be classified according to their origin, from: (i) some species of seaweed, such as agar–agar or carrageenan; (ii) plant tissue extracts, such as pectin, β-glucan or inulin; (iii) plant exudates, such as gum arabic (extracted from the resin of some varieties of acacia); (iv) different viscous plant substances (also called mucilages), such as guar gum or psyllium; (v) exopolysaccharides (EPS) of microbial origin, such as xanthan gum (synthesized by *Xanthomonas campestris*), or gellan gum (synthesized by *Sphingomonas elodea*), brought naturally from the addition of sourdoughs to the GF batter or artificially included on it; and (vi) cellulose-derived molecules, such as methylcellulose (MC), carboxymethylcellulose (CMC), or hydroxypropyl methylcellulose (HPMC) [22,23].

This group of compounds can mimic, to some extent, the viscoelastic properties of gluten in the dough. This is due to its capacity to interact with water and form a network-like structure (gel properties) that increases the viscosity of the mixture, as well as the capacity to retain the CO_2 produced during fermentation. They also stimulate the gelatinization of starches during baking, reducing the crystallization of amylopectin (starch retrogradation), and keeping products fresh for longer periods of time [24].

Hydrocolloids are the most widely used additives in the GF products' industry. Their ability to bind water in doughs (increasing viscosity and providing gel characteristics, which somehow mimics gluten technological properties), was already discovered in the 1950s. In this context, and from a scientific point of view, hydrocolloids are the most studied additives. There is a great number of experimental research studies that have analyzed how these molecules behave in different mixture/dough matrices.

Although the same type of hydrocolloid is used, results published in the literature are divergent, since the added concentration range is another variable to be considered (Table 2). It is usual to employ concentrations ranging from 0.3 to 5%, always selecting the lowest concentration with the best results. Additives are expensive and can provide (based on the concentration they are used), strange and undesirable flavors to the final product. In addition, the relationship between concentration and technological improvement is not directly proportional: once an optimum concentration is achieved (based on each additive and each dough), increasing the amount of additive does not lead to further improvement of the final product, and a collapse of the dough may occur, thus decreasing the improvement/s obtained [11].

Table 2. Hydrocolloids used in gluten-free baked goods.

Food Product	Cereal(s) or Pseudo-Cereal(s) Used in the Product	Main Flour(s)	Hydrocolloids	Technological Outcome	Reference(s)
GF bread	Brown rice	Brown rice	Xanthan gum, guar gum, xanthan-locust bean gum, MC [1], CMC [2], HPMC [3]	↑[4] Porosity, ↑ cohesiveness and elasticity	[25,26]
GF bread	Buckwheat	Buckwheat flour	0.14% xanthan gum	↑ Bread volume ↓[5] Crumb hardness/firmness	[12,27]
GF bread	Buckwheat	Buckwheat flour	Guar gum, HPMC, tragacanth gum	↑ Crumb alveoli resistance, ↑ elasticity	[25]
GF [7] bread	Teff, buckwheat, rice maize	Teff, buckwheat, rice, or maize flours	1.5% HPMC	dov [6]	[12]
	Rice	Rice flour and potato starch	Fructans (such as inulin)	↑ Bread volume ↓ Crumb hardness	[28]
	Maize	Maize starch, potato starch	Inulin (<10 polymerization degrees)	↑ Bread volume ↓ Crumb hardness	
	Maize	Maize starch, zein	HPMC, high β-glucan oat bran	Positive rheology, good crumb structure	
	Maize	Maize starch, potato starch	5% Inulin	↑ Bread volume (4%) ↓ Crumb hardness	[29]
	Maize	Maize starch, potato starch	8% Inulin	↑ Bread volume (9%) ↓ Crumb hardness Wrinkling of the crust	
	Rice	Rice flour and potato starch	4% to 12% ITFs [8] (Raftilose® Synergy1)	↑ Specific volume, darker crust, appealing crust and crumb	
	Rice	Rice flour	Inulin	↑ Volume, delayed staling, improved crumb, smoother crust	
	Rice	Rice flour, potato starch, cassava starch, sour tapioca flour	ITFs (inulin, FOS [9])	Color and porosity improvements Improved texture, taste and flavor	
	Maize, rice	Maize flour, rice flour, inactive soy flour	CMC or xanthan gum	dov	[30,31]
	Rice	Maize flour, carob flour, resistant starch (RS)	Carob flour, resistant starch (RS)	Low crumb firmness and improved porosity values with 15 g carob flour, 10 g RS, 10 g protein and 140 g water/100 g flour	
	Maize	Maize starch, potato starch	Flaxseed mucilage	Improved sensory acceptance	[15]
	Amaranth	Maize starch, amaranth flour, pea isolate	Psyllium	Improved final product quality	

Table 2. *Cont.*

Food Product	Cereal(s) or Pseudo-Cereal(s) Used in the Product	Main Flour(s)	Hydrocolloids	Technological Outcome	Reference(s)
	Rice, quinoa	Rice flour, quinoa flour	Xanthan gum	dov	[32]
	Rice, buckwheat	Rice flour, buckwheat flour			
GF layer cakes	Rice	Rice flour	Inulin, oat fibers, guar gum	Same volume as control ↑ Crumb firmness ↓ Elasticity	[28]
GF cheese bread	Maize	Pre-cooked cornflour, cassava starch	9% FOS	↓ Hydration; ↑ solubility of starch–FOS mixtures	[29]
GF bread	Maize	Maize flour	1.77% HPMC	↑ Bread volume ↓ Crumb hardness/firmness	[12]
"Empanadas" and piecrusts	Maize	Maize starch	Guar gum, HPMC, xanthan gum	↑ Elasticity	[25]
GF bread	Maize	Maize flour, maize starch	Xanthan gum	↑ Specific volume ↓ Crumb hardness	
GF bread	Maize	Maize starch, potato starch	Pectin, whey protein	dov	[32]
GF bread	Maize	Maize flour, maize starch	Guar gum, pectin	↓ Firmness, ↓ crumb hardening	[25]
GF bread	Rice	Rice flour	2.2% HPMC	dov	[12]
GF bread	Rice	Rice flour	HPMC	↑ Elasticity and viscosity	[25]
GF bread	Rice	Rice flour	HPMC	dov	
GF bread	Rice	Rice flour	Xanthan gum, carob gum, guar gum, HPMC	↑ Viscoelasticity	
GF bread	Rice	Rice flour	HPMC	↑ Specific volume	
GF bread	Rice	Rice flour	HPMC, xanthan gum	↑ Specific volume	
GF bread	Rice	Rice flour	HPMC	↓ Crumb firmness	
GF bread	Rice	Rice flour	HPMC	↑ Moisture content Enhanced sensory properties	
GF bread	Rice	Rice flour	HPMC, guar gum, CMC	↑ Specific volume	[25]
GF flat bread	Rice	Rice flour	15 g/kg xanthan gum 10 g/kg CMC 10 g/kg xanthan gum	↑ Crumb alveoli size ↑ Crumb porosity ↑ Dough yield	[15]
GF bread	Rice	Rice flour	HPMC	dov	[32]
GF bread	Rice	Rice flour	HPMC, β-glucan	dov	
GF bread	Rice	Rice flour	Xanthan gum, guar gum, carob gum	dov	
GF cake and muffin products	Rice	Rice flour	Tragacanth gum, xanthan gum	dov	

Table 2. *Cont.*

Food Product	Cereal(s) or Pseudo-Cereal(s) Used in the Product	Main Flour(s)	Hydrocolloids	Technological Outcome	Reference(s)
GF bread	Rice	Rice flour, carob flour, resistant starch	Carob gum, DATEM®, whey protein concentrate, α-amylase, transglutaminase, hemicellulase	dov	
GF bread	Rice, buckwheat	Rice flour, buckwheat flour	Xanthan gum	dov	
GF bread	Rice, maize	Rice flour, maize flour, soy flour	Carrageenan, alginate, xanthan gum, CMC	↑ Consistency, ↑ starch retrogradation, ↑ amylopectin retrogradation	[25]
GF bread	Rice, maize	Rice flour, maize starch	Xanthan gum	dov	
			CMC, pectin, agarose, xanthan gum	↑ Elasticity ↑ Dough strength	
			CMC, xanthan gum	↓ Crumb firmness ↑ Crumb porosity	
Egyptian balady flat bread	Rice, maize	Rice flour, maize starch, potato starch	Xanthan gum, guar gum	↓ Loss of moisture ↓ Hardness/firmness	
GF bread	Brown rice, maize, buckwheat	Brown rice flour, maize starch, soybean flour, buckwheat flour	Xanthan gum, Konjac gum	↓ Elasticity, cohesiveness, and resilience	
	Rice, maize	Rice flour, maize starch, chestnut flour	HPMC, lupine protein, vegetable fiber; guar gum, skimmed milk, cellulose	dov	[32]
	Rice, maize	Rice flour and maize starch	HPMC, skimmed milk, egg powder, soy protein, xanthan gum	dov	
			HPMC, vegetable fiber (bamboo, oat, pea, potato)	dov	
GF bread	Rice, maize	Rice, maize, maize starch	Xanthan gum	↑ Color improvements, ↑ Volume, hydration	[25]
GF bread	Rice, maize, quinoa	Rice flour, maize flour, maize starch, quinoa flour	HPMC, amyloglucosidase, α-amylase	↑ Volume, ↑ firmness	[32]
GF bread	Rice, maize, rice	Rice flour, maize flour, rice starch, rice protein	HPMC, carob gum, guar gum, psyllium, beetroot fiber, amylase	dov	
GF bread	Sorghum, maize	Decorticated sorghum flour, maize starch	Xanthan gum	dov	[25]
GF bread	Teff	Teff flour	0.04% xanthan gum 2% HPMC	↑ Bread volume ↓ Crumb hardness/firmness	[12]

[1] MC: methylcellulose; [2] CMC: carboxymethylcellulose; [3] HPMC: hydroxypropyl methylcellulose; [4] ↑: results in an increase of the mentioned feature; [5] ↓: results in a decrease of the mentioned feature; [6] dov: dependent on variables; [7] GF: gluten-free; [8] ITFs: inulin-type fructans; [9] FOS: fructooligosaccharides.

There are different flours (mixed, or not), hydrocolloids (mixed, or not, at different concentrations), and other substances (such as water, salt, sugar, honey, butter, milk, whey, etc.) that may be present in the dough. Water can be highlighted among all of them, due to its technological impact (it is fundamental in the final product and must be also optimized). GF dough generally requires greater amounts of water, ranging from 50 to 218%, and this proportion has an influence on the other parameters. It even affects baking: more hydrated doughs need baking containers (because of their lower densities), and the size, dimensions, and material of these containers also influence the final baked good.

Longer baking times are also needed to remove this excess of water, which requires lower baking temperatures so that, for example, the crust is not excessively browned. Consequently, there are many parameters to consider when choosing the best ingredients, processes, and additives to obtain the desired result. All this complexity is reflected in the papers selected for this part of the review (Table 2) and, to some extent, it explains the disparity and lack of homogeneity between the obtained results.

The most used hydrocolloids in GF bakery products, and the ones that seem to work better, are HPMC and xanthan gum. Both HPMC and the rest of the cellulose derivatives employed as additives usually come from plant sources, although the so-called bacterial cellulose (BC) is also described, a related molecule synthesized by bacteria of the genus *Gluconacetobacter*, especially *G. sansei*. Recent studies have concluded that the production cost of this BC is so high, and the recovery yield so low, that it cannot be applied at an industrial scale [33].

In the paper by Hager and Arendt [27], included in the review published by Capriles and Arêas in 2014 [12], the use of these two hydrocolloids (HPMC and xanthan gum) was reviewed, reaching the conclusion that HPMC has positive effects on formulations with teff and corn flours, negative effects when rice flour is used, and no changes for buckwheat flour; no conclusions were reached when the effect of xanthan gum was studied.

In the same study [27], a very little amount of hydrocolloid (around 0.14%) was needed when adding xanthan gum to buckwheat flour to obtain optimal results, determined by a higher loaf volume and softer crumb. To obtain the same results in corn flour, a higher concentration of HPMC (1.77%) was needed. During a third part of the same study, to check if the effects of these hydrocolloids were synergistic (potentiated), or could present some antagonism, teff flour and different ratios of HPMC and xanthan gum were tested. To reach the established objectives, it was necessary to slightly increase the concentration of HPMC compared to the one used alone (up to 2%), but the amount of xanthan gum to be added was very small, around 0.04% (70% less than used alone).

Schober et al. [34] obtained an improvement of sorghum bread quality with HPMC (2%) alone, but also showed that a previous sourdough fermentation of the total sorghum flour in combination with HPMC (2%), could solve some technological problems and lead to a superior quality sorghum bread.

Campo et al. [35] worked with GF bread formulas containing different combinations of teff flour (10%) and commercial dried cereal sourdoughs (rice or buckwheat, 15%) or *Lb. helveticus* fresh sourdough (15%), all of the batches including 0.75% HPMC, as a standardized concentration of this hydrocolloid. Bread with a combination of teff (10%) and rice-based sourdough achieved the best sensory results in terms of flavor [35].

Dermirkesen et al. [36] added different hydrocolloids (xanthan gum, carob gum, guar gum, and HPMC) to rice flour and, in their experimental conditions, the best combination was obtained by mixing xanthan and guar gums (paper included in the review published in 2016 by Mir et al. [25]). However, in another study also using rice flour, the highest loaf volumes were showed when CMC and HPMC were combined [37], included in the review published in 2016 by Mir et al. [25].

Keeping in mind that all these additives must be declared on the label and make the final product more expensive, research about other compounds providing more benefits than only technological is being encouraged—for example, those with added nutritional properties, such as inulin or β-glucans [38].

Regarding inulin, the results were again different between studies. Gularte's group employed inulin in GF baked goods, and the results were not satisfactory: compared with control, it did not improve final loaf volume and, in addition, its use was counterproductive, increasing crumb firmness and decreasing elasticity ([39], study included in the review published in 2014 by O'Shea et al. [28]). In contrast, although Korus et al. obtained positive results by adding only 4% inulin, undesirable crust wrinkles appeared when inulin was increased up to 8% ([40], work included in the review published in 2016 by Drabińska et al. [29]).

The conclusion reached after reviewing all these studies, which is not only applicable to hydrocolloids but also to the use of any additive in GF baked goods, is that no correlation between the variables is found, and each case must be analyzed and assessed individually. The effect of additives, or adjuvants in the dough depends on the type and concentration of the additive, its interaction with other additives/ingredients, and any other technological parameter of the process. Besides the scientific literature results, the selection of the best compound/s to achieve a specific technological property should also consider if the substance is previously authorized as a food additive within regulations from the specific regions or countries and the individual restrictions to its use that would apply in every case.

2.2. Proteins

The use of proteins in GF baked products responds to a double objective: firstly, the nutritional value is increased (providing higher levels of protein and essential amino acids) and, secondly, some of these proteins (with the capacity for stabilizing foams and emulsions), can mimic gluten technological properties, improving the organoleptic characteristics, and leading to higher quality products.

The most used proteins come from egg and milk; proteins from soybean and other cereals and/or pseudo-cereals are also widely used:

- Egg proteins (helped by the lecithin present in the yolk), act as foaming and emulsifying agents, and they are capable of stabilizing emulsions. These properties will improve the dough structure and gas retention, providing a softer crumb with more uniformly distributed alveoli. In addition, egg is a food with a very interesting nutritional profile, considered as a good source of high biological value proteins, fats, vitamins, and minerals (especially iron).
- The most-used milk proteins are casein, caseinates, and whey proteins. These proteins have gluten-like functional and technological properties, capable of creating cross-linked networks, and with a high capacity for swelling and water retention. Regarding GF bread, milk proteins contribute to Maillard reaction (between amino acids and reducing sugars), improving texture, roasted flavors and, mainly, both color and aroma crust quality.
- Although soybean is a protein-rich food, it is deficient in sulfur-containing amino acids, such as the essential amino acid methionine. It is used as a functional food to increase the nutritional value of GF bakery products, since it contains bioactive compounds such as isoflavones. Due to its technological properties, soybean also has a positive impact on the quality of the final product, by improving crumb, volume, water retention, and sensory assessment.

When proteins are reviewed (Table 3), each individual case must be studied, analyzing the type of flour, the protein that has been used, and the manufacturing process.

One disadvantage of using proteins is that some of them (such as from milk, egg, and soybean), are classified as allergens, not being well accepted by patients with allergies, intolerances, and/or sensitivities to these foods. On the other hand, if milk-derived proteins are used, they must be low in lactose, since CD patients may have a secondary intolerance to this disaccharide, due to lactase deficiency, because of their villus atrophy [42].

Table 3. Proteins used in gluten-free baked goods.

Food Product	Cereal(s) or Pseudo-Cereal(s) Used in the Product	Main Flour(s) Used in the Product	Protein Supplementation/Additives	Technological Outcome	Reference(s)
GF [1] bread	Rice	Rice flour, cassava starch, soy flour	7.5% soy 7.8% milk powder	↑ [2] Nutritive value, without sensorial changes	[12]
	Rice	Rice flour	Bovine plasma protein	dov [3]	[30]
	Rice	Rice flour	Bovine serum albumin	dov	
	Maize	Maize starch, potato starch	Collagen	dov	
	Variable	Variable	Egg, caseinate, whey protein, milk protein	dov	[30,41]
	Variable	Variable	Egg	Improved structure, stable foaming, and gas retention	[12]
	Variable	Variable	Lactose free milk powder	dov, darkening of the crumb	
Precooked rice pasta	Rice	Rice flour, yellow pea flour, chickpea flour, lentil flour	Legume protein	Increased protein and dietary fiber content	[15]
GF bread	Rice, buckwheat, quinoa	Rice flour, quinoa flour, buckwheat flour, potato starch	Quinoa protein	↑ Elasticity and dietary fiber content improved dough structure	
GF bread	Rice	Rice flour	Rice bran protein concentrate	↑ Elasticity, shear strength, volume, gas retention and shelf life	
GF bread	-	Soy flour	Soy	dov (↑ loaf volume, ↓ [4] crumb hardness)	[12,42,43]
GF bread	-	Soy flour	Soy protein	dov	
GF bread	Variable	Soy flour	Soy protein, milk powder	dov	
GF bread	Variable	Soy flour	Soy, pea	dov	
GF bread	Variable	Starch from different sources	Whey protein	dov	[12,30]
GF bread	Maize	Maize flour	Zein	dov	[12,30]
Egg yolk muffins	Maize	Maize	Egg yolk granulates, apple pectins, gelatine	dov	[32]
GF bread	Maize	Maize starch, carob germ flour	Carob protein, HPMC [5]	dov	[12]
GF doughs	Maize	Unmodified maize starch	Zein, HPMC	dov	
GF muffins	Maize	Maize starch, kidney bean flour, field pea flour, amaranth flour	Protein isolates	dov	[32,41,44]

Table 3. *Cont.*

Food Product	Cereal(s) or Pseudo-Cereal(s) Used in the Product	Main Flour(s) Used in the Product	Protein Supplementation/Additives	Technological Outcome	Reference(s)
GF muffins	Rice		Egg, fructose, inulin, sucralose	dov	
	Rice		Jambolan fruit pulp, soy Protein isolates, glycerol monostearate, xanthan gum	dov	
	Rice		Soya bean protein isolate, pea protein isolate, egg white isolate, casein, xanthan gum	dov	
	Rice		Soy protein isolates, glycerol monostearate, xanthan gum, black carrot dietary fiber concentrate	dov	
GF bread	Buckwheat, rice	Buckwheat flour, rice flour, chickpea flour	Green mussel protein hydrolysates	dov	
GF bread	Wheat	Wheat starch	6% whey protein	Darker crust, white crumb, ↑ volume, improved texture	[12]
	Wheat	Wheat starch	Whey protein	dov	

[1] GF: gluten-free; [2] ↑: results in an increase of the mentioned feature; [3] dov.: dependent on variables; [4] ↓: results in a decrease of the mentioned feature; [5] HPMC: hydroxypropylmethylcellulose.

2.3. Enzymes

Enzymatic technology is widely used in GC bakery for improving dough properties and final quality. Among all the used enzymes, highlights include: (i) amylase, breaking complex carbohydrates into sugars that can be used as substrates; and (ii) proteases, hydrolyzing gluten and being used, for example, in the production of cookies, providing a better malleable dough.

In GF bakery, enzymes are used to modify the proteins present in the dough into others capable of mimicking the action of the gluten proteins they lack (Table 4). The most frequently used enzymes are:

- Enzymes that modify starches, such as amylase and cyclodextrin glycosyltransferase; the latter degrades starch and produces dextrin that has been experimentally proven to increase the solubility of hydrophobic proteins, which in turn increases CO_2 retention, providing a bigger loaf volume and a better texture [12,29,45]. Schober et al. indicated that bacterial α-amylase is used to supply sugars in the sourdough fermentation step, and also exerts an anti-staling effect in GF starch breads, so they included 0.01% of this enzyme in their sorghum sourdough formula [34].
- Enzymes that crosslink, or connect proteins, such as transglutaminase (TGase) and gluco-oxidase (GO). These enzymes, which catalyze protein polymerization and crosslinking reactions, can create a kind of network or mesh, such as the three-dimensional structure provided by gluten, that improves CO_2 retention [12,30,46].
- Proteases that hydrolyze the peptide bonds of the proteins. This property can improve texture and final quality of rice-flour-based breads [12,30]. Additionally, proteolysis that occurs during the sourdough fermentation process could prevent interferences between protein aggregation upon baking and the starch gel, which seems to be desirable in GF sorghum breads [34].

Table 4. Enzymes used in gluten-free baked goods.

Food Product	Cereal(s) or Pseudo-Cereal(s) Used in the Product	Main Flour(s)	Enzymes/Additives	Technological Outcome	Reference(s)
GF [1] bread	Brown rice, buckwheat, maize, oat sorghum or teff	Brown rice, buckwheat, maize, oat, sorghum or teff flours	0.1 or 10 U [2] of TGase [3]/g of protein	Depending on protein source and enzyme dosage	[12,47]
GF bread	Buckwheat, brown rice	Buckwheat flour, brown rice flour	0.1 to 10 U of TGase/g protein	↑ [4] Increased batter pseudoplasticity, ↑ water holding capacity, improved crumb texture and structure	[15]
GF bread	Buckwheat, sorghum, or maize	Buckwheat, sorghum, or maize flours	0.01% or 0.1% proteases	Liquid-like batters, poor viscoelastic behavior, ↓ [5] gas-holding capacity	[12]
GF bread	Buckwheat, rice	Buckwheat flour, rice flour	Amylase	dov [6]	[30]
GF bread	Rice	Rice flour	Cyclodextrinase	dov	[12,30]
GF bread	Rice, sorghum, maize	Rice, sorghum, maize flours	GO [7]	dov	[12,30]
GF bread	Rice	Jasmine rice flour, pregelatinized tapioca starch	TGase	dov, TGase increased loaf volume and softened bread crumb.	[30,46,48]
GF bread	Oat	Oat flour	Tyrosinase, laccase, xylanase	dov, tyrosinase increased firmness of the dough, laccase and xylanase improved specific volume	
GF cake and muffin products	Rice	Rice flour, legume flour, chickpea flour, pea flour, lentil flour, bean flour	α-amylase, amyloglucosidase, trypsin, GO	dov	[32]
GF bread	Rice	Rice flour	0.01% GO 2% HPMC [8]	↑ Final volume, smoother crumb	[12]
GF bread	Rice	Rice flour	1 U TGase/g	Improved crumb texture	[15]
GF bread	Rice	Rice flour	1.35 U of TGase/g rice flour protein 0.67% albumin 0.67% casein	↑ Final volume, less compact crumb	[12]
GF dough and bread	Rice	Rice flour	*Aspergillus oryzae* protease	↑ Viscosity, improved gas-holding capacity, volume improvements	
GF bread	Rice	Rice flour	Glutathione oxidase	↑ Elasticity and volume improved gas-holding capacity	[15]
GF bread	Rice	Rice flour	Microbial TGase HPMC	dov	[12]
GF bread	Rice	Rice flour	Proteases	Depending on protease amount added	

[1] GF: gluten-free; [2] U: units; [3] TGase: transglutaminase; [4] ↑: results in an increase of the mentioned feature; [5] ↓: results in a decrease of the mentioned feature; [6] dov: dependent on variables; [7] GO: glucose oxidase; [8] HPMC: hydroxypropyl methylcellulose.

To deal with the lack of gluten of these GF baked products, enzymes are perhaps the least used additives because, among other reasons, it is a very recent research area. Moreover, enzymes work at very low concentrations and what initially seems to be an advantage makes that slight increase of enzymes produce huge protein changes with unexpected results in the final products (such as loaves with low volume and very hard crumb) [44].

Renzetti et al.'s [43] paper included in the review published in 2017 by Naqash et al. [15] investigated the use of TGase in GF bakery without any other adjuvant addition. Their conclusion was that TGase could improve the functionality of GF flours, obtaining positive results in buckwheat and whole rice breads, also being of interest to continue researching the use of TGase together with other additives. Mohammadi et al.'s [49] paper included in the review published in 2017 by Naqash et al. [15] studied the addition of TGase together with guar gum in rice flour; the combination that better worked in their conditions was 1 U/g of TGase and 20 to 30 g/kg of guar gum (as more TGase was added, the hardness of the crumb was increased).

The use of enzymes in dough is widespread because of its technological potential for modifying proteins. Moore et al.'s [45] paper included in the review published in 2014 by Capriles and Arêas [12] tested increasing concentrations of TGase with the addition of proteins from different sources (egg, milk, soybean, cereals, etc.), without finding a clear correlation. The improvement of the dough was based on the flour, TGase concentration, and type of protein used. However, Storck et al.'s [46] paper included in the review published in 2014 by Capriles and Arêas [12] optimized the use of TGase and protein in their rice-flour-based model. The mixture of 1.35 U of TGase for each gram of protein (albumin+casein), together with 0.67% albumin and 0.67% casein, was the combination that provided the highest volume, and a crumb with more alveoli and less hardness.

However, recent observations have established a possible association between the increased use of microbial TGase in food processing and the surge in incidence of celiac disease [47].

2.4. Emulsifiers

Emulsifiers are substances with an amphiphilic nature, which means that one side of the molecule is hydrophilic (water soluble) and the other side is hydrophobic (water insoluble). This dual nature allows emulsifiers to stand between two immiscible phases, connect them, reduce surface tension, and form a stable, homogeneous, and fluid emulsion. The most frequently used emulsifiers are:

- Soy lecithin, a plant origin product, which is extracted from soybeans. It has a very high concentration of phospholipids that contribute to dough extensibility, and flour hydration properties.
- Mono- and di-glycerides of fatty acids (E–471) [48] have the property of softening the dough, facilitating mixtures at an industrial level, thus achieving a crumb with more alveoli and a larger final volume. They also decrease starch retrogradation, which improves the shelf life of bakery products (especially pastries).
- Esters of mono- and di-glycerides fatty acids (E–472a–E47f) [48], are mainly used in the preparation of bread, since they provide a better "body" to the dough (an excessively liquid dough is an important defect of the GF products); this equates to a firmer dough with greater gas retention, and both texture and final volume improvements. These emulsifiers also contribute to an increased shelf life of bakery products.

There are few studies where emulsifiers are used as a separate category of additives (Table 5). This is because many additives with emulsifying properties are classified as hydrocolloids, proteins, or enzymes (described in the previous subsections). It is worth highlighting those studies where DATEM® (a commercial emulsifier) is investigated.

Table 5. Emulsifiers used in gluten-free baked goods.

Food Product	Cereal(s) or Pseudo-Cereal(s) Used in the Product	Main Flour(s)	Emulsifiers	Technological Outcome	Reference
GF [1] dough	Buckwheat	Buckwheat flour	DATEM®	dov [2]	[12]
GF cheese bread	-	Cassava starch	DATEM®	dov	[50]
GF bread formulas	Rice	Rice flour	0.5% DATEM® 0.5% (xanthan gum/guar)	Improved final product (with highest scores for texture acceptability)	[12]
GF bread	Rice	Rice flour, tigernut flour	DATEM®, xanthan gum, guar gum	dov	[32]
GF cake and muffin products	Rice, maize	Rice flour, maize flour	Lecithin	dov	

[1] GF: gluten-free; [2] dov: dependent on variables.

3. Sourdough Biotechnology

As previously described, sourdough can be considered as a specific ecosystem of LAB and yeasts that coexist in a flour–water matrix. Sourdough biotechnology could have a prehistoric origin, since ancient loaves have been found in Egyptian tombs, and wheat sheaves in human settlements dating from over 8000 years ago [51].

The elaboration of bread with these leavening microorganisms was abandoned in the second half of the 20th century, because of changes in food habits and the availability of commercialized pressed yeast. At that time, the food industry was consolidated, refrigerators arrived for domestic homes, and a boom of processed and ready-to-eat food products started to be sold in supermarkets.

Furthermore, important social changes started to happen, such as female economic independence, changes in eating behaviors (e.g., eating outside the home), etc. that have reduced the available time for cooking. It is important to note that the elaboration of homemade sourdough bread is a long process that requires time and dedication.

Bread is a basic food in the worldwide diet. Although white wheat bread, which is the most frequently sold bread, is usually manufactured without sourdough, it has good organoleptic and technological properties due to gluten proteins. By contrast, artisan bread is more expensive and oriented to specific demographics (and not the general public), although both profiles of consumers are starting to merge.

Actual food research in this field is mainly focused on the improvement of these products by using sourdough. Due to the nature of sourdough, the benefits and technological properties provided to the bakery products by these autochthonous microorganisms can be extended to all types of sourdough (including those made with GF flours). This capacity for improving the baked goods' quality will depend on the microorganisms' capacity to resist environmental stress, and to establish inter-dependent associations that will keep them stable along the entire fermentation process [52].

3.1. Factors Affecting Sourdough Microbiota

3.1.1. Sourdough Fermentation Processes

It is fundamental to know the technological factors that affect and select the sourdough biodiversity, and those out of control, which can be responsible for the variability and dispersion observed in the results of different research articles in this field. Furthermore, it is important to be aware of manageable factors to optimize the process and focus this biotechnology into the final desired bakery product.

Among the non-controllable technological factors are the biochemical composition of the food ingredients (not only between flours from different grains, but also between the same flour type from different origins), and the house microbiota. It has been experimentally demonstrated that house microbiota is different depending on where the elaboration of the sourdough had been taking place (in a bakery, or in a relatively sterile environment, such as in a laboratory) [53].

On the other hand, some of the technological factors that can be controlled by the operator are:

1. Sourdough Type

Depending on the process, four sourdough types can be distinguished (Figure 1) (some authors consider that, depending on certain parameters, there could also be subtypes) [54]:

- Type 0 sourdough is a type of pre-dough, also known as mother sponge, characterized by a short fermentation time at room temperature (RT, <30 °C). This provokes the initial propagation of native and exogenous LAB, with a higher proliferation rate compared to yeast, producing bioactive molecules and organic acids (lactic and acetic acids) that diminish the pH (pH~4). Given the short fermentation time, yeast growing is not enough in the sourdough and it is mandatory to add commercial yeast preparations. The microbiota that can be found in type 0 sourdough is a variety of LAB species; some of them are present in other types of sourdough, and others are not usually isolated and do not contribute to the improvement of the final product. It should be noted that in this type of sourdough there is no time to select those microorganisms with a higher adaptability to sourdough ecosystems, such as the yeast *Saccharomyces cerevisiae*. Typical examples are solid pre-ferments, such as *biga* from Italy and *pâte fermentêe* from France; and hydrated pre-ferments, such as the *levain levure* from north Europe, and *poolish* from Poland.
- Type I sourdough can be considered as the traditional sourdough, probably the one that spontaneously emerged in antiquity. Used in artisan bakeries and domestic settings, it considerably increases the quality of the final baked good. Type I sourdoughs have a long fermentation time at RT and are composed of very few microorganism species with the highest adaptation rates, the highest resistance, being the most competitive, and capable of stablishing solid associations between them. A typical example is the sourdough from San Francisco, mainly fermented by the LAB *Lactobacillus sanfranciscensis* (named because it was first isolated and described in this type of sourdough—reclassified as *Fructilactobacillus sanfranciscensis* [55]—) and the yeast *Candida humilis*. The association between these two microorganisms is very stable, since *Lb. sanfranciscensis* use maltose and *Candida humilis* use glucose, so they do not compete for the carbon source. They are also very competitive, displacing other species [56].
- Type II sourdough is a semiliquid fermented dough that can be bombed and used at an industrial scale. A starter culture is usually added to this type of sourdough, which is composed of LAB species that rapidly acidify the mixture and/or generate compounds that provide the aromas and flavors of traditional sourdough. Long fermentation times are used (two to five days) in only one step and at high temperatures (>30 °C). At these conditions, LAB rapidly proliferate (due to the high temperatures that facilitate their growing), with the consequent production of organic acids, the decrease of pH (pH < 3.4), and the yeast growing inhibition at this pH. This leads to the selection of acid-tolerant and thermophilic LAB (selection that is forced when commercial starter cultures are used) and requires adding industrial yeast. Some examples of *Lactobacillus* species isolated from type II sourdough are *Lactobacillus fermentum* (pro synonymon —pro synon.—*Limosilactobacillus fermentum*), *Lactobacillus plantarum* (pro synon. *Lactiplantibacillus plantarum*) and *Lactobacillus reuteri* (pro synon. *Limosilactobacillus reuteri*) [55]; from rye sourdough, *Lb. amylovorus* is also frequently isolated [54].
- Type III sourdough is a freeze-dried type II sourdough to facilitate its commercialization and later industrial use.

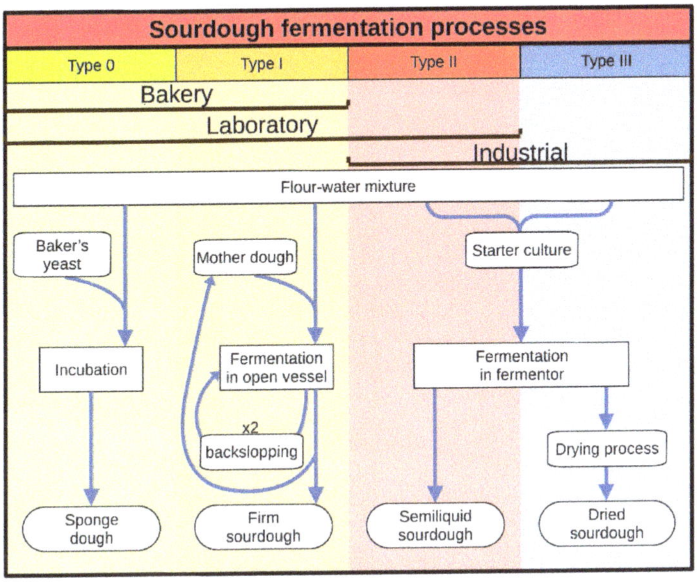

Figure 1. Types of sourdough fermentation processes according to the process technology applied. Adapted from: [54].

2. Temperature of Fermentation

It has been described how the temperature of fermentation is a key factor for classifying the different types of sourdough, but inside the same type of sourdough, temperature is also a decisive factor; for example, the effect over the microbiota composition of a type I sourdough will not be the same if the RT is 20 °C or 35 °C.

The geographic location will determine the selection of the final microbiota. For example, *Lb. sanfranciscensis* (an endemic specie of type I sourdough) is not isolated in tropical climates, since it is a mesophilic species adapted to cold–temperate weathers. When the environmental temperature is high, it stimulates the proliferation of thermophilic species of *Lactobacillus*, such as *Lb. fermentum* (pro synon. *Limosilactobacillus fermentum*), *Lb. casei/paracasei* (pro synon. *Lacticaseibacillus casei*/*L. paracasei*) and *Lb. reuteri* (pro synon. *Limosilactobacillus reuteri*) [55,57].

3. Dough Yield

The dough yield (DY) is the proportion of water and flour in the sourdough. Low DY results in solid doughs, with higher acetic acid and lower lactic acid proportions, because of the inhibition of yeast by acetic acid. Indeed, the velocity of acidification of sourdough is also affected by DY, increasing both values proportionally: high DY results in a higher hydration of the dough and higher acidification velocity, probably due to a better diffusion of acids in a hydrated mixture [58].

4. Other Factors

Some other factors that can affect the sourdough elaboration process are [54,58]:

- The pH of the sourdough, affected by LAB or yeast presence and fermentation stage [58].
- Additional nutrient sources: traditional ingredients added to sourdough final mixes complement the nutrient content of the sourdough—e.g., adding mono- and disaccharides or different amino acid sources, thus affecting the intrinsic parameters for microbial growth [58,59] and the microbial composition itself [60].

- Ash content in the bran fraction of the flour. The bran fraction contains several minerals and micronutrients that can promote the growth of LAB in the sourdough. The ash content also influences the buffering capacity of the sourdough system that makes it possible to reach a higher total titratable activity [58].
- The amount of added salt can promote the presence of osmotolerant microorganisms such as yeast [54,58].
- The redox potential, depending on the oxygen availability, DY, frequency of dough refreshments, etc. [54,58].
- The resting time of the dough and its temperature; if it is performed at cold temperatures, it will favor microorganisms that are resistant to cold stress and to the absence of substratum [54].

3.1.2. Instrumental Techniques for the Isolation and Identification of Microorganisms

Besides all variables that have been previously described, the instrumental techniques can provide new factors that have an impact on the results of the studies about sourdough microbiota; therefore, they should also be considered.

1. Sampling

Sampling is a critical step in all analytical techniques. As the whole sample cannot be analyzed, a representative aliquot must be selected, and the results extrapolated to the whole sample. Since the population of microorganisms varies along time and accordingly with the biotechnological process, the standardization of the sampling methodology, to obtain comparable results, is also required [61].

2. Fermentation Place

It should be considered that the microbiota analysis consists of the isolation and identification of the autochthonous microbiota, which comes not only from the food ingredients but also from the working place (e.g., the table and tools where the sourdough is made) and from the baker's hands [62]. These environmental microorganisms are known as "house microbiota" [63].

If a microorganism is not present in some of the ingredients, and the sourdough is fermented in a relatively sterile environment (such as a laboratory), that microorganism will not be isolated from the sourdough. However, in highly contaminated environments (i.e., bakeries), with the presence of many different types of flour and other ingredients that can provide their own microbiota, it is reasonable to think that different microorganisms will be isolated in comparison to those found when the fermentation is produced in a laboratory [63].

Some authors have investigated whether the daily introduction of a type of flour in a bakery, and the fermentation of the corresponding type I sourdough, could define a house microbiota that could be used afterwards as an inoculum, similarly to the elaboration of wine, or cheese [62].

It has been hypothesized that house microbiota could mainly be responsible for isolating the same microorganisms from a specific sourdough produced in the same region. Nowadays, it has also been postulated that these similarities could also be due to the use of the same flour type, the same environmental conditions, and similar traditional food technological processes [64].

Furthermore, sourdoughs of every region and country are gaining importance as an identity sign, highlighting the need to preserve the biodiversity of each fermentation process. This is the reason why the non-profit initiative, Puratos Sourdough Library, a library of fermented doughs, was created in Belgium in 2013, to maintain sourdoughs worldwide. Currently, 1500 LAB species and 700 yeasts have been isolated from the 84 different sourdoughs collected by this library [65].

3. Isolation and Identification Techniques

During the last years, research about sourdough autochthonous microbiota has shown some variability in the obtained results. This lack of uniformity is mainly due to the different isolation and identification methodologies. Table 6 (Section 3.2.2.) specifies if the microorganism is identified by molecular techniques (based on genotypic factors), or by culturing methods (based on phenotype factors).

Phenotype methods are traditional identification methods of microorganisms, developed by culturing in agar plates. The sample is cultured in a non-selective enriched solid medium to isolate different colonies. Each colony is grown in liquid cultures that allow their rapid proliferation. They are then seeded again in specific and selective media for each type of microorganism. After confirming the isolation of single bacteria or yeast strains, its identity would be checked by different techniques, such as morphology assessment using microscopy methods, carbohydrate metabolism tests, or fermentation tests. With this methodology, it is necessary to know the type of microorganism we are searching for, since selective and differential growing media are used, with concrete substratum that allow the proliferation of only one, or a few species.

Genotype techniques are more recent and are based on molecular biology and the species identification by deoxyribonucleic acid (DNA). In this group, polymerase chain reaction (PCR and real-time PCR), microarray massive sequencing, and pyrosequencing techniques can be found [54,64].

3.2. *Sourdough Autochthonous Microbiota*

3.2.1. Gluten-Containing Sourdough

Studies about the microbiota of GC sourdough are relatively recent. Spicher [66] and a Spanish research group headed by Benedito de Barber [67] were the first ones to investigate the autochthonous microbiota, with the intention of rescuing the sourdough tradition, as well as improving the quality of the mainly produced breads (based on short-time fermentations made by commercial yeast, with the only objective of producing CO_2).

Numerous studies have been published investigating not only the beneficial properties that sourdough can provide to bakery products, but also which microorganisms (among all microbiota) are the responsible ones. Most of these studies are focused on wheat and, to a lesser extent, rye and barley.

The autochthonous microbiota of GC sourdough has been deeply studied during the last years. In a meta-analysis performed by Van Kerrebroeck et al. and published in 2017 [68], 583 sourdoughs were analyzed, and it was concluded that, in these sourdoughs, the most proliferating LAB were heterofermentative (which produce acetic acid, lactic acid, ethanol, and CO_2 from the digestion of monosaccharides), although some homofermentative LAB (which only produce lactic acid) were also found. The isolated LAB species were mainly from the genera *Lactobacillus* [68]: *Lb. sanfranciscensis*, *Lb. plantarum*, *Lb. brevis* (pro synon. *Levilactobacillus brevis)* [55], *Pediococcus pentosaceus*, *L. paralimentarius*, and *L. fermentum* (LAB from the genera *Leuconostoc* and *Weisella* were also isolated, but in a lower proportion).

The main isolated yeast species were *S. cerevisae* (present in almost all bakeries, since it is used as a commercial yeast, and it is part of the house microbiota) and *C. humilis* (reclassified as *Kazachstania humilis*) [68]. In another review published in 2013, where 287 sourdoughs were analyzed, the main isolated yeasts were: *S. cerevisiae*, *C. humilis*, *Wickerhamomyces anomalus*, *Torulaspora delbrueckii*, *Kazachstania exigua*, *Pichia kudriavzevii*, and *Candida glabrata* [64].

3.2.2. Gluten-Free Sourdough

Research about GF sourdough has not evolved in the same way than its GC counterpart. Figure 2 depicts a comparison of articles (published in Scopus during the last 12 years) by using the terms "sourdough", or "sourdough AND gluten-free". Before 2008, the search with "sourdough AND gluten-free" retrieves a scarce number of results, and before 2005, there are no results available in this database for these search terms.

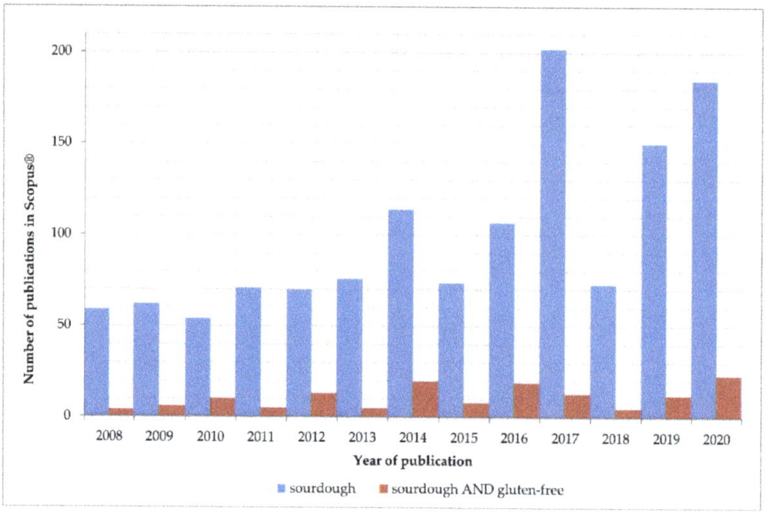

Figure 2. Number of publications retrieved from Scopus® in the last 12 years using the terms "sourdough" or "sourdough AND gluten-free".

The microorganisms (LAB and yeasts) isolated from different GF sourdough are presented in Table 6, according to information retrieved from different works and summarized by reviews from De Vuyst et al. [54] and Gobbetti et al. [59]. The sourdoughs are classified based on its origin (country), type of flour, fermentation method, fermentation place, and identification method. These results are difficult to compare, because of controllable and non-controllable factors that select the sourdough microbiota, including dough yield, propagation temperature, number and frequency of refreshments, use of starters, or the use of other ingredients.

Table 6. Microorganisms isolated from different GF sourdoughs.

Country [1]	**Flour Type** [1]	**Propagation Method** [1]	**Identification Method** [1]	**Microorganisms Reported (LAB** [2]**/Y** [3]**)**	**References(s)**
Argentina	Amaranth	Laboratory	Molecular	LAB: *Lactobacillus plantarum* [4]	[59,69]
	Quinoa	Laboratory	Molecular	LAB: *Lb. brevis* [5], *Lb. plantarum*	[58,59]
		n.i.	n.i.	LAB: *Lb. plantarum*	[17,64]
Belgium	Teff	Bakery	Molecular	LAB: *L. brevis*, *L. helveticus*, *Lb. plantarum*, *L. sanfranciscensis*, *P. pentosaceus*	[70]
				Y: *K. exigua*	
		Laboratory	Molecular	LAB: *L. fermentum*, *Lb. plantarum*, *L. sanfranciscensis*, *W. cibaria*, and *P. pentosaceus*	
				Y: *S. cerevisiae*	
Botswana	Sorghum	n.i.	n.i.	LAB: *Lb. harbinensis* [6], *Lb. parabuchneri* [7], *Lb. plantarum*	[64,71]
China	Rice	Bakery	Molecular	LAB: *Enterococcus durans*, *E. faecium*, *Lb. plantarum*, *Pediococcus pentosaceus*	[59,72]
				Y: *Saccharomyces cerevisiae*, *Saccharomycopsis fibuligera*, *Torulaspora delbrueckii*, *Wickerhamomyces anomalus*	

Table 6. *Cont.*

Country [1]	Flour Type [1]	Propagation Method [1]	Identification Method [1]	Microorganisms Reported (LAB [2]/Y [3])	Reference(s)
	Maize	Bakery	Molecular	LAB: *E. durans*, *Lb. plantarum*, *P. pentosaceus* Y: *S. cerevisiae*, *T. delbrueckii*, *W. anomalus*	
Ethiopia	Teff	Laboratory	Phenotypic	LAB: *E. faecalis*, *Lb. brevis*, *Lb. fermentum* [8], *Lb. plantarum*, *Leuconostoc mesenteroides*	[59,73]
		Laboratory	Molecular + phenotypic	LAB: *Lb. fermentum*, *Lb. graminis* [9], *Lb. parabuchneri*, *Lb. plantarum*	[59,74]
		Laboratory	Phenotypic	LAB: *E. casseliflavus*, *Lb. fermentum*, *Lactococcus piscium*, *Lc. plantarum*, *Lc. raffinolactis*, *Le. mesenteroides*, *P. acidilactici*, *P. pentosaceus* Y: *Candida humilis*, *C. tropicalis*, *Kazachstania exigua*, *Pichia norvegensis*, *S. cerevisiae*	[59,75]
		Laboratory	Molecular + phenotypic	LAB: *Lb. fermentum*, *Lb. graminis*, *Lb. parabuchneri*, *Lb. plantarum*	[64,74]
France	Rice + buckwheat	Laboratory	Molecular	LAB: *Lb. sakei* [10] Y: *C. humilis*	[59,76]
Ghana	Maize	Bakery	Phenotypic	Y: *C. tropicalis*, *Kluyveromyces marxianus*, *P. kudriavzevii*, *S. cerevisiae*	[59,77]
Germany	Buckwheat	Laboratory	Molecular	LAB: *Lb. fermentum*, *Lb. helveticus*, *Lb. paralimentarius*, *Lb. plantarum* Y: not detected	[59,78]
	Amaranth	Laboratory	Molecular	LAB: *Lb. paralimentarius*, *Lb. plantarum*, *Lb. sakei*, *P. pentosaceus*	[59,79]
		Laboratory, use of a starter including all LAB species on the right column	Molecular	LAB: *Lb. plantarum*, *Lb. sakei*, *P. pentosaceus*	
		Laboratory, use of a starter including all LAB species on the right column and *Lb. acetotolerans*, *Lb. brevis*, *Lb. casei*, *Lb. curvatus*, *Lb. sanfranciscensis*, *Lb. spicheri*, *Lc. lactis*, *Le. paramesenteroides* and yeast species *C. humilis*, *W. anomalus*, *P. kudriavzevii*, *S. cerevisiae*, *Torulaspora sp*	Molecular	LAB: *Lb. fermentum*, *Lb. helveticus*, *Lb. paralimentarius*, *Lb. plantarum*, *Lb. spicheri* [11] Y: *C. glabrata*, *S. cerevisiae*	[59,78]
		Laboratory	Molecular	LAB: *Lb. plantarum*, *Lb. sakei*	[64,79]

Table 6. *Cont.*

Country [1]	Flour Type [1]	Propagation Method [1]	Identification Method [1]	Microorganisms Reported (LAB [2]/Y [3])	Reference(s)
	Rice	Laboratory, use of a starter (mother sponge) including underlined species on the right column and *Lb. perolens*	Molecular + phenotypic	LAB: *Lb. paracasei, Lb. paralimentarius,Lb. spicheri*	[59,80]
				Y: *S. cerevisiae*	
		Laboratory, use of a starter including underlined species on the right column and yeast specie *P. membranifaciens.*	Molecular + phenotypic	LAB: *Lb. curvatus, Lb. fermentum, Lb. gallinarum, Lb. kimchii* [12],*Lb. plantarum, Lb. pontis* [13]	
				Y: *P. kudriavzevii, S. cerevisiae*	
		Laboratory	Molecular	LAB: *Lb. fermentum, Lb. helveticus, Lb. plantarum, Lb. pontis*	[59,78]
				Y: *S. cerevisiae*	
		Laboratory	Molecular	LAB: *Lb. kimchii, Lb. paralimentarius, Lb. perolens* [14]	[64,80]
	Maize	Laboratory, use of a starter including all species on the right column and *Lb. acetotolerans, Lb. brevis, Lb. casei, Lb. curvatus, Lb. sanfranciscensis, Lb. spicheri, Lc. lactis, Le. paramesenteroides* and yeast species *C. humilis, W. anomalus, Torulaspora sp.*	Molecular	LAB: *Lb. fermentum, Lb. helveticus, Lb. paralimentarius, Lb. pontis*	[59,78]
				Y: *P. kudriavzevii, S. cerevisiae*	
	Millet		Molecular	LAB: *Lb. fermentum, Lb. helveticus, Lb. pontis*	
				Y: *S. cerevisiae*	
	Quinoa		Molecular	LAB: *Lb. fermentum, Lb. helveticus, Lb. paralimentarius, Lb. plantarum, Lb. pontis*	
				Y: *P. kudriavzevii, S. cerevisiae*	
Italy	Quinoa	Laboratory	Molecular	LAB: *Lb. plantarum*	[17,59]
	Teff	Laboratory	Molecular	LAB: *Lb. plantarum, Lb. fermentum.*	[81]
				Y: *S. cerevisiae*	
Ireland	Buckwheat	Laboratory use of a starter use of a starter including all LAB species on the right column and *Lb. helveticus, Lb. paracasei, Lb. pontis, Lb. reuteri,* and yeast species *C. humilis* and *S. pastorianus*	Molecular	LAB: *Lb. amylovorus, Lb. brevis, Lb. fermentum, Lb, frumenti* [15], *Lb. paralimentarius, Lb. plantarum, Lb. sanfranciscensis* [16], *Leuconostoc argentinum* [17], *Weissella cibaria*	[59,82]
				Y: not detected	
		Laboratory	Molecular + phenotypic	LAB: *Lb. acidophilus, Lb. amylovorus, Lb. crispatus, Lb. fermentum, Lb. gallinarum, Lb. graminis, Lb. helveticus, Lb. plantarum, Lb.sakei, Lb. vaginalis*	[64,83]

Table 6. *Cont.*

Country [1]	Flour Type [1]	Propagation Method [1]	Identification Method [1]	Microorganisms Reported (LAB [2]/Y [3])	Reference(s)
	Teff	Laboratory	Molecular	LAB: *Lb. crispatus, Lb. fermentum, Lb. gallinarum, Lb. graminis, Lb. plantarum, Lb. sakei, Lb. vaginalis, Le. holzapfelii, P. pentosaceus, W. cibaria*	[59,83]
				Y: *K. barnetti*	
		Laboratory, use of a starter use of a starter including all LAB species on the right column and *Lb. helveticus, Lb. paracasei, Lb. pontis, Lb. reuteri*, and yeast species *C. humilis* and *S. pastorianus*	Molecular	LAB: *Lb. amylovorus, Lb. brevis, Lb. fermentum, Lb. frumenti, Lb. paralimentarius, Lb. plantarum, Lb. pontis, Lb. reuteri* [18], *Lb. sanfranciscensis, P. acidilactici*	
				Y: *K. barnettii, S. cerevisiae*	
		Laboratory	Molecular + phenotypic	LAB: *Lb. amylovorus, Lb. fermentum, Lb. gallinarum, Lb. plantarum, Lb. vaginalis* [19]	[64,83]
		Laboratory	Molecular	LAB: *Lb. fermentum, Lb. gallinarum, Lb. pontis, Lb. vaginalis, Le. holzapfelii, P. pentosaceus*	[59,83]
				Y: *C. glabrata, S. cerevisiae*	
Morocco	Maize	n.i.	n.i.	LAB: *Lb. alimentarius, Lb. casei* [20]	[64,84]
Nigeria	Maize	Laboratory	Molecular	LAB: *Lb. brevis, Lb. casei, Lb. fermentum, Lb. plantarum, Le. mesenteroides, P. acidilactici*	[59,85]
				Y: *C. albicans, S. cerevisiae, Schizosaccharomyces pombe*	
		Laboratory	Phenotypic	LAB: *Lb. brevis, Lb. casei, Lb. fermentum, P. acidilactici, P. pentosaceus*	[59,86]
		Laboratory	Molecular	LAB: *Lb. acidophilus, Lb. brevis, Lb. casei, Lb. fermentum, Lb. plantarum*	[64,85]
Portugal	Maize	Bakery	Phenotypic	LAB: *E. casseliflavus, E. durans, E. faecium, Lb. brevis, Lb. curvatus, Lc. lactis subsp. lactis, Leuconostoc* spp., *Streptococcus constellatus, S. equinus*	[59,64,87]
				Y: *S. cerevisiae, T. delbrueckii, W. anomalus*	
Saudi Arabia	Sorghum	Bakery	Phenotypic	LAB: *Lb. brevis, Lb. cellobiosus* [21], *Lb. lactis, P. pentosaceus*	[59,64,88]
				Y: *C. norvegensis, C. parapsilosis, Rhodotorula glutinis*	

Table 6. *Cont.*

Country [1]	Flour Type [1]	Propagation Method [1]	Identification Method [1]	Microorganisms Reported (LAB [2]/Y [3])	Reference(s)
Sudan	Sorghum	Laboratory	Phenotypic	LAB: *Lb. brevis*, *Lb. confusus* [22], *Lactobacillus* spp., *P. pentosaceus* Y: *C. intermedia*, *Debaromyces hansenni*	[59,89]
		Laboratory	Phenotypic	LAB: *Lb. amylovorus*, *Lb. fermentum*, *Lb. reuteri* Y: *P. kudriavzevii*	[59,90]
		Laboratory	Molecular + phenotypic	LAB: *E. faecalis*, *Lb. fermentum*, *Lb. helveticus*, *Lb. reuteri*, *Lb. vaginalis*, *Lc. lactis*	[59,64, 90,91]

[1] The sourdoughs are classified depending on the origin of the country, the type of flour, the propagation place (laboratory or bakery), and the identification method (molecular or phenotypic). Each row corresponds to an independent experiment. [2] LAB: lactic acid bacteria species; [3] Y: yeast species; n.i.: not indicated. [4] *Lactobacillus plantarum* (Orla-Jensen 1919) Bergey et al. 1923 pro synonymon (pro synon.) *Lactiplantibacillus plantarum* (Orla-Jensen 1919) Zheng et al. *2020*. [5] *Lactobacillus brevis* (Orla-Jensen 1919) Bergey et al. 1934 pro synon. *Levilactobacillus brevis* (Orla-Jensen 1919) Zheng et al. 2020. [6] *Lactobacillus harbinensis* Miyamoto et al. 2006 pro synon. *Schleiferilactobacillus harbinensis* (Miyamoto et al. 2006) Zheng et al. 2020. [7] *Lactobacillus parabuchneri* pro synon. *Lentilactobacillus parabuchneri* (Farrow et al. 1989) Zheng et al. 2020. [8] *Lactobacillus fermentum* Beijerinck 1901 pro synon. *Limosilactobacillus fermentum* (Beijerinck 1901) Zheng et al. 2020. [9] *Lactobacillus graminis* Beck et al. 1989 pro synon. *Latilactobacillus graminis* (Beck et al. 1989) Zheng et al. 2020. [10] *Lactobacillus sakei* Katagiri et al. 1934 pro synon. *Latilactobacillus sakei* (Katagiri et al. 1934) Zheng et al. 2020. [11] *Lactobacillus spicheri* Meroth et al. 2004 pro synon. *Levilactobacillus spicheri* (Meroth et al. 2004) Zheng et al. 2020. [12] *Lactobacillus kimchii* Yoon et al. 2000 pro synon. *Companilactobacillus kimchii* (Yoon et al. 2000) Zheng et al. 2020. [13] *Lactobacillus pontis* Vogel et al. 1994 pro synon. *Limosilactobacillus pontis* (Vogel et al. 1994) Zheng et al. 2020. [14] *Lactobacillus perolens* Back et al. 2000 pro synon. *Schleiferilactobacillus perolens* (Back et al. 2000) Zheng et al. 2020. [15] *Lactobacillus frumenti* Müller et al. 2000 pro synon. *Limosilactobacillus frumenti* (Müller et al. 2000) Zheng et al. 2020. [16] *Lactobacillus sanfranciscensis* corrig. (ex Kline and Sugihara 1971) Weiss and Schillinger 1984 pro synon. *Fructilactobacillus sanfranciscensis* (Weiss and Schillinger 1984) Zheng et al. 2020. [17] *Leuconostoc argentinum* Dicks et al. 1993 pro synon. *Leuconostoc lactis* Garvie 1960. [18] *Lactobacillus reuteri* Kandler et al. 1982 pro synon. *Limosilactobacillus reuteri* (Kandler et al. 1982) Zheng et al. 2020. [19] *Lactobacillus vaginalis* Embley et al. 1989 pro synon. *Limosilactobacillus vaginalis* (Embley et al. 1989) Zheng et al. 2020. [20] *Lactobacillus casei* (Orla-Jensen 1916) Hansen and Lessel 1971 pro synon. *Lacticaseibacillus casei* (Orla-Jensen 1916) Zheng et al. 2020. [21] *Lactobacillus cellobiosus* (Rogosa et al. 1953) pro synon. *Limosilactobacillus fermentum* (Beijerinck 1901) Zheng et al. 2020. [22] *Lactobacillus confusus* (Holzapfel and Kandler 1969) Sharpe et al. 1972 pro synon. *Weissella confusa* corrig. (Holzapfel and Kandler 1969) Collins et al. 1994.

The type of sourdough determines the microorganisms that will proliferate. Studies included in both reviews [54,59] are mainly focused on type 0 and type I sourdoughs, the most interesting ones.

Selecting the same type of sourdough (made from corn), Vogelmann et al. ([84] included in the review published by Luc De Vuyst et al. in 2017 [54]) isolated different species when it was fermented in Germany, or in China, with the only exception of *S. cerevisiae*. Considering that type I sourdough is fermented at RT, this value fluctuates between countries, and could be a main determinant for the selection of microorganisms. Besides that, the corn sourdough from China used a traditional starter culture, named *Jiaozi*, which could have addressed the selection of the final microbiota composition [92].

Figures 3 and 4 show heat maps depicting the frequency of isolation of different yeast and LAB species from different GF sourdoughs, based on the findings of the present review.

A similar scenario than the one described for GC sourdoughs is observed in Figure 3, where the frequencies of different yeast species are shown. *S. cerevisiae*, being used as a commercial starter culture, is part of the bakery's environment and can be isolated from most of the GF sourdoughs. If we compare these results with the ones presented in Table 6, the absence of *S. cerevisiae* in the sourdoughs is related to a fermentation performed in the laboratory.

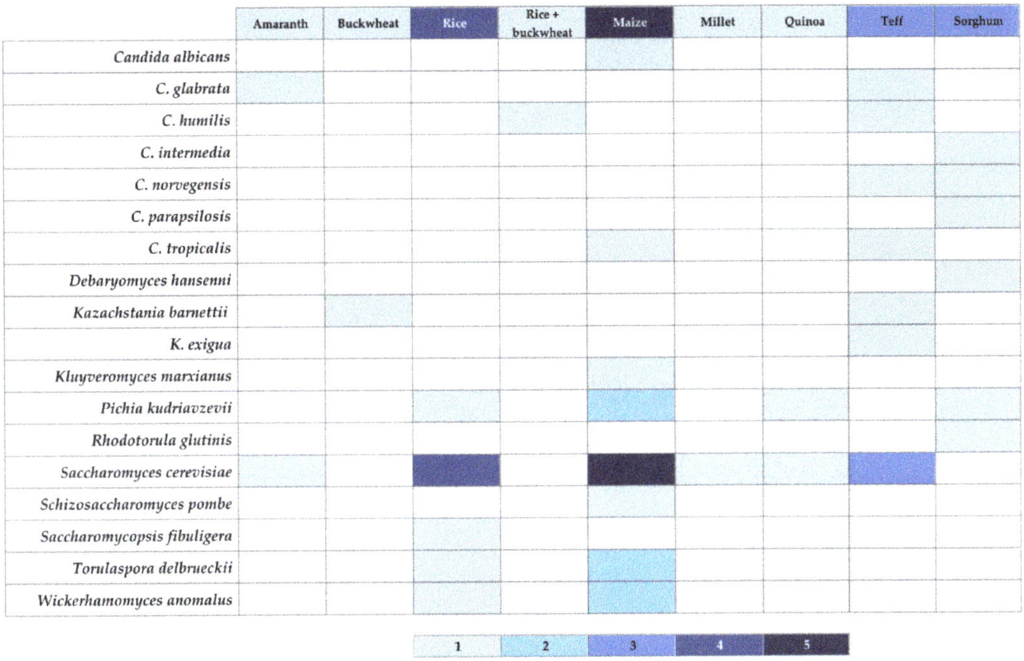

Figure 3. Heat map for yeast species. The presence of certain yeast species, isolated from the specific GF sourdough indicated in table header is described with colored cells. The intensity of blue color, as shown in the scale at the bottom, represents the least (1) and the most (5) frequent isolations, within the findings of this review. The intensity of blue color in the table header cells represents the least (1) and the most (5) analyzed type of GF sourdough within the examined results. Authors' own elaboration based on the findings of the present review.

In Figure 4, where the frequencies of different LAB species are shown, there are some recurrent bacteria that can be widely isolated due to their colonization ability. For example, *Lb. fermentum* has been isolated from practically all sourdoughs, indicating that this microorganism should be specially considered in sourdough biotechnology. The following ones, in decreasing order of frequency, are *Lb. plantarum* and *P. pentosaceus*. However, *Lb. sanfranciscensis*, considered as an endemic bacteria of type I GC sourdoughs, has only been isolated in two types of GF flours (buckwheat and teff), and not in all cases.

According to the Spanish bread quality standard [86], it can be indicated that a bread is made with sourdough as long as it is in a proportion equal or superior to 5% of the total weight of the flour of the final dough. The most-used proportion of gluten-free sourdough is usually around 20% [34,87], since it seems to give better results. However, it has been observed that this amount depends on the type of flour used to make the gluten-free sourdough. For example, in the elaboration of GF bread with sourdough from chestnut flour, good results were observed with concentrations between 30 and 50% [88]. Using both fresh and freeze-dried rice sourdough flour, the best sensory results were obtained with 10 to 20% of added sourdough [89]. In another work, the best results were obtained using amounts of 20 to 30% with both fresh and freeze-dried sourdough from buckwheat flour [90]. In a similar research using both fresh and freeze-dried sourdough from amaranth, the best sensory results were obtained with an amount of 10%, GF bread being sensorially rejected if the concentrations added were of 20% [91].

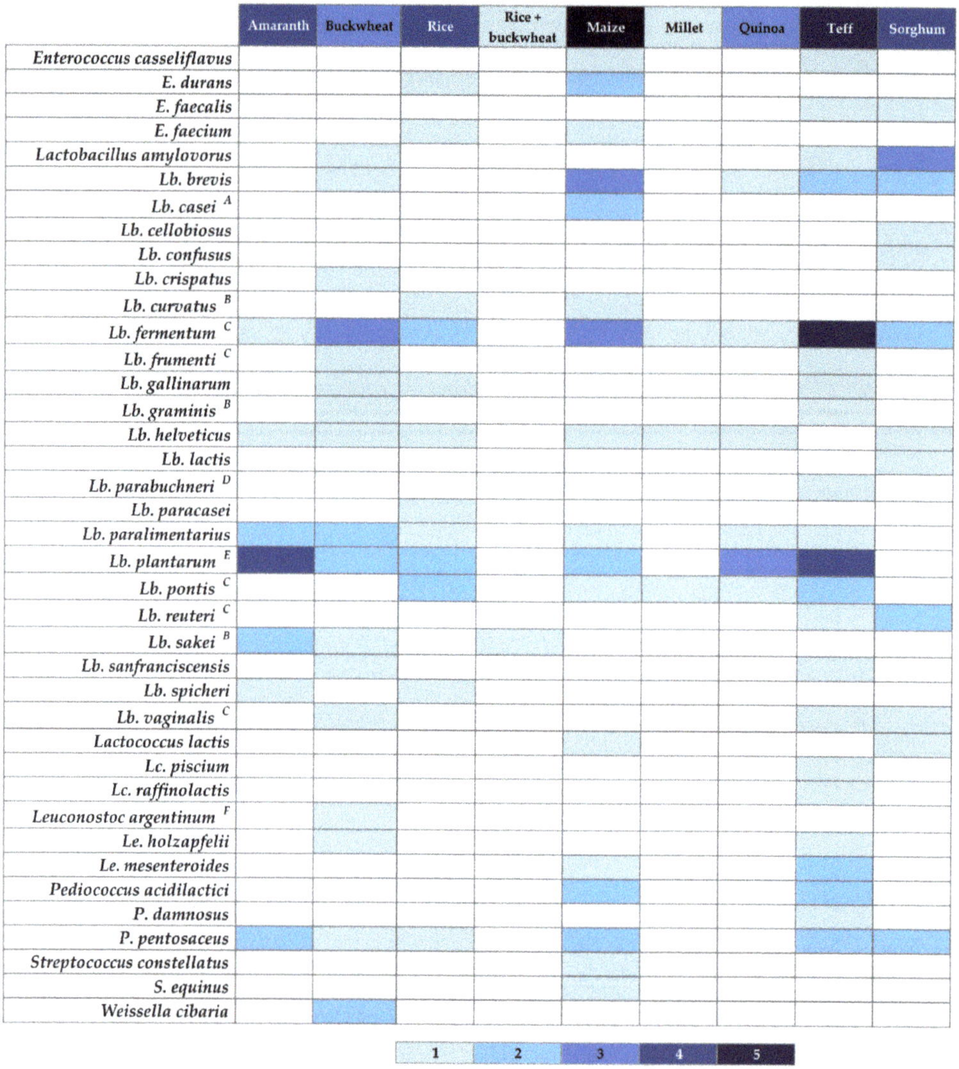

[A] *Lactobacillus casei* (Orla-Jensen 1916) Hansen and Lessel 1971 pro synonymon (pro synon.) *Lacticaseibacillus casei* (Orla-Jensen 1916) Zheng et al. 2020.

[B] *Lactobacillus curvatus* Troili-Petersson 1903 pro synon. *Latilactobacillus curvatus* (Troili-Petersson 1903) Zheng et al. 2020; *Lactobacillus graminis* Beck et al. 1989 pro synon. *Latilactobacillus graminis* (Beck et al. 1989) Zheng et al. 2020; *Lactobacillus sakei* Katagiri et al. 1934 pro. synon. *Latilactobacillus sakei* (Katagiri et al. 1934) Zheng et al. 2020.

[C] *Lactobacillus fermentum* Beijerinck 1901 pro synon. *Limosilactobacillus fermentum* (Beijerinck 1901) Zheng et al. 2020; *Lactobacillus frumenti* Müller et al. 2000 pro synon. *Limosilactobacillus frumenti* (Müller et al. 2000) Zheng et al. 2020; *Lactobacillus pontis* Vogel et al. 1994 pro synon. *Limosilactobacillus pontis* (Vogel et al. 1994) Zheng et al. 2020; *Lactobacillus reuteri* Kandler et al. 1982 pro synon. *Limosilactobacillus reuteri* (Kandler et al. 1982) Zheng et al. 2020; *Lactobacillus vaginalis* Embley et al. 1989 pro synon. *Limosilactobacillus vaginalis* (Embley et al. 1989) Zheng et al. 2020.

[D] *Lactobacillus parabuchneri* (Farrow et al. 1989) pro synon. *Lentilactobacillus parabuchneri* (Farrow et al. 1989) Zheng et al. 2020.

[E] *Lactobacillus plantarum* (Orla-Jensen 1919) Bergey et al. 1923 pro synon. *Lactiplantibacillus plantarum* (Orla-Jensen 1919) Zheng et al. 2020.

[F] *Leuconostoc argentinum* Dicks et al. 1993 pro synon. *Leuconostoc lactis* Garvie 1960.

Figure 4. Heat map for LAB species. The presence of certain LAB species, isolated from the specific GF sourdough indicated in table header, is described with colored cells. The intensity of blue color, as shown in the scale at the bottom, represents the least (1) and most (5) frequent isolations, within the findings of this review. The intensity of blue color in the table header cells represents the least (1) and most (5) analyzed type of GF sourdough within the examined results. The bacterial nomenclature was revised according to Zheng et al. [55] and Parte et al. [85]. Authors' own elaboration based on the findings of the present review.

4. Identification of Microorganisms Capable of Producing Hydrocolloid-Like Compounds

The overall benefits that sourdough provides to bakery products have already been described: improvements at organoleptic (taste, texture, and aroma) and nutritional (hydrolysis of anti-nutrients, such as phytic acid) levels, the extension of shelf life, and synthesis of functional molecules (prebiotics, antioxidants, antifungals, peptidases that degrade immunogenic peptides, etc.).

All these properties are mainly attributable to the microbiota (bacteria and yeasts) that proliferates and is established in the sourdough. As a result of the metabolic processes, these microorganisms synthesize and release molecules with diverse properties and functionalities. Within this biodiversity, bacterial contributions are the most relevant. The main function of yeasts is the CO_2 production, although they also contribute to the synthesis of metabolites, such as alcohols and derived esters, and the characteristic flavor and aromas of the crumb of fermented products [93].

Analyzing the published literature, it has been observed that bacteria are the microorganisms that contribute most to these technological improvements by synthesizing a diverse group of molecules, called EPS. These molecules are long-chain carbohydrates (polysaccharides) that widely differ among them in terms of their molecular characteristics, composition, structure, and even mechanisms by which they are synthesized [94–96]. In sourdough, EPS can improve technological properties, avoiding the addition of other hydrocolloids. Moreover, they can present other properties, such as prebiotic, immunomodulatory, antioxidant, pathogen inhibition, etc. [97–99].

There are two types of EPS—heteropolysaccharides (HePS) and homopolysaccharides (HoPS) [96,99–103]:

- HePS are described as such because the sugar polymer chain is made of different monosaccharides, usually D-galactose, D-glucose, R-rhamnose and, to a lesser extent, other N-acetylated monosaccharides, varying from two to eight different monomers, and with a molecular weight up to 10^6 Da. A large variety of HePS can be synthesized by LAB, depending on the type of monosaccharides, bonds between these monosaccharides, and spatial configurations (linear vs. branched). As an example, Suzuki et al. studied how *Lactococcus lactis* can synthesize a high number of different HePS [102]. HePS are synthesized from sugar–nucleotide precursors, intracellularly (in the cytoplasm), and in small quantities, usually between 10 and 166 mg/L. The yield of this synthesis depends on several factors: by optimizing some culture parameters of *Lb. plantarum*, Ismail and Nampoothiri achieved a final EPS concentration of 1.2 g/L [103]. Xanthan and gellan gums are HePS synthesized by bacteria belonging to phylum "*Proteobacteria*".
- HoPS are polymers based on a single type of monosaccharide (glucose or fructose), and, because of this, they are recognized as glucans or fructans (also designated as fructooligosaccharides or FOS) [96,100]. Its synthesis is extracellular, from sucrose, by the action of enzymes (glycosyl hydroxylases), and with a molecular weight greater than HePS ($>10^6$ Da). For the polymerization of glucose or fructose, these enzymes employ the energy of the glycosidic bond. HoPS are synthesized by different genera of LAB (mainly, *Lactobacillus, Streptococcus, Leuconostoc, Oenococcus* and *Weissella*) and in an amount greater than HePS, reaching up to 10 g/L. In addition to this first classification of HoPS (in glucans and fructans), these compounds are also classified based on the carbons involved in the glycosidic linkages of the backbone chain of the polymer.
    - Within the group of glucans, the following types are recognized: *dextrans, mutans, reuterans*, and *alternans*. Dextrans are the HoPS with the most technological relevance, being the only EPS synthesized at an industrial level, widely used as, for example, a thickener for jams and ice cream: they reduce crystallization, increase moisture retention, and do not affect taste.

  - Two types of fructans can be distinguished: *inulin* and *levan*. As its prebiotic properties, inulin is acquiring a greater role in the current market. Recently, it has been reported that fructans can induce gastrointestinal symptoms in individuals with self-reported non-celiac gluten sensitivity [104].

Normally, LAB species that synthesize HoPS only produce a single glycosyl hydroxylase enzyme and, consequently, a single type of EPS. There are some exceptions, such as *Leuconostoc mesenteroides*, which produces dextran, alternan and levan [96].

Once the EPS types are exposed, and which LAB are related to their synthesis have been identified, the next step will be the study and physical–chemical characterization of each EPS, to determine its activity and technological properties which are contributing or could contribute to the doughs [100]. From the point of view of the GF bakery industry, the most important property of certain EPS is to aid to resemble texture and appearance of GF baked goods to wheat-based baked products.

At this point, it is essential to remember that because EPS are a very heterogeneous group of compounds, not all of them have the same properties; therefore, not all of them can emulate the functions of gluten molecules in doughs.

Current research is focused on the study of each type of EPS and on the identification of those with technological potential as substitutes of gluten. This will allow three approaches, based on sourdough and LAB, to try to solve the problem of low sensory quality of gluten-free products [21,81,105,106]:

- Using mixtures of GF flours, where each flour supplies a type of bacteria that produces the EPS that we are looking for.
- Using controlled fermentation processes oriented to the development of the microbiota of interest.
- Using commercial starters based on bacteria strains selected because of their technological potential.

The technological and functional properties of EPS is due to its ability to act as hydrocolloids in the dough [58,100]: (i) increasing water absorption, (ii) improving rheology, (iii) increasing the final volume, (iv) increasing the softness of the crumb, and (v) increasing the shelf life by avoiding starch retrogradation.

We have already seen that in the GF products' industry, the use of hydrocolloids is widely employed, HPMC and xanthan gum (which is the only microbial EPS with relevance as an additive) being the most widely used [101]. The characterization of certain EPS confirms that, in the dough, they behave in a similar way to these exogenous additives. They are also capable of interacting with water molecules and forming a mesh-like structure with gel properties, which increases CO_2 retention (although the exact mechanisms of this behavior are still unknown) [101].

The EPS that are most used for this purpose are the HoPS because they are synthesized extracellularly, reaching higher concentrations that are relevant at a functional level. It is estimated that the amount of HoPS synthesized can reach values around 0.8% w/v, and considering that hydrocolloids are usually added in dough at 0.3% w/v, it is logical to think that they could be used as potential substitutes of these additives [21,101].

Zannini et al. presented a brief classification of HoPS, the corresponding LAB that are involved in their synthesis and the main food industrial applications of HoPS in an interesting mini-review [96]. The EPS synthesized by different LAB, and the properties attributed to them in experimental tests on specific sourdoughs has also been reviewed by Lynch et al. [101].

The conditions of EPS production by sourdough lactobacilli depend on several factors, such as sourdough composition (available carbon sources, mainly sugars, and their concentration, nitrogen sources, content of other nutrients), fermentation conditions (time, temperature, oxygen, pH), *Lactobacillus* species, and the type of flour used, among others [100,101,107–109]. The concentration of fermentable sugars present in the dough affects the EPS microbial synthesis [110]. Sucrose concentration is of particular relevance for some species, such as *Weissella cibaria* [96,110,111].

Considering this information, we could think that it is as simple as selecting some LAB and designing a starter culture with technological properties. This selection would be made based on its ability to synthesize EPS and other properties of interest, such as its growth kinetics, its acidification capacity, its fermentation quotient (ratio between acetic acid and lactic acid), its release of amino acids involved in the formation of aroma and flavor, or its ability to hydrolyze immunogenic gluten peptides (eliminating possible cross contamination and making safer products for CD patients) [111].

However, considering what a sourdough is, the inherent complexity and the variability factors that affect this ecosystem, it is logical to think that the development of these starters is somewhat more complex.

Experimental tests suggest that the selection of these LAB should be carried out on the endemic bacteria of each sourdough; that is, they should be isolated in that specific process, in such a way that we can ensure that they will be adapted to that substrate and fermentation conditions and be competitive enough to outperform the rest of the present microorganisms [112].

Again, we find that research on GF sourdoughs is scarce, and the use of commercial starters tested (with good results) in GC doughs is not useful in GF flours. Moroni et al. investigated two commercial starters for GC doughs in buckwheat and teff flours, with negative results. In fact, both *Lb. helveticus* as *Lb. paracasei*, which were both part of this starter, were not isolated from the mature sourdoughs [77]. Galle et al., using *Lb. buchneri* (producer of HePS) in sorghum sourdough, also obtained loaves with a loss of elasticity with respect to the control, a phenomenon that did not occur in doughs made with wheat [113].

Therefore, it is important to select bacteria strains within the native microbiota with desirable properties that allow rapid adaptation, intense acidification, and a positive influence at both a technological and nutritional level [114].

As some examples of positive experimental results, Galle et al. showed that sorghum sourdoughs were improved with the addition of *W. cibaria* and *Lb. reuteri* by producing dextran and fructan, respectively [105]. Wolter et al. also optimized the use of *W. cibaria* in their bread model made with buckwheat, quinoa, sorghum, and teff flours. They also verified how the type of flour influenced the amount of dextran synthesized by this bacterium [87]. In a research study developed by Nami et al., the use of sourdoughs with starters based on combinations of four LAB species improved the quality and shelf-life of GF pearl millet bread, with starters based on *L. brevis* and *L. paralimentarius* being the most successful ones [106]. Dingeo et al. achieved good nutritional values in gluten-free muffins baked with a teff Type-I sourdough, dominated mainly by *Lb. plantarum*, *Lb. fermentum* and *S. cerevisiae* [76]. The interpretation we can give is that further investigation is necessary for each particular case. Starting from bacteria present in the sourdoughs of each type of flour and specific process, those most interesting (from a technological point of view), could be selected.

On the other hand, the use of starters provides additional benefits to the use of sourdoughs since it directs the selection of microorganisms in some way [57]. In addition, it can be very useful in type II sourdoughs, so that not only acidification occurs, but also benefits attributable to the use of sourdough.

5. Concluding Remarks and Future Perspectives

Once the main functional and technological properties of the most commonly used additives and adjuvants in GF bakery have been described, the reviews selected to develop Section 2 of this paper are presented in Tables 2–5. The descriptors depicted in these tables are: (i) the type of flour used in the preparation; (ii) the additive or mixture of additives, and their concentration (if it was mentioned in the article); and (iii) both the positive and negative technological properties described in the final product. Most of the studies refer to GF bread and, in almost all cases, the type or types of flours used in the preparation are also indicated (when the study refers to another type of product, it is also indicated in the tables). The overall conclusion of Section 2 is that it is complicated to establish beneficial or harmful properties (from a technological point of view), of any additive, since they are based on a set of variables (e.g., food matrix, type of additive, concentration at which it will be used, or interactions between the different ingredients and the subsequent processing). As with any other ingredient, additives make the final product more expensive, and need to be tested for every specific condition, since their technological contribution depends on the characteristics of each dough. In addition, additives must be declared on the label, which is a problem for some consumers who are reluctant to use food additives.

From Section 3, it can be concluded that there is a high variability of microorganisms present in GF sourdough. The papers analyzed suggest that, similarly to GC flours, their GF counterparts have endemic LAB that can be isolated in practically all GF sourdoughs. Therefore, the study of autochthonous microbiota highlights that there are some species strong enough and adapted to the ecosystem that can be considered as endemic in these sourdoughs, and able to compete and proliferate independently of the process. However, more studies are needed to compare the results and to correctly identify autochthonous microbiota in GF sourdough.

It can be postulated from Section 4 that each sourdough contains at least one EPS-producing *Lactobacillus* strain, so the use of fermentation could replace additives as functional ingredients. From the knowledge of the microbiota present in the GF sourdoughs and the EPS synthesized by these microorganisms, the best species could be selected (based on their technological and nutritional potential) as starter cultures. These starters, formed by bacteria and yeasts selected for their technological characteristics, could improve bakery processes (including products fermented at industrial level). Further research is necessary in this field to develop the full potential of an economic and ecological biotechnology, such as the use of sourdough, which is capable of positively influencing all the parameters with which we measure the final quality of GF products.

Author Contributions: Conceptualization, L.R. and J.R.M.; validation, A.A.-H. and J.R.M.; investigation, L.R.; writing—original draft preparation, L.R.; writing—review and editing, L.R., A.A.-H. and J.R.M.; visualization, A.A.-H., L.R., J.R.M., J.A.M.-P., M.M.-C., A.P.-M., P.C.-L., E.T.-J.; supervision, A.A.-H. and J.R.M.; funding acquisition, A.A.-H. and J.R.M. All authors have read and agreed to the published version of the manuscript.

Funding: The research leading to these results has received funding from "la Caixa" Foundation and Caja Burgos Foundation, grant number FUi1-PI001. The corresponding author wishes to acknowledge the financial support granted to the project FUi1-PI001 "*Caracterización de la microbiota indígena de harinas sin gluten (Proyecto MADRE)*" by these entities.

Data Availability Statement: No new data were created or analyzed in this study. Data sharing is not applicable to this article.

Conflicts of Interest: The authors declare no conflict of interest. The funders had no role in the design of the study; in the collection, analyses, or interpretation of data; in the writing of the manuscript, or in the decision to publish the results.

References

1. Ribeiro, M.; Nunes, F.M.; Rodriguez-Quijano, M.; Carrillo, J.M.; Branlard, G.; Igrejas, G. Next-generation therapies for celiac disease: The gluten-targeted approaches. *Trends Food Sci. Technol.* **2018**, *75*, 56–71. [CrossRef]
2. Xhakollari, V.; Canavari, M.; Osman, M. Factors affecting consumers' adherence to gluten-free diet, a systematic review. *Trends Food Sci. Technol.* **2019**, *85*, 23–33. [CrossRef]
3. Scherf, K.A.; Catassi, C.; Chirdo, F.; Ciclitira, P.J.; Feighery, C.; Gianfrani, C.; Koning, F.; Lundin, K.E.A.; Schuppan, D.; Smulders, M.J.M.; et al. Recent Progress and Recommendations on Celiac Disease from the Working Group on Prolamin Analysis and Toxicity. *Front. Nutr.* **2020**, *7*, 14. [CrossRef]
4. Lundin, K.E.A.; Wijmenga, C. Coeliac disease and autoimmune disease—genetic overlap and screening. *Nat. Rev. Gastroenterol. Hepatol.* **2015**, *12*, 507–515. [CrossRef]
5. Singh, P.; Arora, A.; Strand, T.A.; Leffler, D.A.; Catassi, C.; Green, P.H.; Kelly, C.P.; Ahuja, V.; Makharia, G.K. Global Prevalence of Celiac Disease: Systematic Review and Meta-analysis. *Clin. Gastroenterol. Hepatol.* **2018**, *16*, 823–836.e2. [CrossRef] [PubMed]

6. Ciccocioppo, R.; Kruzliak, P.; Cangemi, G.C.; Pohanka, M.; Betti, E.; Lauret, E.; Rodrigo, L. The Spectrum of Differences between Childhood and Adulthood Celiac Disease. *Nutrients* **2015**, *7*, 8733–8751. [CrossRef] [PubMed]
7. Lionetti, E.; Gatti, S.; Pulvirenti, A.; Catassi, C. Celiac disease from a global perspective. *Best Pract. Res. Clin. Gastroenterol.* **2015**, *29*, 365–379. [CrossRef] [PubMed]
8. Fedorak, R.N.; Switzer, C.M.; Bridges, R.J. Canadian Digestive Health Foundation Public Impact Series 4: Celiac disease in Canada. Incidence, prevalence, and direct and indirect economic impact. *Can. J. Gastroenterol.* **2012**, *26*, 350–352. [CrossRef] [PubMed]
9. Catassi, C.; Fabiani, E.; Ratsch, I.M.; Coppa, G.V.; Giorgi, P.L.; Pierdomenico, R.; Alessandrini, S.; Iwanejko, G.; Domenici, R.; Mei, E.; et al. The coeliac iceberg in Italy. A multicentre antigliadin antibodies screening for coeliac disease in school-age subjects. *Acta Paediatr. Suppl.* **1996**, *412*, 29–35. [CrossRef]
10. Capriles, V.D.; dos Santos, F.G.; Arêas, J.A.G. Gluten-free breadmaking: Improving nutritional and bioactive compounds. *J. Cereal Sci.* **2016**, *67*, 83–91. [CrossRef]
11. Masure, H.G.; Fierens, E.; Delcour, J.A. Current and forward looking experimental approaches in gluten-free bread making research. *J. Cereal Sci.* **2016**, *67*, 92–111. [CrossRef]
12. Capriles, V.D.; Arêas, J.A.G. Novel Approaches in Gluten-Free Breadmaking: Interface between Food Science, Nutrition, and Health. *Compr. Rev. Food Sci. Food Saf.* **2014**, *13*, 871–890. [CrossRef]
13. do Nascimento, A.B.; Fiates, G.M.R.; Dos Anjos, A.; Teixeira, E. Analysis of ingredient lists of commercially available gluten-free and gluten-containing food products using the text mining technique. *Int. J. Food Sci. Nutr.* **2013**, *64*, 217–222. [CrossRef] [PubMed]
14. Hüttner, E.K.; Arendt, E.K. Recent advances in gluten-free baking and the current status of oats. *Trends Food Sci. Technol.* **2010**, *21*, 303–312. [CrossRef]
15. Naqash, F.; Gani, A.; Gani, A.; Masoodi, F.A. Gluten-free baking: Combating the challenges - A review. *Trends Food Sci. Technol.* **2017**, *66*, 98–107. [CrossRef]
16. Moroni, A.V.; Dal Bello, F.; Arendt, E.K. Sourdough in gluten-free bread-making: An ancient technology to solve a novel issue? *Food Microbiol.* **2009**, *26*, 676–684. [CrossRef]
17. Rizzello, C.G.; Lorusso, A.; Montemurro, M.; Gobbetti, M. Use of sourdough made with quinoa (Chenopodium quinoa) flour and autochthonous selected lactic acid bacteria for enhancing the nutritional, textural and sensory features of white bread. *Food Microbiol.* **2016**, *56*, 1–13. [CrossRef] [PubMed]
18. Novotni, D.; Čukelj, N.; Smerdel, B.; Bituh, M.; Dujmić, F.; Ćurić, D. Glycemic index and firming kinetics of partially baked frozen gluten-free bread with sourdough. *J. Cereal Sci.* **2012**, *55*, 120–125. [CrossRef]
19. Polanco Allué, I. *Libro Blanco de la Enfermedad Celíaca*, 1st ed.; ICM: Madrid, Spain, 2008; ISBN 978-84-936109-4-4.
20. Nionelli, L.; Rizzello, C.G. Sourdough-Based Biotechnologies for the Production of Gluten-Free Foods. *Foods* **2016**, *5*, 65. [CrossRef]
21. Galle, S.; Schwab, C.; Arendt, E.; Ganzle, M. Exopolysaccharide-forming Weissella strains as starter cultures for sorghum and wheat sourdoughs. *J. Agric. Food Chem.* **2010**, *58*, 5834–5841. [CrossRef]
22. BeMiller, J.N. 9-Hydrocolloids. In *Gluten-Free Cereal Products and Beverages*; Arendt, E.K., Dal Bello, F., Eds.; Food Science and Technology; Academic Press: San Diego, CA, USA, 2008; pp. 203–215. ISBN 978-0-12-373739-7.
23. Sabanis, D.; Tzia, C. Effect of hydrocolloids on selected properties of gluten-free dough and bread. *Food Sci. Technol. Int.* **2011**, *17*, 279–291. [CrossRef]
24. Zannini, E.; Jones, J.M.; Renzetti, S.; Arendt, E.K. Functional Replacements for Gluten. *Annu. Rev. Food Sci. Technol.* **2012**, *3*, 227–245. [CrossRef]
25. Mir, S.A.; Shah, M.A.; Naik, H.R.; Zargar, I.A. Influence of hydrocolloids on dough handling and technological properties of gluten-free breads. *Trends Food Sci. Technol.* **2016**, *51*, 49–57. [CrossRef]
26. Demirkesen, I.; Kelkar, S.; Campanella, O.H.; Sumnu, G.; Sahin, S.; Okos, M. Characterization of structure of gluten-free breads by using X-ray microtomography. *Food Hydrocoll.* **2014**, *36*, 37–44. [CrossRef]
27. Hager, A.-S.; Arendt, E.K. Influence of hydroxypropylmethylcellulose (HPMC), xanthan gum and their combination on loaf specific volume, crumb hardness and crumb grain characteristics of gluten-free breads based on rice, maize, teff and buckwheat. *Food Hydrocoll.* **2013**, *32*, 195–203. [CrossRef]
28. O'Shea, N.; Arendt, E.; Gallagher, E. State of the art in gluten-free research. *J. Food Sci.* **2014**, *79*, R1067-76. [CrossRef] [PubMed]
29. Drabińska, N.; Zieliński, H.; Krupa-Kozak, U. Technological benefits of inulin-type fructans application in gluten-free products—A review. *Trends Food Sci. Technol.* **2016**, *56*, 149–157. [CrossRef]
30. Wang, K.; Lu, F.; Li, Z.; Zhao, L.; Han, C. Recent developments in gluten-free bread baking approaches: A review. *Food Sci. Technol.* **2017**, *37*, 1–9. [CrossRef]
31. Tsatsaragkou, K.; Gounaropoulos, G.; Mandala, I. Development of gluten free bread containing carob flour and resistant starch. *LWT - Food Sci. Technol.* **2014**, *58*, 124–129. [CrossRef]
32. Gao, Y.; Janes, M.E.; Chaiya, B.; Brennan, M.A.; Brennan, C.S.; Prinyawiwatkul, W. Gluten-free bakery and pasta products: Prevalence and quality improvement. *Int. J. Food Sci. Technol.* **2017**, *53*, 19–32. [CrossRef]
33. Costa, A.F.S.; Almeida, F.C.G.; Vinhas, G.M.; Sarubbo, L.A. Production of Bacterial Cellulose by Gluconacetobacter hansenii Using Corn Steep Liquor As Nutrient Sources. *Front. Microbiol.* **2017**, *8*, 2027. [CrossRef]

34. Schober, T.J.; Bean, S.R.; Boyle, D.L. Gluten-Free Sorghum Bread Improved by Sourdough Fermentation: Biochemical, Rheological, and Microstructural Background. *J. Agric. Food Chem.* **2007**, *55*, 5137–5146. [CrossRef]
35. Campo, E.; del Arco, L.; Urtasun, L.; Oria, R.; Ferrer-Mairal, A. Impact of sourdough on sensory properties and consumers' preference of gluten-free breads enriched with teff flour. *J. Cereal Sci.* **2016**, *67*, 75–82. [CrossRef]
36. Demirkesen, I.; Mert, B.; Sumnu, G.; Sahin, S. Rheological properties of gluten-free bread formulations. *J. Food Eng.* **2010**, *96*, 295–303. [CrossRef]
37. Cato, L.; Gan, J.J.; Rafael, L.G.B.; Small, D.M. Gluten free breads using rice flour and hydrocolloid gums. *Food Aust.* **2004**, *56*, 75–78.
38. Perez-Quirce, S.; Caballero, P.A.; Vela, A.J.; Villanueva, M.; Ronda, F. Impact of yeast and fungi $(1 \rightarrow 3)(1 \rightarrow 6)$-β-glucan concentrates on viscoelastic behavior and bread making performance of gluten-free rice-based doughs. *Food Hydrocoll.* **2018**, *79*, 382–390. [CrossRef]
39. Gularte, M.A.; de la Hera, E.; Gómez, M.; Rosell, C.M. Effect of different fibers on batter and gluten-free layer cake properties. *LWT Food Sci. Technol.* **2012**, *48*, 209–214. [CrossRef]
40. Korus, J.; Grzelak, K.; Achremowicz, K.; Sabat, R. Influence of prebiotic additions on the quality of gluten-free bread and on the content of inulin and fructooligosaccharides. *Food Sci. Technol. Int.* **2006**, *12*, 489–495. [CrossRef]
41. Deora, N.S.; Deswal, A.; Mishra, H.N. Functionality of alternative protein in gluten-free product development. *Food Sci. Technol. Int.* **2015**, *21*, 364–379. [CrossRef] [PubMed]
42. Fasano, A.; Catassi, C. Celiac Disease. *N. Engl. J. Med.* **2012**, *367*, 2419–2426. [CrossRef]
43. Renzetti, S.; Dal Bello, F.; Arendt, E.K. Microstructure, fundamental rheology and baking characteristics of batters and breads from different gluten-free flours treated with a microbial transglutaminase. *J. Cereal Sci.* **2008**, *48*, 33–45. [CrossRef]
44. Steffolani, M.E.; Ribotta, P.D.; Pérez, G.T.; León, A.E. Effect of glucose oxidase, transglutaminase, and pentosanase on wheat proteins: Relationship with dough properties and bread-making quality. *J. Cereal Sci.* **2010**, *51*, 366–373. [CrossRef]
45. Moore, M.M.; Heinbockel, M.; Dockery, P.; Ulmer, H.M.; Arendt, E.K. Network formation in gluten-free bread with application of transglutaminase. *Cereal Chem.* **2006**, *83*, 28–36. [CrossRef]
46. Storck, C.R.; da Rosa Zavareze, E.; Gularte, M.A.; Elias, M.C.; Rosell, C.M.; Guerra Dias, A.R. Protein enrichment and its effects on gluten-free bread characteristics. *LWT Food Sci. Technol.* **2013**, *53*, 346–354. [CrossRef]
47. Lerner, A.; Matthias, T. Possible association between celiac disease and bacterial transglutaminase in food processing: A hypothesis. *Nutr. Rev.* **2015**, *73*, 544–552. [CrossRef]
48. European Commission Commission Regulation (EU). No 1129/2011 of 11 November 2011 amending Annex II to Regulation (EC) No 1333/2008 of the European Parliament and of the Council by establishing a Union list of food additives. *Off. J. Eur. Union* **2011**, *L*, 1–177. [CrossRef]
49. Mohammadi, M.; Azizi, M.H.; Neyestani, T.R.; Hosseini, H.; Mortazavian, A.M. Development of gluten-free bread using guar gum and transglutaminase. *J. Ind. Eng. Chem.* **2015**, *21*, 1398–1402. [CrossRef]
50. López-Tenorio, J.A.; Rodríguez-Sandoval, E.; Sepúlveda-Valencia, J.U. The Influence of Different Emulsifiers on the Physical and Textural Characteristics of Gluten-Free Cheese Bread. *J. Texture Stud.* **2015**, *46*, 227–239. [CrossRef]
51. Rubel, W. *Bread: A Global History*; Reaktion Books: London, UK, 2011; ISBN 9781861898548.
52. Gänzle, M.; Ripari, V. Composition and function of sourdough microbiota: From ecological theory to bread quality. *Int. J. Food Microbiol.* **2016**, *239*, 19–25. [CrossRef] [PubMed]
53. Ruiz Rodríguez, L.; Vera Pingitore, E.; Rollan, G.; Cocconcelli, P.S.; Fontana, C.; Saavedra, L.; Vignolo, G.; Hebert, E.M. Biodiversity and technological-functional potential of lactic acid bacteria isolated from spontaneously fermented quinoa sourdoughs. *J. Appl. Microbiol.* **2016**, *120*, 1289–1301. [CrossRef]
54. De Vuyst, L.; Van Kerrebroeck, S.; Leroy, F. Chapter Two - Microbial Ecology and Process Technology of Sourdough Fermentation. *Adv. Appl. Microbiol.* **2017**, *100*, 49–160. [CrossRef]
55. Zheng, J.; Wittouck, S.; Salvetti, E.; Franz, C.M.A.P.; Harris, H.M.B.; Mattarelli, P.; O'toole, P.W.; Pot, B.; Vandamme, P.; Walter, J.; et al. A taxonomic note on the genus Lactobacillus: Description of 23 novel genera, emended description of the genus Lactobacillus beijerinck 1901, and union of Lactobacillaceae and Leuconostocaceae. *Int. J. Syst. Evol. Microbiol.* **2020**, *70*, 2782–2858. [CrossRef]
56. Michel, E.; Monfort, C.; Deffrasnes, M.; Guezenec, S.; Lhomme, E.; Barret, M.; Sicard, D.; Dousset, X.; Onno, B. Characterization of relative abundance of lactic acid bacteria species in French organic sourdough by cultural, qPCR and MiSeq high-throughput sequencing methods. *Int. J. Food Microbiol.* **2016**, *239*, 35–43. [CrossRef]
57. Coda, R.; Di Cagno, R.; Gobbetti, M.; Rizzello, C.G. Sourdough lactic acid bacteria: Exploration of non-wheat cereal-based fermentation. *Food Microbiol.* **2014**, *37*, 51–58. [CrossRef]
58. Chavan, R.; Chavan, R. Sourdough Technology—A Traditional Way for Wholesome Foods: A Review. *Compr. Rev. Food Sci. Food Saf.* **2011**, *10*, 169–182. [CrossRef]
59. Gobbetti, M.; Minervini, F.; Pontonio, E.; Di Cagno, R.; De Angelis, M. Drivers for the establishment and composition of the sourdough lactic acid bacteria biota. *Int. J. Food Microbiol.* **2016**, *239*, 3–18. [CrossRef] [PubMed]
60. Minervini, F.; De Angelis, M.; Di Cagno, R.; Gobbetti, M. Ecological parameters influencing microbial diversity and stability of traditional sourdough. *Int. J. Food Microbiol.* **2014**, *171*, 136–146. [CrossRef]

61. De Vuyst, L.; Van Kerrebroeck, S.; Harth, H.; Huys, G.; Daniel, H.-M.; Weckx, S. Microbial ecology of sourdough fermentations: Diverse or uniform? *Food Microbiol.* **2014**, *37*, 11–29. [CrossRef]
62. Scheirlinck, I.; Van der Meulen, R.; De Vuyst, L.; Vandamme, P.; Huys, G. Molecular source tracking of predominant lactic acid bacteria in traditional Belgian sourdoughs and their production environments. *J. Appl. Microbiol.* **2009**, *106*, 1081–1092. [CrossRef] [PubMed]
63. Minervini, F.; Lattanzi, A.; De Angelis, M.; Celano, G.; Gobbetti, M. House microbiotas as sources of lactic acid bacteria and yeasts in traditional Italian sourdoughs. *Food Microbiol.* **2015**, *52*, 66–76. [CrossRef]
64. Huys, G.; Daniel, H.-M.; De Vuyst, L. Taxonomy and Biodiversity of Sourdough Yeasts and Lactic Acid Bacteria. In *Handbook on Sourdough Biotechnology*; Gobbetti, M., Gänzle, M., Eds.; Springer: Boston, MA, USA, 2013; pp. 105–154. ISBN 978-1-4614-5425-0.
65. Puratos Sourdough Library—A Unique Collection of Natural Ferments Providing a Range of Flavours. Available online: https://www.puratos.com/commitments/next-generation/product-heritage/sourdough-library (accessed on 30 May 2018).
66. Spicher, G. Die Mikroflora des Sauerteiges. I. Mitteilung: Untersuchungen über die Art der in Sauerteigen anzutreffenden stäbchenförmigen Milchsäurebakterien (Genus Lactobacillus Beijerinck). *Zeitblatt fur Bakteriol. II Abt* **1959**, *113*, 80–106.
67. Barber, S.; Torner, M.J.; Martínez-Anaya, M.A.; de Barber, C.B. Microflora of the sour dough of wheat flour bread IX. Biochemical characteristics and baking performance of wheat doughs elaborated with mixtures of pure microorganisms. *Z. Lebensm. Unters. Forsch.* **1989**, *189*, 6–11. [CrossRef]
68. Van Kerrebroeck, S.; Maes, D.; De Vuyst, L. Sourdoughs as a function of their species diversity and process conditions, a meta-analysis. *Trends Food Sci. Technol.* **2017**, *68*, 152–159. [CrossRef]
69. Ruiz Rodríguez, L.; Vera Pingitore, E.; Rollan, G.; Martos, G.; Saavedra, L.; Fontana, C.; Hebert, E.M.; Vignolo, G. Biodiversity and technological potential of lactic acid bacteria isolated from spontaneously fermented amaranth sourdough. *Lett. Appl. Microbiol.* **2016**, *63*, 147–154. [CrossRef]
70. Harth, H.; Van Kerrebroeck, S.; De Vuyst, L. Impact of process conditions on the microbial community dynamics and metabolite production kinetics of teff sourdough fermentations under bakery and laboratory conditions. *Food Sci. Nutr.* **2018**, *6*, 1438–1455. [CrossRef]
71. Sekwati-Monang, B.; Gänzle, M.G. Microbiological and chemical characterisation of ting, a sorghum-based sourdough product from Botswana. *Int. J. Food Microbiol.* **2011**, *150*, 115–121. [CrossRef]
72. Obiri-Danso, K. Microbiological studies on corn dough fermentation. *Ceral Chem.* **1994**, *71*, 186–188.
73. Gashe, B.A. Involvement of Lactic Acid Bacteria in the Fermentation of TEF (Eragrosfis tef), an Ethiopian Fermented Food. *J. Food Sci.* **1985**, *50*, 800–801. [CrossRef]
74. Sterr, Y.; Weiss, A.; Schmidt, H. Evaluation of lactic acid bacteria for sourdough fermentation of amaranth. *Int. J. Food Microbiol.* **2009**, *136*, 75–82. [CrossRef]
75. Meroth, C.B.; Hammes, W.P.; Hertel, C. Characterisation of the Microbiota of Rice Sourdoughs and Description of Lactobacillus spicheri sp. nov. *Syst. Appl. Microbiol.* **2004**, *27*, 151–159. [CrossRef]
76. Dingeo, C.; Difonzo, G.; Paradiso, V.M.; Rizzello, C.G.; Pontonio, E. Teff type-i sourdough to produce gluten-free muffin. *Microorganisms* **2020**, *8*, 1149. [CrossRef] [PubMed]
77. Moroni, A.V.; Arendt, E.K.; Morrissey, J.P.; Bello, F.D. Development of buckwheat and teff sourdoughs with the use of commercial starters. *Int. J. Food Microbiol.* **2010**, *142*, 142–148. [CrossRef] [PubMed]
78. Moroni, A.V.; Arendt, E.K.; Bello, F.D. Biodiversity of lactic acid bacteria and yeasts in spontaneously-fermented buckwheat and teff sourdoughs. *Food Microbiol.* **2011**, *28*, 497–502. [CrossRef] [PubMed]
79. Boraam, F.; Faid, M.; Larpent, J.; Breton, A. Lactic acid bacteria and yeast associated with traditional Moroccan sour-dough bread fermentation. *Sci. Aliments* **1993**, *13*, 501–509.
80. Edema, M.; Sanni, A.I. Micro-population of fermenting maize meal for sour maize bread production in Nigeria. *Niger. J. Microbiol.* **2006**, *20*, 937–946.
81. Sanni, A.I.; Onilude, A.A.; Fatungase, M.O. Production of sour maize bread using starter-cultures. *World J. Microbiol. Biotechnol.* **1997**, *14*, 101–106. [CrossRef]
82. Mohammed, S.I.; Steenson, L.R.; Kirleis, A.W. Isolation and characterization of microorganisms associated with the traditional sorghum fermentation for production of Sudanese Kisra. *Appl. Environ. Microbiol.* **1991**, *57*, 2529–2533. [CrossRef]
83. Hamad, S.H.; Böcker, G.; Vogel, R.F.; Hammes, W.P. Microbiological and chemical analysis of fermented sorghum dough for Kisra production. *Appl. Microbiol. Biotechnol.* **1992**, *37*, 728–731. [CrossRef]
84. Vogelmann, S.A.; Hertel, C. Impact of ecological factors on the stability of microbial associations in sourdough fermentation. *Food Microbiol.* **2011**, *28*, 583–589. [CrossRef]
85. Parte, A.C.; Carbasse, J.S.; Meier-Kolthoff, J.P.; Reimer, L.C.; Göker, M. List of prokaryotic names with standing in nomenclature (LPSN) moves to the DSMZ. *Int. J. Syst. Evol. Microbiol.* **2020**, *70*, 5607–5612. [CrossRef]
86. Real Decreto 308/2019, de 26 de abril, por el que se aprueba la norma de calidad para el pan. Available online: https://www.boe.es/eli/es/rd/2019/04/26/308 (accessed on 28 June 2021).
87. Wolter, A.; Hager, A.-S.; Zannini, E.; Czerny, M.; Arendt, E.K. Influence of dextran-producing Weissella cibaria on baking properties and sensory profile of gluten-free and wheat breads. *Int. J. Food Microbiol.* **2014**, *172*, 83–91. [CrossRef]
88. Aguilar, N.; Albanell, E.; Miñarro, B.; Capellas, M. Chestnut flour sourdough for gluten-free bread making. *Eur. Food Res. Technol.* **2016**, *242*, 1795–1802. [CrossRef]

89. Różyło, R.; Rudy, S.; Krzykowski, A.; Dziki, D.; Siastała, M.; Polak, R. Gluten-Free Bread Prepared with Fresh and Freeze-Dried Rice Sourdough-Texture and Sensory Evaluation. *J. Texture Stud.* **2016**, *47*, 443–453. [CrossRef]
90. Różyło, R.; Rudy, S.; Krzykowski, A.; Dziki, D.; Gawlik-Dziki, U.; Różyło, K.; Skonecki, S. Effect of adding fresh and freeze-dried buckwheat sourdough on gluten-free bread quality. *Int. J. Food Sci. Technol.* **2015**, *50*, 313–322. [CrossRef]
91. Różyło, R.; Rudy, S.; Krzykowski, A.; Dziki, D. Novel Application of Freeze-Dried Amaranth Sourdough in Gluten-Free Bread Production. *J. Food Process Eng.* **2015**, *38*, 135–143. [CrossRef]
92. Li, Z.; Li, H.; Bian, K. Microbiological characterization of traditional dough fermentation starter (Jiaozi) for steamed bread making by culture-dependent and culture-independent methods. *Int. J. Food Microbiol.* **2016**, *234*, 9–14. [CrossRef] [PubMed]
93. De Vuyst, L.; Harth, H.; Van Kerrebroeck, S.; Leroy, F. Yeast diversity of sourdoughs and associated metabolic properties and functionalities. *Int. J. Food Microbiol.* **2016**, *239*, 26–34. [CrossRef]
94. Arendt, E.K.; Moroni, A.; Zannini, E. Medical nutrition therapy: Use of sourdough lactic acid bacteria as a cell factory for delivering functional biomolecules and food ingredients in gluten free bread. *Microb. Cell Fact.* **2011**, *10*, S15. [CrossRef]
95. Nácher-Vázquez, M. Dextranos de Bacterias Lácticas Aisladas de Productos Cárnicos: Caracterización y Aplicaciones. Ph.D. Thesis, Universitat de València, Valencia, Spain, 2015.
96. Zannini, E.; Waters, D.M.; Coffey, A.; Arendt, E.K. Production, properties, and industrial food application of lactic acid bacteria-derived exopolysaccharides. *Appl. Microbiol. Biotechnol.* **2016**, *100*, 1121–1135. [CrossRef] [PubMed]
97. Laiño, J.; Villena, J.; Kanmani, P.; Kitazawa, H. Immunoregulatory Effects Triggered by Lactic Acid Bacteria Exopolysaccharides: New Insights into Molecular Interactions with Host Cells. *Microorganisms* **2016**, *4*, 27. [CrossRef]
98. Wang, K.; Li, W.; Rui, X.; Chen, X.; Jiang, M.; Dong, M. Characterization of a novel exopolysaccharide with antitumor activity from Lactobacillus plantarum 70810. *Int. J. Biol. Macromol.* **2014**, *63*, 133–139. [CrossRef] [PubMed]
99. Wang, Y.; Ganzle, M.G.; Schwab, C. Exopolysaccharide synthesized by Lactobacillus reuteri decreases the ability of enterotoxigenic Escherichia coli to bind to porcine erythrocytes. *Appl. Environ. Microbiol.* **2010**, *76*, 4863–4866. [CrossRef] [PubMed]
100. Galle, S.; Arendt, E.K. Exopolysaccharides from sourdough lactic acid bacteria. *Crit. Rev. Food Sci. Nutr.* **2014**, *54*, 891–901. [CrossRef]
101. Lynch, K.M.; Coffey, A.; Arendt, E.K. Exopolysaccharide producing lactic acid bacteria: Their techno-functional role and potential application in gluten-free bread products. *Food Res. Int.* **2017**. [CrossRef] [PubMed]
102. Suzuki, C.; Kobayashi, M.; Kimoto-Nira, H. Novel exopolysaccharides produced by Lactococcus lactis subsp. lactis, and the diversity of epsE genes in the exopolysaccharide biosynthesis gene clusters. *Biosci. Biotechnol. Biochem.* **2013**, *77*, 2013–2018. [CrossRef]
103. Ismail, B.; Nampoothiri, K.M. Production, purification and structural characterization of an exopolysaccharide produced by a probiotic Lactobacillus plantarum MTCC 9510. *Arch. Microbiol.* **2010**, *192*, 1049–1057. [CrossRef]
104. Verbeke, K. Nonceliac Gluten Sensitivity: What Is the Culprit? *Gastroenterology* **2018**, *154*, 471–473. [CrossRef]
105. Galle, S.; Schwab, C.; Dal Bello, F.; Coffey, A.; Gänzle, M.G.; Arendt, E.K. Influence of in-situ synthesized exopolysaccharides on the quality of gluten-free sorghum sourdough bread. *Int. J. Food Microbiol.* **2012**, *155*, 105–112. [CrossRef]
106. Nami, Y.; Gharekhani, M.; Aalami, M.; Hejazi, M.A. Lactobacillus-fermented sourdoughs improve the quality of gluten-free bread made from pearl millet flour. *J. Food Sci. Technol.* **2019**, *56*, 4057–4067. [CrossRef]
107. Dertli, E.; Mercan, E.; Arıcı, M.; Yılmaz, M.T.; Sağdıç, O. Characterisation of lactic acid bacteria from Turkish sourdough and determination of their exopolysaccharide (EPS) production characteristics. *LWT Food Sci. Technol.* **2016**, *71*, 116–124. [CrossRef]
108. Franco, W.; Pérez-Díaz, I.M.; Connelly, L.; Diaz, J.T. Isolation of exopolysaccharide-producing yeast and lactic acid bacteria from quinoa (Chenopodium quinoa) sourdough fermentation. *Foods* **2020**, *9*, 337. [CrossRef]
109. Rühmkorf, C.; Jungkunz, S.; Wagner, M.; Vogel, R.F. Optimization of homoexopolysaccharide formation by lactobacilli in gluten-free sourdoughs. *Food Microbiol.* **2012**, *32*, 286–294. [CrossRef] [PubMed]
110. Montemurro, M.; Pontonio, E.; Rizzello, C.G. Design of a "clean-label" gluten-free bread to meet consumers demand. *Foods* **2021**, *10*, 462. [CrossRef] [PubMed]
111. Coda, R.; Nionelli, L.; Rizzello, C.G.; De Angelis, M.; Tossut, P.; Gobbetti, M. Spelt and emmer flours: Characterization of the lactic acid bacteria microbiota and selection of mixed starters for bread making. *J. Appl. Microbiol.* **2010**, *108*, 925–935. [CrossRef]
112. Nionelli, L.; Curri, N.; Curiel, J.A.; Di Cagno, R.; Pontonio, E.; Cavoski, I.; Gobbetti, M.; Rizzello, C.G. Exploitation of Albanian wheat cultivars: Characterization of the flours and lactic acid bacteria microbiota, and selection of starters for sourdough fermentation. *Food Microbiol.* **2014**, *44*, 96–107. [CrossRef] [PubMed]
113. Galle, S.; Schwab, C.; Arendt, E.K.; Gänzle, M.G. Structural and rheological characterisation of heteropolysaccharides produced by lactic acid bacteria in wheat and sorghum sourdough. *Food Microbiol.* **2011**, *28*, 547–553. [CrossRef] [PubMed]
114. Mamhoud, A.; Nionelli, L.; Bouzaine, T.; Hamdi, M.; Gobbetti, M.; Rizzello, C.G. Selection of lactic acid bacteria isolated from Tunisian cereals and exploitation of the use as starters for sourdough fermentation. *Int. J. Food Microbiol.* **2016**, *225*, 9–19. [CrossRef]

foods

MDPI

Article

Amorphophallus konjac: A Novel Alternative Flour on Gluten-Free Bread

Fernanda Laignier [1], Rita de Cássia Coelho de Almeida Akutsu [1], Iriani Rodrigues Maldonade [2], Maria Teresa Bertoldo Pacheco [3], Vera Sônia Nunes Silva [3], Marcio Antônio Mendonça [4], Renata Puppin Zandonadi [1], António Raposo [5,*] and Raquel Braz Assunção Botelho [1,*]

1 Department of Nutrition, College of Health Sciences, University of Brasília, Brasília 70910900, Brazil; felaignier@hotmail.com (F.L.); rita.akutsu@gmail.com (R.d.C.C.d.A.A.); renatapz@unb.br (R.P.Z.)
2 Brazilian Agricultural Research Corporation (EMBRAPA-Hortaliças), Brasília 70359970, Brazil; irianimaldonade@embrapa.br
3 Institute of Food Technology, São Paulo 13070178, Brazil; mtb@ital.sp.gov.br (M.T.B.P.); vera.silva@ital.sp.gov.br (V.S.N.S.)
4 College of Agronomy and Veterinary Medicine, University of Brasília, Brasília 70910900, Brazil; marcioamen@gmail.com
5 CBIOS (Research Center for Biosciences and Health Technologies), Universidade Lusófona de Humanidades e Tecnologias, Campo Grande 376, 1749-024 Lisboa, Portugal
* Correspondence: antonio.raposo@ulusofona.pt (A.R.); raquelbotelho@unb.br (R.B.A.B.); Tel.: +55-61-981378620 (R.B.A.B.)

Citation: Laignier, F.; Akutsu, R.d.C.C.d.A.; Maldonade, I.R.; Bertoldo Pacheco, M.T.; Silva, V.S.N.; Mendonça, M.A.; Zandonadi, R.P.; Raposo, A.; Botelho, R.B.A. *Amorphophallus konjac*: A Novel Alternative Flour on Gluten-Free Bread. *Foods* **2021**, *10*, 1206. https://doi.org/10.3390/foods10061206

Academic Editors: Maria Papageorgiou and Theodoros Varzakas

Received: 7 May 2021
Accepted: 24 May 2021
Published: 27 May 2021

Abstract: The demand for gluten-free products is rising, but their production with similar quality as their gluten counterparts is challenging. This study aimed to develop gluten-free bread samples using different concentrations of *Amorphophallus konjac* flour (0%, 12.5%, 25%, 37.5%, and 50% of the total flour content) and to evaluate their nutritional and physicochemical properties. Proteins, lipids, carbohydrates, moisture, ash content, fibers, resistant starch, firmness, specific volume, and color were evaluated using official methods. Protein varied from 2.95% to 4.94%, the energy value from 347.93 to 133.55 kcal/100 g, dietary fiber from 8.19 to 17.90%, and resistant starch from 0.67% to 0.75% on wet basis. The addition of konjac flour positively influenced the specific volume. Higher concentrations of konjac flour in the formulations led to lower calories of the bread due to the significant addition of water to the dough. The bread samples with konjac showed high fiber content due to the composition of the flour. They had lower levels of carbohydrates, which can positively influence the glycemic index. Konjac flour provided dough mold, growth, and better texture for gluten-free bread. The best formulations were prepared in concentrations up to 37.5% konjac. The 50% konjac bread showed slightly reduced specific volume and pale color.

Keywords: gluten-free; bread; *Amorphophallus konjac*; baking

1. Introduction

Due to the growing trend in consuming gluten-free products, the food industry has sought to expand and diversify its production to meet this increasing demand [1]. The gluten-free diet (GFD) has become popular since it is the only treatment for those who suffer from gluten-related disorders (GRD) [2,3], and their relatives consume gluten-free products to support the treatment and avoid food cross-contamination. Moreover, some individuals without GRD have adhered to GFD, believing in GFD's potential health benefits, despite the lack of scientific evidence on it [2,4–7]. Therefore, about 10% of the world population has adopted a GFD [8–11]. In this sense, the gluten-free food market is expected to grow between 2019 and 2025, from US$ 3.73 billion to US$ 6.43 billion worldwide [12].

Among the gluten-free foods that make up the market's largest share are bakery products [12], bread being the most desired product by GRD people. However, producing gluten-free bread with similar characteristics to the traditional product becomes a technological challenge due to the absence of proteins forming the gluten network, with elasticity and extensibility [8,13,14]. The absence of gluten in bread formulations, for example, results in some qualitative problems, low volume, brittle texture, undesirable taste, and short shelf life [15].

Foste et al. [16] claim that gluten-free products are high in starch and low in some nutrients, and fiber and strategies are needed to balance them. Soluble fibers as beta-glucan, chitosan, psyllium, and glucomannan have been studied as potential application in the gluten-free bakery market with potentially positive effects on consumers' health [17,18]. Glucomannan is extracted from the tuber *Amorphophallus konjac* (a perennial plant from the subtropical regions of South East Asia and Africa). It is used in Chinese medicine for detoxification, cancer suppression, stasis of blood, treatment of asthma, cough, hernia, breast pain, and hematological and skin disorders [19]. Due to its water absorption capacity and stability, it is considered a source of hydrocolloidal dietary fiber. It has been used as a supplement to treat and prevent excess weight and diabetes and dermatological conditions [20].

As a food additive, glucomannan was tested in bakery products, drinks, bread, and pasta [19,21], but the flour of *Amorphophallus konjac* was not broadly studied. The flour from *Amorphophallus konjac* is considered a functional ingredient [22] containing about 1.4–3.4% of proteins, 78–80% of fibers, 8% of starch, and 1.7–2.1% of ash content [23–25]. This flour shows important health benefits in reducing cholesterol and triglycerides, improving blood sugar levels and promoting intestinal activity and human immune function. These health benefits potentially contribute to GRD individuals' health and can be considered a potential healthy ingredient in gluten-free products [22]. Until now, no study used konjac flour as a substitute for wheat flour but as a food additive to improve gluten-free bread characteristics. Nakamura et al. [26] used *Amorphophallus konjac* flour in concentrations of 0.25%, 0.50%, and 0.75% as a thickener in gluten-free bread, and Moore et al. [27] used konjac flour at 1.5% added to 0.9% of xanthan in bread production.

Considering the potential application of *Amorphophallus konjac* flour and its potential benefits to individuals suffering from GRD, this study aimed to develop gluten-free bread with *Amorphophallus konjac* flour and to evaluate the nutritional and physicochemical properties of the formulations with different concentrations of the flour.

2. Materials and Methods

This experimental study took place in the Dietetic Laboratory from the University of Brasilia (Brazil) for bread production. For chemical composition analysis, color, texture, and specific volume, research was conducted in the Food Analysis laboratory (UnB), EMBRAPA, and ITAL (Food Technology Institute). All samples were developed and analyzed in triplicate.

2.1. Bread Preparation

Amorphophallus konjac flour was purchased from a Brazilian pharmaceutical company (SM pharmaceutical enterprises) in three different lots (lots: 18F13-B022-034221, 18F13-B022-034226, MW20171011-1750) imported from China in packages of 1000 g each. These three flours were mixed in the exact amounts to prepare the formulations. The used flours were composed of 70% glucomannan konjac, according to the labels. They were previously microbiologically tested (total count of bacteria, mold and yeasts, *Escherichia coli*, and *Salmonella*) and approved by the Brazilian sanitary legislation.

The control gluten-free bread (GFB) was composed of potato starch (30%) and rice flour (70%) as flour basis, added sucrose (12 g/100 g of control flour basis); salt (3 g/100 g of control flour basis); water (34.5 g/100 g of control flour basis); soy oil (16.5 g/100 g of control flour basis); whole egg (29.5 g/100 g of control flour basis); and yeast (1.5 g/100 g of control flour basis). This formulation was adjusted from the bread formulation studied by Aguiar [28], and no additives were used. The same ingredients were used in the modified GFB samples with the addition of konjac flour (Table S1 in Supplementary

File). For the other samples, *Amorphophallus konjac* flour was added in the proportion of 12.5%, 25%, 37.5%, and 50% of the flour quantity. According to the literature, konjac flour presents almost 80% of fibers and only 8% of starch, making it unfeasible to use it above the maximum percentage applied in this study. Preliminary tests were performed with percentages superior to 50%, and bread samples presented a strong odor and "taste of fish". For the samples added of konjac flour, the amount of sucrose, salt, soy oil, and eggs was the same as the control sample (considering the weight of the control flour basis). The amount of water was adjusted due to konjac flour's fiber content that can absorb 200 times their weight in water [29]. The amount of water was previously tested (every 5 mL until a moldable dough was achieved), and the final water amount was determined in each sample based on the dough's characteristics. Therefore, in the konjac flour samples, we used water in a concentration of 131%, 228%, 297%, and 406%, respectively, based on the weight of the control flour basis. The ingredients besides the konjac flour were purchased from local stores in the Federal District, Brazil.

After weighing ingredients, the yeast was pre-activated in sucrose and warm water (38 °C) for 10 min. Separately, rice flour, potato starch, konjac flour, and salt were mixed. Eggs, water, and oil were added to the dry ingredients. Finally, activated fresh biological yeast was added, and the dough was homogenized. This dough was kneaded and rested for 50 min, and then, it went through a second kneading and modeling (40 g spheres). We used a preheated gas oven Brastemp®-Brazil, at 180 °C to bake all the samples. On the day of the analysis, samples were prepared and baked.

The cooking factor for each sample of bread was determined using the formula proposed by Araujo et al. [30].

$$\text{Cooking Factor} = \frac{\text{Baked bread (g)}}{\text{Bread dough}} \tag{1}$$

Moreover, we calculated the weight loss percentage [31] after baking using the formula:

$$\text{Weight loss after baking} = \frac{\text{Dough (g)} \times 100}{\text{Baked bread (g)} - 100} \tag{2}$$

2.2. Chemical Characterization

The Adolfo Lutz Institute's analytical standards [32] were used to determine the moisture by the loss of water by drying, direct drying in the oven at 105 °C, followed by weighing the dry sample until constant weight. The determination of crude protein was performed using the official Kjeldahl method with adaptations. The samples went through the stages of digestion of organic matter, the distillation of nitrogen with the formation of ammonium hydroxide, and titrated directly with HCl 0.1 N. This resulted in a percentage of nitrogen calculated according to the volume spent on HCl, multiplied by the general factor of protein following the Association of Official Analytical Chemists (AOAC 991.22) [33]. The Am 5-04 method was used to determine lipid content, using the XT15 extractor (Ankom, Macedon, NY, US) from Ankom Technology, carried out by extraction with petroleum ether, by dragging, under pressure [34]. The ash content of the dry samples was determined by the incineration residue obtained from heating in a muffle furnace at 600 °C using a heating ramp of 240 min, according to method 945.45 [33]. Total dietary fiber (TDF) was evaluated by the enzymatic-gravimetric method, which consists of gelatinization and partial hydrolysis of starch, followed by hydrolysis of part of proteins and residual starch. Its value is expressed after subtracting the analytical blank (AB) and the protein and mineral content determined in the residues [33]. The total carbohydrate content was determined by difference, subtracting from 100 the values found for moisture, protein content, lipids, ash, and total fibers, according to method 986.25 [33]. According to the AOAC 2002.02 method and the American Association of Cereal Chemists (AACC 32-40.01) method, the resistant starch content was determined using a commercially test kit (Megazyme International Ireland Ltd., Wicklow, Ireland).

2.3. Color Evaluation

The evaluation of the color of the crust, crumb, and bottom of the bread samples was carried out in a spectrophotometer ColorQuestXE (HunterLab, Reston, VA, USA). We obtained from the Hunter system the values for the coordinates L* (measurable in terms of white to black intensity), a* (measurable in terms of intensity from red to green), and b* (measurable in terms of yellow and blue intensity). It was possible to obtain hue angle h* (Equation (3)), color saturation or chroma C* (Equation (4)), and color difference ΔE (Equation (5)) [35–38]. L_0, a_0 and b_0 are the coordinates obtained for the control sample.

$$h* = \text{arctang}\ (b * /a*) \tag{3}$$

$$C* = \sqrt{(a *^2 + b*^2)} \tag{4}$$

$$\Delta E = \sqrt{[(L * - L*_0)^2 + (a * - a_0)^2 + (b * - b_0)^2]} \tag{5}$$

2.4. Texture Analysis

The texture profile (TPA) analysis of the bread samples was performed using the method 74-09.01—*Measurement of Bread Firmness by Universal Testing Machine* [39]. The equipment used was the TA.XTplus connected to the Software Exponent (version 6.1.4., Stable Micro System, Surrey, UK) A 36 mm cylindrical probe (Stable Micro System, Surrey, UK) was used, test speed 1.7 mm/s; deformation level of 40%, trigger load 5 g. This probe is usually used for bread samples. Bread samples were evaluated after baking and standing two hours out of the oven for cooling. They were tested as baked in small spheres (balls) like brioche bread. Data are expressed as the force necessary to deform the product as the cylindrical probe enters in contact with the bread.

2.5. Specific Volume

Specific volume was measured by the rapeseed displacement method [40] through the ratio between volume (cm^3) and mass (g) of each sample.

2.6. Total Energy Value

The total energy value was reached with the macronutrients from the proximate composition analysis and the Atwater factors multiplying fats by 9 kcal/g, proteins by 4 kcal/g, and carbohydrates by 4 kcal/g [41]. As mentioned in Section 2.2, carbohydrates were reached by subtracting from 100 the values found for moisture, protein content, lipids, ash, and total fibers.

2.7. Statistical Analysis

The results were subjected to one-way ANOVA followed by Tukey's post hoc test, with the level of $p < 0.05$ considered significant. Statistical analyses were performed using software SPSS-IBM (24.0, IBM, Armonk, NY, USA) All analyses were conducted in triplicate.

3. Results

3.1. Preparation of Bread Samples and Cooking Quality

Table 1 presents some characteristics of the formulations in the cooking process. Cooking times and weight losses during baking were different depending on the moisture of the dough. Table S1 presents the different formulations.

Table 1. Cooking characteristics of different formulations of gluten-free bread samples.

	Percentage of Konjac Flour in Bread Formulations				
	0	12.5	25	37.5	50
% of cooking weigh loss	17	20	27.70	32.50	32.50
Cooking factor	0.83	0.80	0.72	0.68	0.68
Cooking time (Minutes)	24	24	30	30	33

3.2. Chemical Characterization

Table 2 presents the chemical composition of the different gluten-free bread formulations. All samples were baked simultaneously and taken to the food analysis laboratory after one hour to start all chemical analysis. Protein content was slightly reduced with konjac flour increase with the lowest and significant value for the 50% konjac flour bread ($p = 0.008$). Lipids reduced 26.1% when comparing the control bread and the 25% konjac, and 51.3% compared to the 50% konjac. There was a reduction of 38.1% between the control sample and the 50% konjac sample for carbohydrates. The 50% konjac bread presented 34.5 more times fiber than the control bread for dietary fibers. The total energy value (TEV) decreased by 61.6% between control and 50% konjac samples.

Table 2. Chemical composition of different formulations of gluten-free bread with and without konjac flour addition on a wet basis.

	Konjac Flour Percentage in Bread Formulations				
	0	12.5	25	37.5	50
Protein (g/100 g)	5.9 ± 0.37 ^{a}	4.94 ± 0.27 ^{b}	4.0 ± 0.41 ^{c}	3.88 ± 0.16 ^{c}	2.95 ± 0.41 ^{d}
Lipid (g/100 g)	10.84 ± 0.19 ^{a}	8.13 ± 0.14 ^{b}	5.59 ± 0.12 ^{c}	3.89 ± 0.64 ^{d}	3.59 ± 0.09 ^{d}
Carbohydrates (g/100 g)	56.70 ± 0.54 ^{a}	37.21 ± 0.69 ^{b}	30.73 ± 0.27 ^{c}	31.81 ± 0.59 ^{c}	22.37 ± 0.27 ^{d}
Moisture (g/100 g)	23.90 ± 0.28 ^{e}	39.98 ± 0.56 ^{d}	46.88 ± 0.48 ^{b}	43.97 ± 0.42 ^{c}	51.54 ± 0.86 ^{a}
Ash content (g/100 g)	1.85 ± 0.03 ^{a}	1.56 ± 0.01 ^{bcd}	1.64 ± 0.04 ^{bcd}	1.54 ± 0.08 ^{cd}	1.66 ± 0.00 ^{bc}
Dietary Fibers (g/100 g)	0.82 ± 0.02 ^{e}	8.19 ± 0.01 ^{d}	11.16 ± 0.09 ^{c}	14.92 ± 0.06 ^{b}	17.90 ± 0.32 ^{a}
Resistant Starch (g/100 g)	0.64 ± 0.09 ^{c}	0.67 ± 0.04 ^{c}	0.70 ± 0.02 ^{b}	0.75 ± 0.07 ^{a}	0.70 ± 0.01 ^{b}
Energy value (kcal/100 g)	347.93 ± 1.6 ^{a}	241.73 ± 2.8 ^{b}	189.19 ± 2.1 ^{c}	177.76 ± 3.7 ^{d}	133.55 ± 1.8 ^{e}

Means followed by the same letter within lines do not differ statistically $p > 0.05$. All the analysis were performed in triplicate.

3.3. Specific Volume, Firmness and Color

The specific volume (SV) is a measure to verify the dough's ability to expand and retain the gas during baking [42]. Table 3 presents the data of the SV and firmness of the control and konjac samples. A statistically significant difference ($p < 0.05$) was obtained for SV from the control and all konjac samples, with a higher volume with the increase of konjac. However, only 12.5% bread is different from other formulations, demonstrating that the increase in konjac above 25% did not affect the volume. Konjac flour provided more significant bread expansion, contributing to its texture. It was impossible to measure control bread's firmness in the same conditions as the other bread samples. Control bread presented a very hard texture after baking since it was only prepared with potato starch and rice flour and no additives to improve texture. The cylindrical probe did not penetrate the sample because of its hardness; therefore, the equipment did not provide reading parameters.

The average values of Chroma-C*, hue angle-h*, and color difference for the different samples are in Table 4. The chroma is related to the color's purity, and higher values indicate more intense colors [43].

Table 3. Specific volume and texture of bread formulations prepared with different concentrations of konjac flour.

	Konjac Flour Percentage in Bread				
	0	12.5	25	37.5	50
Specific Volume	1.44 ± 0.06 ^{e}	1.61 ± 0.06 ^{d}	1.96 ± 0.06 ^{b}	2.10 ± 0.12 ^{a}	2.05 ± 0.10 ^{ab}
Firmness (g)	*	4505.00 ± 343.97 ^{a}	2508.31 ± 40.94 ^{b}	1469.53 ± 39.91 ^{c}	2334.90 ± 77.05 ^{b}

Means followed by the same letter within lines do not differ statistically $p > 0.05$. * It was not possible to read the control bread due to its firmness.

Table 4. Mean values of chroma (C*), tone color (h*), and color difference (ΔE*) of bread formulations prepared with different concentrations of konjac flour.

Konjac Flour (%)	Crust Color		
	C*	h*	ΔE*
0	34.39 ± 2.37 ^{a}	78.14 ± 2.25 ^{a}	— **
12.5	33.22 ± 0.29 ^{a}	69.65 ± 5.75 ^{d}	5.86 ± 3.14 ^{a}
25	24.53 ± 0.58 ^{b}	70.52 ± 2.65 ^{c}	7.17 ± 1.49 ^{a}
37.5	23.10 ± 0.70 ^{b}	74.43 ± 2.73 ^{b}	7.77 ± 0.95 ^{a}
50	21.85 ± 0.45 ^{b}	73.76 ± 2.78 ^{c}	8.54 ± 0.46 ^{a}
	Crumb color		
	C*	h*	ΔE*
0	26.55 ± 0.56 ^{a}	85.83 ± 0.24 ^{a}	— **
12.5	22.37 ± 0.15 ^{b}	85.06 ± 0.23 ^{b}	5.95 ± 0.44 ^{c}
25	16.75 ± 0.86 ^{c}	86.03 ± 0.29 ^{a}	10.23 ± 0.71 ^{b}
37.5	14.12 ± 0.70 ^{d}	86.26 ± 0.30 ^{a}	13.24 ± 0.60 ^{a}
50	14.61 ± 0.51 ^{d}	84.76 ± 0.12 ^{b}	14.14 ± 0.93 ^{a}
	Bottom color		
	C*	h*	ΔE*
0	32.56 ± 4.10 ^{a}	60.35 ± 6.11 ^{a}	— **
12.5	31.32 ± 2.75 ^{a}	60.25 ± 2.95 ^{a}	3.88 ± 0.80 ^{b}
25	24.84 ± 2.97 ^{ab}	56.93 ± 4.59 ^{a}	8.86 ± 2.87 ^{ab}
37.5	20.41 ± 1.77 ^{b}	59.71 ± 1.43 ^{a}	11.35 ± 1.92 ^{a}
50	20.44 ± 1.16 ^{b}	57.46 ± 0.81 ^{a}	11.06 ± 1.64 ^{a}

Means followed by the same letter within columns do not differ statistically, $p > 0.05$. All the analysis were performed in triplicate. ** There is not a ΔE* for the control bread sample.

Analyzing the results of color saturation for the bread crust, the control bread obtained the highest average value indicating a more intense color. The bread crumb from the control also obtained the highest average for color saturation, differing statistically ($p < 0.05$) from the others. It indicates that as the konjac flour increases, a change in the crumbs' color also increased, decreasing the degree of saturation and, consequently, the loss in color purity. The highest color saturation value was found in the control bread's crust, and the lowest in the breadcrumbs 37.5 and 50%.

4. Discussion

Bread is one of the most popular items in the customer's purchase basket [44], reaching the worldwide average consumption of 18 kg/year per capita [45,46]. To provide similar products for GRD individuals who follow a GFD is until now a challenge, mainly in terms of physicochemical properties. Weight loss when baking is very pronounced in gluten-free bread due to the absence of gluten's protein network. Cooking factor expresses the dough's ability to retain the water added to it [42].

In our study, the weight loss varied between 17 (control bread) and 32.5% (37.5% and 50% konjac flour bread). Moore et al. [27] using konjac flour at 1.5% showed a weight cooking loss of 9.20%, lower than our results. Turkut et al. [47] obtained losses from

14.4% to 15.4%, and Zelada et al. [31] observed a loss ranging from 11.9 to 15.0%, both lower than our findings. Weight loss during cooking provides information mainly on the amount of evaporated water. However, it also represents the loss of organic material, such as fermented sugars released in the form of CO_2 [48]. The format in which the konjac bread was shaped is different from the bread samples in the mentioned studies. Konjac bread samples were molded into small spheres, and in other studies, they were shaped as loaves. According to Horstmann, Foschia, and Arendt [48], it is possible that bread samples with a larger surface area present high cooking loss. Our results point to higher losses as konjac flour is added to the formulation. However, for the doughs' shape before baking, the amount of water added to the formulations was 980% higher than the control bread. The network formed in these doughs did not allow all the added water to be retained, even with the high fiber content. It was observed that bread samples with more added water were kept longer in the oven to present the crunchiest crust. Therefore, longer baking time led to higher water losses.

The konjac bread samples had lower protein levels than the bread studied by other authors [28,49]. Wang et al. [50] stated that the incorporation of protein ingredients in gluten-free doughs could improve the sensory and nutritional quality of gluten-free bread, in addition to an increase in flavor. The addition of proteins helps in forming a network similar to gluten in wheat bread [13]. However, in this study, the only variation was the konjac flour and water content to evaluate the use of a product rich in fiber, forming a barrier to maintain volume and texture.

The protein content found in 12.5% konjac bread (4.9%) was higher than the other konjac gluten-free bread samples (Table 2) and slightly higher than the average protein found in gluten-free bread evaluated by Cornicelli et al. (4.29%, wet basis) [51] and by Roman, Belorio and Gomez (3.91%, wet basis) [52]. The highest amount of protein in the 12.5% konjac bread could be explained by the lower amount of water necessary to achieve moldable dough than the other konjac GFB samples (Supplementary File—Table S1), as confirmed by the higher moisture content in this sample. Roman, Belorio, and Gomez [52] claim that 81% of commercial gluten-free bread in their study had added proteins and that even so, the protein content of these bread samples was lower than their gluten-free counterparts.

Bread samples with 37.5% and 50% konjac had the lowest levels of lipids. When evaluating the lipid content on formulations based on rice flour, Saueressig, Kaminski, and Escobar [49] observed that the highest average was 3.80%, similar to bread with a higher concentration of konjac flour (3.59%) in this study. According to Brandão and Lucena [53], the fats added to the formulations improve the dough's quality, increase its extensibility, and provide the softness of the crumb and a more pleasant flavor. Jamieson, Weir, and Gougeon [54] observed that industrialized gluten-free products had up to 1.3 times more fat than their gluten counterparts, significantly increasing the consumption of calories. The 12.5% and 25% konjac GFB samples presented lipid content of 8.13% and 5.59%, respectively, similar to the GFB produced by Jamieson, Weir, and Gougeon (6.8%) [54].

All konjac bread samples obtained lower average carbohydrates than studies of bread with and without gluten by Cornicelli et al. [51]; Jamieson, Weir, and Gougeon [54]; and Roman, Belorio, and Gomez [52]. According to Jamieson, Weir, and Gougeon [54], gluten-free products generally have higher sugar content than gluten bread.

A significant difference ($p < 0.05$) was obtained for the moisture content comparing all samples. The highest content was for 50% konjac bread, making up for more than half of the baked bread's weight, which justifies the lower energy value. Aguiar [28] produced gluten-free bread with sorghum flour, and the highest value for moisture was 53.24%. The moisture content of a product influences the choice of packaging, the form of storage, and its processing [32]. Parry [55] reports that the use of konjac flour in concentrations of 0.1% to 0.5% influenced the release of moisture in bread, sweets, and bakery products. Horstmann, Foschia, and Arendt [48] affirm that bread moisture reflects in crumb softness after baking. This moisture difference resulted in softer crumbs for the konjac bread samples, as presented by the texture profile data.

The higher the proportion of konjac flour, the lower the protein content, lipids, and carbohydrates. Conversely, the total dietary fiber increases, demonstrating that konjac flour provides 21.8 times more fiber than control bread. According to Parry [55], the fibers in konjac flour can reach 90% of its composition. This fiber (Glucomannan) has beneficial properties such as prebiotic action [25], reducing cholesterol, improving blood sugar levels, and promoting immune function that are essential health benefits to GRD individuals [22].

Regarding fiber, gluten-free bread with konjac obtained better results than those by Saueressig, Kaminski, and Escobar [49], with soluble (inulin) and insoluble (rice bran) fibers. In Saueressig, Kaminski, and Escobar study [49], the formulation that contained the highest percentage of fibers showed an average of 4.88%, lower than our study bread samples with konjac in which dietary fibers ranged from 8.19% to 17.9%. Thus, according to Brazilian legislation, bread with konjac can be classified as food with high fiber content (>6 g/100 g) [56].

The recommendation for daily fiber intake is from 30 to 38 g for men and 21 and 25 g for women [25], a challenging amount to achieve in a gluten-free diet that features low-fiber food. The amount of fiber in labels of gluten-free bread by Nascimento et al. [14] and Lerma et al. [57] was less than the average for the same products with gluten. The average values in gluten-free bread found by these authors were 0.7% and 3.61%, respectively, with low values considering the recommendation.

In the preparation of gluten-free bread, corn, rice, and potato starches are often used to replace wheat flour. However, these products are low in fiber, micronutrients, proteins, and generally have a higher glycemic index [58]. The glycemic response of carbohydrates may increase in gluten-free foods because the gluten protein network surrounds the starch granule, being difficult for amylase action, thus inhibiting its hydrolysis in the lumen of the small gut [2]. Pellegrini and Agostoni [2] and Foste et al. [16] suggest supplementing gluten-free bread with soluble fibers so that there is a reduction in the glycemic index in these bread samples.

Samples with 25%, 37.5%, and 50% konjac had higher resistance starch levels than control bread. Compared to the results obtained in the analysis of commercial gluten-free bread performed by Larretxi et al. [59], bread with konjac flour had a low content of resistant starch. Konjac bread 50% presented 0.70 g/100 g, less than the commercial bread with 3.6% [59]. Resistant starch has a physiological behavior similar to soluble fiber. Its positive effects range from the formation of short-chain fatty acids, due to the prebiotic effect, to the decrease in postprandial glycemia and insulinemia [25].

The energy value ranged between 347 kcal/100 g in control and 133.55 kcal/100 g in the 50% konjac. The average energy value of gluten-free bread by Cornicelli et al. [51] and Roman, Belorio, and Gomez [52] is superior to all bread prepared with konjac flour. A large amount of fiber can explain the low energy value of konjac bread.

The higher was the proportion of konjac flour added to the formulations, the higher were the average moisture, total dietary fibers, and resistant starch. Inversely to this, smaller were the averages for proteins, carbohydrates, lipids, and energy value. These results show that the konjac flour influenced positively the formulations of bread, improving the amount of micronutrients and fibers. The humidity was higher among the konjac bread samples favoring the texture; however, the macronutrients' averages were low due to the increase of fibers.

The bread samples of Moore et al. [27], using 1.5% konjac flour registered 2.08 cm^3/g of SV, similar to bread samples above 25% konjac flour in this study. Hager and Arendt [60] obtained values of 1.78 and 1.63 cm^3/g in their rice and cornbread, respectively. Sandri et al. [61] obtained SV between 1.22 and 1.70 cm^3/g in their bread based on rice flour. Djordjevic et al. (2019) obtained a variation between 1.52 to 3.97 cm^3/g of bread prepared with corn flour with added fibers. Zelada et al. [31] obtained results from 2.41 to 2.92 cm^3/g in bread samples of corn and rice flour. However, the bread samples with higher SV were not precisely those that presented lower firmness values, as stated by Moore et al. [62] and Sandri et al. [61], in which there is a direct relationship between low specific volume and bread hardness. The values found for konjac bread samples are close to the average values presented in other studies.

Djordjevic et al. [63] report that dietary fibers can interfere with the quality of gluten-free bread by improving viscosity, texture, volume, sensory characteristics, and shelf life due to their water-binding ability, gel-forming ability, effects fat mimetics, textural, and thickeners.

In formulations of gluten-free bread prepared with rice flour, Nakamura et al. [26] used konjac flour in concentrations of 0.25%, 0.5%, and 0.75% as a thickener. The bread samples' SV increased with increasing amounts of konjac from 0.25% to 0.50%, but it decreased by 0.75%. These authors also observed that konjac significantly reduced the bread's hardness resulting in softer bread than those only with rice flour.

Texture can be defined as the mechanical, geometric, and surface attributes of a perceptible product using instruments and sensory means [64]. The taste of food is the most observed attribute for its acceptance. However, the texture is the main attribute considered to reject it. Gluten-free bread is characterized by a low volume, crumbly texture, and cracked crust [58], making them unattractive. Table 3 shows that 12.5% bread had the highest average of firmness. According to Giannou and Tzia [65], hardness is the maximum force necessary to compress food between teeth. Bread with higher konjac content is softer and probably easier to chew. Turkut et al. [47] observed that bread with 25% quinoa flour obtained the lowest average instrumental hardness. The results of the instrumental hardness analysis carried out by Arcanjo [66] on his gluten-free rice bread ranged from 1830.28 g to 4587.56 g. Zelada et al. [31], for their gluten-free bread prepared with different hydrocolloids, presented values from 1717 g to 3868 g. Gluten-free bread well evaluated made with fibers from the coffee husk [67] obtained average hardness varying between 1560.75 g and 5585 g.

According to Foste et al. [16], the structure of the gluten-free dough requires higher amounts of water that resembles cake dough. Due to the amount of konjac flour, the dough became consistent, and it was possible to mold it into spheres without difficulty. The bread samples added konjac flour obtained a more significant water addition in their formulations and obtained the greatest losses during baking. The losses did not compromise the bread samples' specific volume, as an increase of volume was observed as the proportion of flour increased.

According to Turkut et al. [47], bread color is the result of chemical reactions between proteins and carbohydrates during the baking process. The Maillard reaction is a way of darkening food that occurs with these two components, high temperatures and under ideal pH conditions [68]. During the heating of foods, reducing amino acids and sugars trigger a complex cascade of reactions that results in the formation of brown substances called melanoidins that provide a more attractive color to these foods [55]. An insufficient amount of reducing sugars and low protein content collaborates for pale color that often occurs in gluten-free bread [69].

When analyzing the results of color saturation for the bread crust, the control bread obtained the highest average value, indicating a more intense color. The control bread crumb also obtained the highest average color saturation. This indicates that, as konjac flour was added to the formulations, the change in the crumb color also increased, decreasing the degree of saturation and, consequently, the loss in color purity. The highest color saturation value was found in the crust of the control bread, and the lowest color saturation in the crumb of the bread 37.5 and 50%.

The results for bread crust tonality in this study corroborate the results found in the bread studied by Messa et al. [1], which also showed a tendency towards yellow. At the bottom of the bread, there was no statistical difference for color shade. However, the bottoms from 25%, 37.5%, and 50% konjac had lower averages indicating a color trend towards red. When analyzing the color difference variable in the bread crust, it was observed that there was no statistical difference between the samples of konjac bread.

5. Conclusions

The weight loss during the baking of the different formulations was lower in the control bread. The moisture content varied between 23.9% and 51.54%. The ash content on konjac bread samples is similar to those found in gluten-free bread from other studies. However, the ash content has increased according to konjac flour addition. It shows that the flour has contributed to the increase of minerals in the composition of the bread. The bread samples with konjac flour showed low caloric values and high fiber contents due to the konjac flour composition.

Additionally, they had lower carbohydrate levels, which can positively influence these samples' glycemic index, but more studies are necessary to evaluate it. Considering the color analysis, the most intense color was obtained in the control bread. As the konjac flour was added to the formulations, the purity of the color was reduced. Konjac flour can be a promising alternative in preparing gluten-free bread because it provided dough mold, growth, and better texture when used in gluten-free bread. The best formulations were prepared in concentrations of up to 37.5% konjac. The 50% konjac bread showed low values for macronutrients, but it was observed that its specific volume was slightly reduced. A limitation of our study is the lack of sensory analysis of the developed bread samples, and further studies are necessary to evaluate their acceptability by consumers.

Supplementary Materials: The following are available online at https://www.mdpi.com/article/10.3390/foods10061206/s1, Table S1: Different formulations of gluten-free bread with and without the addition of konjac flour.

Author Contributions: Conceptualization, F.L., R.d.C.C.d.A.A., and R.B.A.B.; methodology, F.L., M.A.M., I.R.M., M.T.B.P., V.S.N.S., R.P.Z., and R.B.A.B.; formal analysis, R.d.C.C.d.A.A.; investigation, F.L., M.A.M., I.R.M., M.T.B.P., V.S.N.S., and R.B.A.B.; resources, F.L.; writing—original draft preparation, F.L., R.d.C.C.d.A.A., M.A.M., R.P.Z., and R.B.A.B.; writing—review and editing, R.P.Z., A.R., and R.B.A.B.; supervision, R.d.C.C.d.A.A. and R.B.A.B.; project administration, F.L. All authors have read and agreed to the published version of the manuscript.

Funding: This research received no external funding.

Institutional Review Board Statement: Not applicable.

Informed Consent Statement: Not applicable.

Data Availability Statement: The study did not report any data.

Acknowledgments: The authors acknowledge PPGNH/UnB and CAPES for their support.

Conflicts of Interest: The authors declare no conflict of interest.

References

1. Messa, S.; Nabeshima, E.H.; Montenegro, F.M.; Cruz, C.L.d.C.V. Estudo da obtenção de pães de forma sem glúten à base de derivados de mandioca. In Proceedings of the 11th Congresso Interinstitucional de Iniciação Científica—CIIC 2017, Campinas, Brazil, 2–4 August 2017; pp. 1–9.
2. Pellegrini, N.; Agostoni, C. Nutritional aspects of gluten-free products. *J. Sci. Food Agric.* **2015**, *95*, 2380–2385. [CrossRef]
3. Menga, V.; Amato, M.; Phillips, T.D.; Angelino, D.; Morreale, F.; Fares, C. Gluten-free pasta incorporating chia (*Salvia hispanica* L.) as thickening agent: An approach to naturally improve the nutritional profile and the in vitro carbohydrate digestibility. *Food Chem.* **2017**, *221*, 1954–1961. [CrossRef]
4. Missbach, B.; Schwingshackl, L.; Billmann, A.; Mystek, A.; Hickelsberger, M.; Bauer, G.; König, J. Gluten-free food database: The nutritional quality and cost of packaged gluten-free foods. *PeerJ* **2015**, *3*, e1337. [CrossRef]
5. Falcomer, A.L.; Santos Araújo, L.; Farage, P.; Santos Monteiro, J.; Yoshio Nakano, E.; Puppin Zandonadi, R. Gluten contamination in food services and industry: A systematic review. *Crit. Rev. Food Sci. Nutr.* **2018**, *60*, 479–493. [CrossRef]
6. Machado, J.; Gandolfi, L.; Coutinho De Almeida, F.; Malta Almeida, L.; Puppin Zandonadi, R.; Pratesi, R. Gluten-free dietary compliance in Brazilian celiac patients: Questionnaire versus serological test. *Nutr. Clin. Diet. Hosp.* **2013**, *33*, 46–49. [CrossRef]

7. Farage, P.; Zandonadi, R.P.; Ginani, V.C.; Gandolfi, L.; Nakano, E.Y.; Pratesi, R. Gluten-free diet: From development to assessment of a check-list designed for the prevention of gluten cross-contamination in food services. *Nutrients* **2018**, *10*, 1274. [CrossRef] [PubMed]
8. Cross, C. Gluten-free industry is healthy, but is the food? *CMAJ* **2013**, *185*, E610. [CrossRef] [PubMed]
9. Melini, V.; Melini, F. Gluten-free diet: Gaps and needs for a healthier diet. *Nutrients* **2019**, *11*, 170. [CrossRef]
10. Sapone, A.; Bai, J.C.; Ciacci, C.; Dolinsek, J.; Green, P.H.R.; Hadjivassiliou, M.; Kaukinen, K.; Rostami, K.; Sanders, D.S.; Schumann, M.; et al. Spectrum of gluten-related disorders: Consensus on new nomenclature and classification. *BMC Med.* **2012**, *10*, 13. [CrossRef] [PubMed]
11. Gadelha de Mattos, Y.; Puppin Zandonadi, R.; Gandolfi, L.; Pratesi, R.; Yoshio Nakano, E.; Pratesi, C. Self-Reported Non-Celiac Gluten Sensitivity in Brazil: Translation, Cultural Adaptation, and Validation of Italian Questionnaire. *Nutrients* **2019**, *11*, 781. [CrossRef]
12. STATISTA. Global Gluten-Free Food Market Value from 2019 to 2025 (in Billion U.S. Dollars). 2018. Available online: https://www.statista.com/statistics/248467/global-gluten-free-food-market-size/#:~{}:text=The%20global%20market%20for%20gluten,to%208.3%20billion%20U.S.%20dollars (accessed on 20 January 2021).
13. Houben, A.; Höchstötter, A.; Becker, T. Possibilities to increase the quality in gluten-free bread production: An overview. *Eur. Food Res. Technol.* **2012**, *235*, 195–208. [CrossRef]
14. Bagolin do Nascimento, A.; Fiates Rataichesck Medeiros, G.; Teixeira, E. Availbility, cost and nutitional composition of gluten-free products. *Br. Food J.* **2013**, *116*, 1842–1852. [CrossRef]
15. Pszczola, D.E. The rise of gluten-free. *Food Technol.* **2012**, *66*, 55–66.
16. Foste, M.; Verheyen, C.; Jekle, M.; Becker, T. Fibres of milling and fruit processing by-products in gluten-free bread making: A review of hydration properties, dough formation and quality-improving strategies. *Food Chem.* **2019**, *306*, 125451. [CrossRef]
17. Queiroz, M.B.; Nabeshima, E.H. Naturalidade e Autenticidade. In *Brasil Bakery & Confectionery Trends*; ITAL: Campinas, Brazil, 2020; pp. 158–195.
18. Zandonadi, R.P.; Botelho, R.B.A.; Araújo, W.M.C. Psyllium as a Substitute for Gluten in Bread. *J. Am. Diet. Assoc.* **2009**, *109*, 1781–1784. [CrossRef]
19. Chua, M.; Baldwin, T.C.; Hocking, T.J.; Chan, K. Traditional uses and potential health benefits of *Amorphophallus konjac* K. Koch ex N.E.Br. *J. Ethnopharmacol.* **2010**, *128*, 268–278. [CrossRef]
20. Choi, K.H.; Kim, S.T.; Bin, B.H.; Park, P.J. Effect of *konjac glucomannan* (Kgm) on the reconstitution of the dermal environment against uvb-induced condition. *Nutrients* **2020**, *12*, 2779. [CrossRef] [PubMed]
21. Impaprasert, R.; Piyarat, S.; Sophontanakij, N.; Sakulnate, N.; Paengkanya, S.; Borompichaichartkul, C.; Srzednicki, G. Rehydration and textural properties of dried konjac noodles: Effect of alkaline and some gelling agents. *Horticulturae* **2017**, *3*, 20. [CrossRef]
22. Zannini, E.; Jones, J.M.; Renzetti, S.; Arendt, E.K. Functional Replacements for Gluten. *Annu. Rev. Food Sci. Technol.* **2012**, *3*, 227–245. [CrossRef]
23. Li, B.; Xia, J.; Wang, Y.; Xie, B. Grain-Size Effect on the Structure and Antiobesity Activity of Konjac Flour. *J. Agric. Food Chem.* **2005**, *53*, 7404–7407. [CrossRef]
24. Chen, H.-L.; Cheng, H.-C.; Liu, Y.-J.; Liu, S.-Y.; Wu, W.-T. Konjac acts as a natural laxative by increasing stool bulk and improving colonic ecology in healthy adults. *Nutrition* **2006**, *22*, 1112–1119. [CrossRef]
25. Li, B.; Shah, B.R.; Wang, L.; Liu, S.; Li, Y.; Wei, X.; Jin, W.; Li, Z. Health benefits of konjac glucomannan with special focus on diabetes. *Bioact. Carbohydr. Diet. Fibre* **2015**, *5*, 179–187. [CrossRef]
26. Nakamura, R.; Teshima, Y.; Miura, M.; Konishi, F. Effect of Adding Glucomannan on the Rheological Properties, Sensory Characteristics and Staling of Gluten-free Rice Bread. *J. Home Econ. Jpn.* **2016**, *67*, 141–150. [CrossRef]
27. Moore, M.M.; Schober, T.J.; Dockery, P.; Arendt, E.K. Textural comparison of gluten-free and wheat-based doughs, batters and breads. *Cereal Chem.* **2004**, *81*, 567–575. [CrossRef]
28. Andrade de Aguiar, L.; Bobrowski Rodrigues, D.; Vieira Queiroz, V.A.; Melo, L.; de Lacerde de Oliviera Pineli, L. Comparison of two rapid descriptive sensory techniques for profiling and screening of drivers of liking of sorghum breads. *Food Res. Int.* **2020**, *106*, 892–900. [CrossRef]
29. Canga, A.G.; Martínez, N.F.; Sahagún, A.M.; Vieitez, J.J.G.; Liébana, M.J.D.; Pardo, Á.P.C.; Robles, L.J.C.; Sierra, M. Glucomanano: Propiedades y aplicaciones terapéuticas. *Nutr. Hosp.* **2004**, *19*, 45–50.
30. Araújo, W.M.C.; Botelho, R.B.A.; Pilla, N.M.; Borgo, L.A. *Alquimia dos Alimentos*, 2nd ed.; Senac: Brasilia, Brazil, 2014.
31. Encina-Zelada, C.R.; Cadavez, V.; Teixeira, J.A. Bread by a Mixture Design of Xanthan, Guar, and Hydroxypropyl Methyl Cellulose Gums. *Foods* **2019**, *8*, 156. [CrossRef]
32. Izidoro, D.R.; Scheer, A.d.P.; Negre, M.F.d.O.; Haminiuk, C.W.I.; Sierakowski, M.-R. Avaliação físico-química, colorimétrica e aceitação sensorial de emulsão estabilizada com polpa de banana verde. *Rev. Inst. Adolfo Lutz* **2008**, *67*, 167–176.
33. AOAC. *Official Methods of Analysis of the Association of Official Analytical Chemists*, 19th ed.; AOAC International: Rockville, MD, USA, 2012.
34. AOCS. *Approved Procedure Am 5-04 Rapid Determination of Oil/Fat Utilizing High Temperature Solvent Extraction*; AOCS: Urbana, IL, USA, 2004; pp. 4–6.
35. Little, A. A research note off on a tangent. *J. Food Sci.* **1975**, *40*, 410–411. [CrossRef]

36. Mclellan, M.R.; Lind, L.R.; Kime, R.W. Hue Angle Determinations and Statistical. *J. Food Qual.* **1994**, *18*, 235–240. [CrossRef]
37. Maskan, M. Drying, shrinkage and rehydration characteristics of kiwifruits during hot air and microwave drying. *J. Food Eng.* **2001**, *48*, 177–182. [CrossRef]
38. Francis, F.J. The origin of tan-1 a/b. *J. Food Sci.* **1975**, *40*, 412. [CrossRef]
39. American Association of Cereal Chemists. *AACC 74-09.01: Measurement of Bread Firmness by Universal Testing Machine*; AACC: St. Paul, MN, USA, 1999.
40. AACC. *Approved Methods of the American Association of Cereal Chemists*, 10th ed.; AACC: St. Paul, MN, USA, 2000.
41. Ribeiro, P.; de Morais, T.B.; Colugnati, F.A.B.; Sigulem, D.M. Tabelas de composição química de alimentos: Análise comparativa com resultados laboratoriais. *Rev. Saude Publica* **2003**, *37*, 216–225. [CrossRef] [PubMed]
42. Evangelho, J.A.; Pinto, V.Z.; Zavareze, R.E.; Vanier, L.N.; Dias, A.R.G.; Barbosa, L.M.P.B. Propriedades tecnológicas e nutricionais de pães preparados com diferentes proporções de farinha de arroz e farinha de arroz extrusada. *Rev. Bras. Agrocienc.* **2012**, *18*, 264–282.
43. Jacomino, A.P.; Mendonça, K.; Kluge, R.A. Armazenamento refrigerado de limões "Siciliano" tratados com etileno. *Rev. Bras. Frutic.* **2003**, *25*, 45–48. [CrossRef]
44. FMCG NEWS GLOBAL. Global Bread and Bakery Consumption Continues to Experience Modest Growth. 2018. Available online: https://www.bizcommunity.com/Article/1/162/176273.html (accessed on 28 April 2019).
45. Eglite, A.; Kunkulberga, D. Foodbalt 2017 Bread Choice and Consumption Trends. In Proceedings of the 11th Baltic Conference on Food Science and Technology "Food Science and Technology in a Changing World", Jelgava, Latvia, 27–28 April 2017; pp. 178–182. [CrossRef]
46. The World Bank Global Consumption Database: Bread. Available online: http://datatopics.worldbank.org/consumption/product/Bread (accessed on 28 April 2019).
47. Turkut, G.M.; Cakmak, H.; Kumcuoglu, S.; Tavman, S. Effect of quinoa flour on gluten-free bread batter rheology and bread quality. *J. Cereal Sci.* **2016**, *69*, 174–181. [CrossRef]
48. Horstmann, S.W.; Foschia, M.; Arendt, E.K. Correlation analysis of protein quality characteristics with gluten-free bread properties. *Food Funct.* **2017**, *8*, 2465–2474. [CrossRef] [PubMed]
49. Saueressig, A.L.C.; Kaminski, T.A.; Escobar, T.D. Inclusion of dietary fiber in gluten-free breads. *Braz. J. Food Technol.* **2016**, *19*, e2014045.
50. Wang, K.; Lu, F.; Zhao, L.; Han, C. Recent developments in gluten-free bread baking approaches: A review. *Food Sci. Technol.* **2017**, *37* (Suppl. 1), 1–9. [CrossRef]
51. Cornicelli, M.; Saba, M.; Machello, N.; Silano, M.; Neuhold, S. Nutritional composition of gluten-free food versus regular food sold in the Italian market. *Dig. Liver Dis.* **2018**, *50*, 1305–1308. [CrossRef]
52. Roman, L.; Belorio, M.; Gomez, M. *Gluten-Free Breads: The Gap Between Research and Commercial Reality*; Blackwell Publishing Inc.: Hoboken, NJ, USA, 2019; Volume 18, pp. 690–702.
53. Brandão, S.S.; Lira, H.d.L. *Tecnologia de Panificação e Confeitaria*; EDUFRPE: Recife, Brazil, 2011; ISBN 9788579460852.
54. Jamieson, J.A.; Weir, M.; Gougeon, L. Canadian packaged gluten-free foods are less nutritious than their regular gluten-containing counterparts. *PeerJ* **2018**, *6*, e5875. [CrossRef] [PubMed]
55. Parry, J.M. Konjac Glucomannan. In *Food Stabilizers, Thickeners and Gelling Agents*; Wiley-Blackwell: Hoboken, NJ, USA, 2010; pp. 216–234.
56. Ministério da Saúde—Agência Nacional de Vigilância Sanitária. *RDC No 54, de 12 de Novembro de 2012*; Ministério da Saúde—Agência Nacional de Vigilância Sanitária: Brasília, Brazil, 2012.
57. Calvo-Lerma, J.; Crespo-Escobar, P.; Martínez-Barona, S.; Fornés-Ferrer, V.; Donat, E.; Ribes-Koninckx, C. Differences in the macronutrient and dietary fibre profile of gluten-free products as compared to their gluten-containing counterparts. *Eur. J. Clin. Nutr.* **2019**, *73*, 930–936. [CrossRef]
58. Capriles, V.D.; Arêas, J.A.G. Novel Approaches in Gluten-Free Breadmaking: Interface between Food Science, Nutrition, and Health. *Compr. Rev. Food Sci. Food Saf.* **2014**, *13*, 871–890. [CrossRef]
59. Larretxi, I.; Churruca, I.; Navarro, V.; Miranda, J.; Lasa, A.; Bustamante, M.A.; Simon, E. Impact of analytically measured fiber and resistant starch from gluten free products on celiac peoples diet. *J. Pre Proof* **2019**, *70*, 110586.
60. Hager, A.S.; Arendt, E.K. Influence of hydroxypropylmethylcellulose (HPMC), xanthan gum and their combination on loaf specific volume, crumb hardness and crumb grain characteristics of gluten-free breads based on rice, maize, teff and buckwheat. *Food Hydrocoll.* **2013**, *32*, 195–203. [CrossRef]
61. Sandri, L.T.B.; Santos, F.G.; Fratelli, C.; Capriles, V.D. Development of gluten-free bread formulations containing whole chia flour with acceptable sensory properties. *Food Sci. Nutr.* **2017**, *5*, 1021–1028. [CrossRef]
62. Moore, M.M.; Heinbockel, M.; Dockery, P.; Ulmer, H.M.; Arendt, E.K. Network formation in gluten-free bread with application of transglutaminase. *Cereal Chem.* **2006**, *83*, 28–36. [CrossRef]
63. Djordjević, M.; Šoronja-Simović, D.; Nikolić, I.; Djordjević, M.; Šereš, Z.; Milašinović-Šeremešić, M. Sugar beet and apple fibres coupled with hydroxypropylmethylcellulose as functional ingredients in gluten-free formulations: Rheological, technological and sensory aspects. *Food Chem.* **2019**, *295*, 189–197. [CrossRef] [PubMed]
64. Esteller, M.S.; Amaral, R.L.; Da Silva Lannes, S.C. Effect of sugar and FAT replacers on the texture of baked goods. *J. Texture Stud.* **2004**, *35*, 383–393. [CrossRef]

65. Giannou, V.; Tzia, C. Frozen dough bread: Quality and textural behavior during prolonged storage—Prediction of final product characteristics. *J. Food Eng.* **2007**, *79*, 929–934. [CrossRef]
66. Arcanjo, F.M. Development and Characterization of Gluten-Free Bread Enriched with Maca (*Lepidium meyenii*), Canary Seed (*Phanaris canariensis*) and Niger (*Guizotia abyssinica*) flours. Master's Thesis, Universidade Tecnológica Federal do Paraná, Medianeira, Brazil, 2017.
67. Rios, M.B.; Iriondo-Dehond, A.; Iriondo-Dehond, M.; Herrera, T.; Dolores, M. Effect of Coffee Cascara Dietary Fiber on the Physicochemical, Nutritional and Sensory Properties of a Gluten-Free Bread Formulation. *Molecules* **2020**, *25*, 1358. [CrossRef] [PubMed]
68. de Oliveira, F.C.; Coimbra, J.S.d.R.; de Oliveira, E.B.; Zuñiga, A.D.G.; Rojas, E.E.G. Food Protein-polysaccharide Conjugates Obtained via the Maillard Reaction: A Review. *Crit. Rev. Food Sci. Nutr.* **2016**, *56*, 1108–1125. [CrossRef] [PubMed]
69. van Riemsdijk, L.E.; Van der Goot, A.J.; Hamer, R.J.; Boom, R.M. Preparation of gluten-free bread using a meso-structured whey protein particle system. *J. Cereal Sci.* **2011**, *53*, 355–361. [CrossRef]

MDPI
St. Alban-Anlage 66
4052 Basel
Switzerland
Tel. +41 61 683 77 34
Fax +41 61 302 89 18
www.mdpi.com

Foods Editorial Office
E-mail: foods@mdpi.com
www.mdpi.com/journal/foods

www.ingramcontent.com/pod-product-compliance
Lightning Source LLC
Chambersburg PA
CBHW061442210926
51850CB00008B/1584
* 9 7 8 3 0 3 6 5 6 5 6 5 1 *